MECHANICAL ENGINEERING THEORY AND APPLICATIONS

GRID ELECTRIFIED VEHICLES

PERFORMANCE, DESIGN AND ENVIRONMENTAL IMPACTS

MECHANICAL ENGINEERING THEORY AND APPLICATIONS

Additional books in this series can be found on Nova's website under the Series tab.

Additional e-books in this series can be found on Nova's website under the e-book tab.

MECHANICAL ENGINEERING THEORY AND APPLICATIONS

GRID ELECTRIFIED VEHICLES

PERFORMANCE, DESIGN AND ENVIRONMENTAL IMPACTS

CARLA ALEXANDRA MONTEIRO DA SILVA, PH.D.
EDITOR

nova publishers
New York

For permission to use material from this book please contact us:
Telephone 631-231-7269; Fax 631-231-8175
Web Site: http://www.novapublishers.com

Library of Congress Cataloging-in-Publication Data

Grid electrified vehicles : performance, design and environmental impacts / editor, Carla Alexandra Monteiro da Silva.
pages cm
Includes biblliographical references and index.
ISBN 978-1-62808-839-7 (hardcover)
1. Electric vehicles. 2. Electric vehicles--Power supply. 3. Electric vehicles--Design and construction. 4. Electric vehicles--Environmental aspects. I. Silva, Carla Alexandra Monteiro da.
TL220.G74 2013
629.22'93--dc23
2013030605

Published by Nova Science Publishers, Inc. † New York

To Miriam and Carina

CONTENTS

PREFACE

The world is mainly dependent on fossil fuels across all major sectors of activity, as fuel is burned industrially, domestically and in transportation and electricity derives from burning fuel at central power plants. The hydrocarbon content of fuels and its interaction with oxygen from the air inevitably produces CO_2 and other pollutants such as HC, CO, NOx and particles hazardous to health. According to the International Energy Agency fossil fuels represented 90% of total primary energy supply in 2010. Greenhouse gas emissions in Europe are currently almost 18% below 1990 levels. Global emissions, however, keep on rising, particularly because of rapid economic growth in countries like China, India and Brazil.

Electricity generation is seen as a flexible mechanism whereby renewable energy may replace fossil fuels. Renewables represent roughly 10% of the electricity generation mix worldwide (in 2010).

The industry sector saw a pronounced decrease of 30% in its energy consumption in the period 1973-2010. The transport sector, being more difficult to control in terms of efficiency and emissions, is the sector which increased oil consumption and GHG emissions by 115% between 1973 and 2010. To reverse this tendency, mobility management policies and more efficient alternative drivetrains running on low carbon fuels should be introduced into the market.

Mobility management engineers focus on changing society's dependence on private transportation and increasing its preference for public transportation and soft modes (walking, cycling). On the other hand, OEMs are advertising more efficient alternative powertrains based on electric drives that can eventually run on electricity. This is the case with the pure electric vehicle (EV), e.g., Nissan Leaf, the hybrid electric vehicle (HEV), e.g., Toyota Prius, the fuel cell vehicle (FCHEV), e.g., Honda FCX prototype, and the plug-in hybrid vehicle (PHEV), e.g., Toyota Prius Plug-in. Changing society's dependence on private transportation is a difficult task as it requires human behavioural changes. Despite the campaigns undertaken, for example the European climate action plan, there is still much resistance to changing to efficient ways of life. This book discusses grid electric vehicles from a holistic perspective, looking at the technology in terms of: control strategy design and environmental life cycle rating; the business model behind 'electricity charging'; the way charging behavior interacts with local electricity grid voltage drop under dumb or smart grid scenarios; the prediction of grid electric vehicle movement and charging periods; the monitoring apparatus needed to acquire real data on both private car trips and bicycle sharing systems; and the obtained 'real' usage of those technologies' results.

The book is composed of 11 chapters. Chapter 1 addresses the differences between pure or battery electric (EV) and hybrid (PHEV) vehicles and discusses seamless transition from the current social and energy infrastructure.

Chapter 2 provides an extensive overview of selected aspects of automotive control system design with a focus on simulation-based development methods. It offers a methodology for simulation-based control network design for implementation of the Data Time Flow simulator, a simulation environment for design and performance analysis for vehicle electronic control systems.

Chapter 3 compares three EV business models: battery leasing, vehicle leasing and charging lease.

Chapter 4 explores how consumer charging behaviour can affect gasoline consumption over the life of a plug-in hybrid electric vehicle. The total fuel savings are calculated as a function of dwell time and infrastructure capability across a range of plug-in hybrid vehicles (PHEVs).

Chapter 5 discusses observable variations in PEV charging behaviours, whether PEV charging is more frequent and variable than has commonly been assumed, and the implications for estimates of social, economic, and environmental impacts.

Chapter 6 offers an analysis of the replacement of a large share of the traditional ICE vehicles with EVs in a low voltage network by analysing 10% to 60% (six to 36 vehicles) replacement in a residential area. Dumb versus smart charging influence on voltage drop and wind power and energy storage are discussed.

Chapter 7 describes a model that generates individual EV movement patterns, tracking them throughout the simulation period, and output indicators such as individual and global energy needs, greenhouse gas emissions impacts and impacts on electricity grids. The model is illustrated through a case study application.

Chapter 8 reviews vehicle labelling and environmental rating schemes.

The three last chapters focus on real monitoring of PEV in light-duty and soft-mode applications. Toyota Prius plug-in and Renault Fluence ZE are analysed. Bike-sharing systems worldwide, the number of electrical bikes and an electric bicycle, QWIC Trend2 by Prio Energy, are reviewed.

In: Grid Electrified Vehicles
Editor: Carla Alexandra Monteiro da Silva

ISBN: 978-1-62808-839-7

Chapter 1

THE PLUG-IN HYBRID ELECTRIC VEHICLE PHEV: DIFFERENCES WITH THE BATTERY ELECTRIC VEHICLE EV

Andrew A. Frank*
CTO Efficient Drivetrains Inc and
Professor Emeritus, University of California - Davis, Davis, CA, US

ABSTRACT

This chapter discusses plug-in hybrid electric vehicles (PHEV) and pure electric vehicles (EV) for the transition from Fossil Fuel Based to a Zero CO_2 society and is broken into the following sections:

1) Introduction of the PHEV and The objective of Vehicle Electrification
2) The limitations of Electric Drive now and into the foreseeable future.
3) The urgency of conversion to renewable energy to replace fossil fuel use in transportation and society.
4) The need for dual or triple fuel systems until new technology is created.
5) Using the PHEV to improve the existing Electric generation and distribution system
6) A road Map for Integrating the PHEV into a no fossil fuel, completely renewable energy society with no loss of function or convenience and yet lower costs.
7) Conclusions.

Keywords: Global warming, fossil fuel, renewable energy, PHEV (Plug-In Hybrid Electric Vehicle), Pure Battery Electric Vehicle (EV), sustainable energy supply, minimizing CO_2 and greenhouse gases, technology solution, Natural Gas-NG, Corporate and private benefits of new energy use.

---

* Corresponding Author address: Email: afrank@efficientdrivetrains.com

INTRODUCTION

The Electric Vehicle, or EV, is a focus of this book but this chapter looks at an alternative but parallel path to allow us to solve the most pressing problem on the Planet—Global Warming from creating CO_2 – in the shortest time frame with a minimum of new technology. This chapter shows alternative emphasis in the near-term and midterm until new, currently un-imagined technology is created. Technology is limited to the known physical elements, physics, and chemical concepts developed and discovered in the last 500 years. Yes, there is probably more to be discovered but we may not have the time to discover and develop these new technologies for mass use in the next 50 to 100 years.

Based on this focus, what can we do now to transition to a Renewable Energy Society in the shortest possible time? Part of the answer can be found by reviewing the evolution of energy development and use.

We began in earnest to transition from an agricultural society based on animal and human power to a machine oriented society without the use of biological slaves about 200 years ago with the creation of the steam engine. This Steam Machine could use cellulose or fossil fuel. It was a true multi-fuel machine that was used to transition from biologically renewable energy sources to a more energy dense material found in the ground but created by the sun thousands-to-millions of years ago and concentrated by the earth—fossil fuel. This provided the beginning of the fossil fuel age, but the transition was possible because of the multi-fuel capability of the steam engine. The transition was un-noticeable even as new power-developing sources were discovered. Otto, Diesel, Atkinson and others spearheaded these discoveries in the late 1800's. All these inventors created engines that used a further concentrated energy-dense material derived from oil or coal to provide transportation machines that were light weight and cost effective. The steam engine continued until recently for long haul and freight transportation where weight and performance was not as important. Coal steam is still used to generate stationary electricity for most of the world today. These large stationary machines are also multi-fuel, being easily switched from coal to natural gas and even garbage.

Now let's get back to vehicles and converting them from gasoline and Diesel to using electricity. We currently use electricity for many transportation systems "successfully" on fixed routes. These are the trains and trolley systems that exist around the world. The word "successfully" is questioned because we have not looked at the efficiency and resource expenditure for these systems. If we consider the amount of conducting material needed for an electric train and the number of people carried per day, it is large because the electricity is used in peak pulses when a train needs it as it passes by a source. Since these trains go by infrequently the energy distribution system needs to be designed for the peak power but the average power is much less, maybe as low as 10%. Thus they are an inefficient use of energy supply resources. To improve this efficiency then it is necessary to find a way to level the load so that electric energy can be delivered at a low power but can supply a high power when needed as the train or vehicle goes by. This can be done with Electric Energy Storage by capacitors or batteries. The storage can be stationary on the ground or mobile in the vehicle. Either way, energy storage is needed to be able to use electricity more efficiently to drive our machines instead of liquid fuel. One of the questions addressed is, "Is it wise for us to move toward an All Electric society?" For example, it makes little sense with our current knowledge of physics and chemistry to consider a 5000 passenger oceangoing craft being powered by batteries. The same goes for aircraft. Thus, society, as we know it, must have some liquid fuel to be competitive to what we have become dependent. The only issue is to stop using fossil-based liquid fuels.

Thus, the objective of electrification can be listed with the following attributes:

1) Improve energy efficiency from renewable sources to the end use in transportation and industry.
2) Reduce liquid forms of energy use as much as possible, but in a coordinated transition so as not to disturb global economic health.
3) Create as much renewable energy as possible, using solar, wind, water and biofuels.
4) Create a transition road map that allows the world's general public to seamlessly transform from fossil fuel consumption to 100% renewable energy consumption in the shortest possible time, beginning with the existing fossil fuel and electric distribution infrastructure.
5) Create transition products that are cheaper and more functional than fossil fuel-only vehicles, and provide for lower running costs.

These attributes can be achieved entirely by the PHEV, but there is certainly space for battery-only vehicles, or EVs. In addition, as we reduce the use of fossil fuel we will need to address the fossil fuel industry that currently has a financial and physical monopoly on our oil dependent world.

LIMITATIONS OF EVs

The advantage of gasoline- and Diesel-fueled vehicles lies in the energy density of the fuel. Part of that density comes from the fact that some of the energy comes from the atmosphere in the form of oxygen. When electricity is created from fossil fuel it can only be created at less than 50% and, when transported to the end use by the distribution system, it could be as low as 30% to 35%. If the fossil fuel is consumed in today's gasoline or Diesel vehicles, it could be as efficient as 30% to 35%. But for passenger cars it is much lower at around 20% to 25% due to the demanded performance and acceleration. If EVs are designed to have the same performance as conventional gasoline and Diesel cars then they also suffer from low energy efficiency from the electric plug to the wheels of the car. Thus, many people have argued that the EV fueled by fossil fuel electric generation is a lot of expense for little or no gain in energy efficiency. In addition, to achieve a comparable range requires a lot of batteries with a real price to be paid in weight and costs using today's battery technology. Of course this could be improved, but that is only a small portion of the problem!

The more important issue currently ignored is how to transfer the equivalent energy required to travel 350 miles (500 km) in 5 to 10 minutes? A simple calculation illustrates the concern. If a vehicle requires 0.3 kWh/mile travelled and it goes 350 miles then it requires 105 kWh of batteries. If we wish to put this much energy into the batteries in 10 minutes, or $1/6^{th}$ hour, then it will take 630 kW of power. But the batteries at this charge level are currently less than 50% efficient, meaning that we need more than 1000 kW of power to feed the batteries in 10 minutes, or over 1 mega-Watt or more for one vehicle for 5 to 10 minutes. Such an electrical system would require about 20% of a typical electric power plant for 10 minutes with the electric wires and transformers necessary to get that power to the vehicle charge station to be used only for 5 or 10 minutes out of perhaps an hour or more. Think of how frequently our gasoline stations are used on the average and yet we have almost every car using gasoline today. It is less than 50% and, in the beginning, 75 years ago, it was less than 5%. Another difference is that a tanker truck can put a store of energy into the ground at the gas station to be used for 2 weeks to a month. Thus, an obvious solution is to put a large battery at the filling station that can supply this megawatt of power for five minutes, or so, and then be charged slowly over many hours at low power. If such a battery were to be developed where such a charge rate can be done with an efficiency of 90% instead of 50% or less, then the idea is to put in a large set of batteries that can be charged from the electric grid at a low power and can be used to discharge quickly into the vehicle's battery. No such technology or delivery system exists today that is comparable to our current system of filling stations that utilize tanker trucks to fill underground storage tanks.

Another fundamental problem is the way electric energy is stored in batteries (Electrochemically, which has internal resistance where the transfer loss goes up proportional to the square of the rate) and the way liquid energy can be stored in bulk with energy transferred to a vehicle with little or no loss. Until technology of electric energy transfer is solved there will always be a loss. This makes electric drive energy much less efficient than liquid fuel in all Fast- or Quick-Charge situations. To improve this situation requires technology we currently do not have, and no one is attacking the problem. The current Quick Charge Stations being installed around the US and Europe are not really good investments and could not happen economically without direct government support. Such support is being made available because people are pushing for a solution without a full understanding of the consequences. Money is being made available for such experiments because politicians do not understand the consequences of Quick Charge. The manufacturers have no reason to tell the truth about overall efficiency, and the electric power companies care less about efficiency because they make more money the more energy they sell! A possible solution would be to have a truly independent technology monitoring body, made up of technology people, who are paid and protected enough to not be influenced by special interest lobbyist from all sectors of the energy industry.

A bigger issue of the EV is range, or how far it can go. Current batteries are all affected by ambient temperature variations. This one parameter will vary the range by as much as 30% or more. Another parameter that affects range is internal temperature caused by use and quick charging and discharging. This could also affect battery longevity. Battery cell balancing is another matter that may further discount range capability when rapidly charging and discharging. Thus at any one time the actual range available can vary by a factor of 2 or more. Compare this to gasoline or Diesel vehicles where the maximum variation for the worst conditions may be $\pm$10%. Anyone owning an EV now knows about these wide variations in capability and if he can live with and understand the limitations then he could be happy. But what of the average person who owns a conventional liquid fuel vehicle with the current convenience and availability of the fueling infrastructure? There is no comparison today. Regardless of the money thrown at the issue to make it acceptable, while ignoring physics and efficiency, the truth will eventually come out and the current experiments will be supplanted by as yet unknown technology, or abandoned due simply to the logic of the amount of energy required to charge batteries and the cost of the infrastructure. The future of the world will depend on the average energy consumption per person. It is our objective to find the best way to minimize this consumption and make sure it all comes from renewable sources as our finite world continues.

Now let's look at the EV's internal technology. All EVs constructed by the major car companies are well designed and use the best available technology, yet they hardly compare to the liquid fuel alternatives that are a small fraction of the cost. Of course there are applications where the EV is 100% satisfactory and those applications and the people who use them should be commended for spending the money and time to drive such vehicles. However, for the average consumer the EV is far from satisfactory since it lacks so many of the essential technical features of the existing, much lower-cost liquid fueled vehicles. Thus there will be a market for EVs but it cannot be anywhere near a double-digit percentage of the overall vehicle market until major breakthroughs discussed above are completely addressed and become competitive with liquid fuel counterparts. To put things in perspective, the Tesla long-range vehicle has 85 kWh of batteries. Yes these batteries can drive the vehicle 300 miles but under very carefully controlled conditions. We must remember that 85 kWh of energy is equivalent to about 2.5 gallons of gasoline. So if you had a gasoline car with such a tank size, would you attempt a 300 mile trip?

The Urgent Need to Stop Using Fossil Fuel for Transportation, Industry and Social Applications

Regardless of the down sides of the EV, we must continue to try new technology because the world is finite and may not last forever. Mankind, using fossil fuel, or stored energy from the sun millions of years in the past, is changing the climate quicker than at any time in the history as we know it. Recent Public Broadcasting Service (PBS) documentaries and books by Climatologists the world around, all confirm the findings. So it is clear that we must stop using fossil fuel without disturbing society, as we now know it. Dire consequences in heat and violent weather will disrupt the planet to conditions not tolerable by our species. The point is that we humans cannot evolve as fast as the climate is changing, so we will become the dinosaurs of the past. But the destruction will come from within rather than from meteors from outside our planet. But how can we do this without large disruption in society. By the way, the richest segment of our society is the oil industry because the world is currently completely dependent on it to support our convenient and productive lifestyle. Thus we need to find a solution that addresses every part of current society so that everyone benefits and can continue to profit equally.

Oil Depletion Causes Global Warming-PHEVs Can Be Pivotal in the Solution with No New Technology Needed

The best solution would be to stop using fossil material for fuel, which is creating CO_2 and other greenhouse gases contributing to the rapid warming of the planet. If the fossil material were developed into recyclable structural materials for the construction industry, replacing wood and other materials, we could begin to restore some of our forests to absorb the excess CO2 being generated and start to use recyclable fossil materials to create new buildings and houses. Then after the products become obsolete or worn out they can be recycled simply by adding primarily concentrated heat energy from the sun through solar, wind or water sources.

We have been using fossil material now for roads and plastics of all kinds already. What is suggested here is simply that this segment of the fossil extraction industry be greatly expanded with the potential to become more valuable than selling the fossil material for destruction of itself and the planet by using it for energy. Policies to make this happen would be needed and it must be shown to the energy industry that making this change is good for their long term future, and could insure them of a continued business model for more years than if they were to use their current extracted material for energy. Their business model may have to change slightly but they have the resources to do that, they only need the thrust and motivation to move in such a direction.

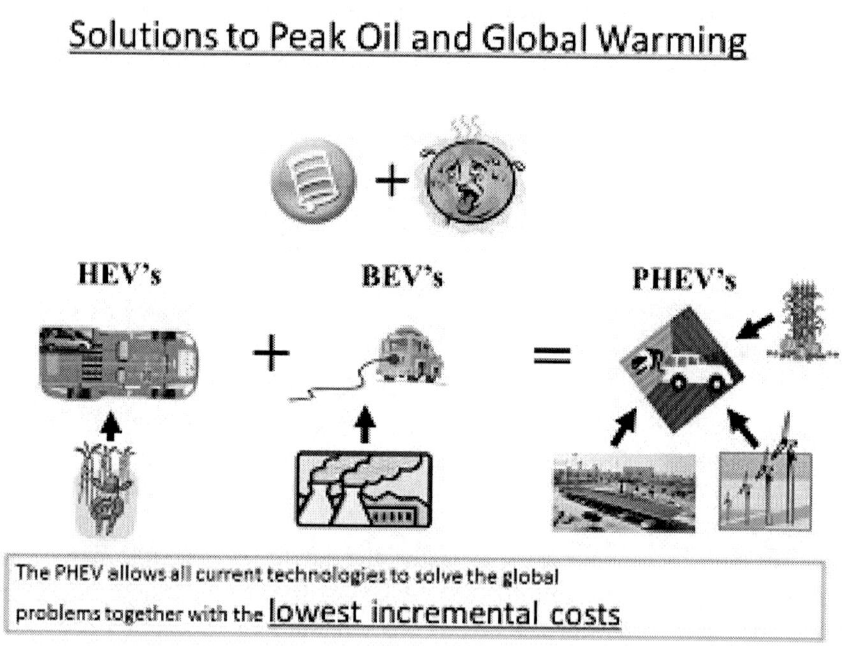

Figure 1. The PHEV can use renewable energy with zero CO_2 emissions.

Since the world is driven by profits, rules and regulations must be created *to* allow this to occur. Also, new material technology is needed to allow mass production of standard shapes used in the construction business.

Refinement of the fossil material will take energy but ultimately this should come from renewable solar, wind, or water, and maybe nuclear energy. This way the current fossil fuel industry can continue its profit making, while benefiting society by creating planet-saving materials and reducing the greenhouse gas emissions gradually, in a controlled rate, to essentially zero, over time.

THE NEED FOR DUAL OR TRIPLE ENERGY USE MACHINES FOR A TRANSITIONING SOCIETY

As we discussed in the Introduction, the evolution from steam to gasoline and Diesel was essentially seamless because the steam engine was a multi-fuel machine. The gasoline engine was not. Diesel engines were designed by Rudolf Diesel to initially burn essentially any combustible material from coal to wood pulp. But it was soon discovered that a liquid fuel contained much more energy and could be handled and distributed much easier than coal or wood. The steam engine, however, when converted to Diesel fuel or oil also became better and more efficient. So the various forms of power production continued with gradual improvements until relatively recently. Thus, the same holds true today–flexible fueled systems allow the transition to new technology in a seamless fashion.

To transition to the use of renewable energy from gasoline and Diesel machines dependent on liquid fossil fuels, could be done in a number of ways. People have suggested a transition directly to biofuels. The problem is that the current liquid energy consumption rate is so high that we could not grow enough fuel to satisfy the energy currently consumed by society. Biofuels have been shown to be producible at a reasonable price competitive with current oil-derived gasoline and Diesel fuel. The main problem is developing enough to satisfy demand. Currently, gasoline in California, due to emissions regulations, is mixed, with up to 10% ethanol. Thus, already 10% of the gasoline sold there is biofuel. So this is likely sustainable. But to make it 100% biofuel would not be possible today or into the future because fuel growth will compete with food growth.

If, however, we create a vehicle that can use 90% electricity and 10% liquid or gaseous fuel, we can transition to a renewable energy society seamlessly without having to create more biofuel. The Chevrolet Volt, the new Ford C-Max and Ford Fusion Energi models, and Toyota Prius Plug-In are PHEVs. Other companies are beginning to follow suit. These vehicles are all designed to use as much electric energy as possible. Their control policy is to assume a fully charged vehicle at the beginning of a trip then deplete the electric batteries to a minimum when a gasoline or Diesel engine is asked to take over carrying the load. The electric battery and motor then are used to provide extra performance to the vehicle. This philosophy allows the liquid-fueled engine to be downsized and run much more efficiently. The amount of down size is dependent on the size of the battery and thus the power of the electric motor.

The author has designed and constructed over ten such vehicles based on various technologies in the powertrain. Some of the vehicles had liquid-fueled engines that were downsized by a factor of 5 (from 3.0 liters to 0.6 liters) and yet had comparable performance to the original vehicle because the electric motor and the gasoline engine powers were designed to be additive.

The result was an All Electric Range (AER) of about 60 miles (100 km) and a performance equal to the conventional car. When the batteries are depleted to the Minimum State of Charge (SOC) after traveling 60 miles (100 km), the engine comes on to maintain the battery SOC, but does not charge the battery above the Minimum SOC. Thus allowing the batteries to be filled from an external source when the vehicle comes to a stop.

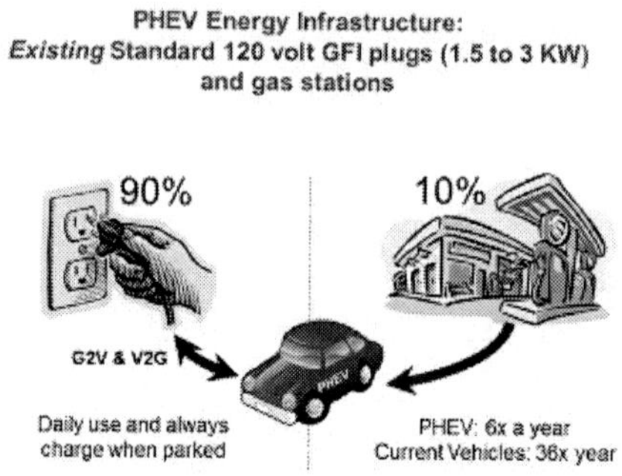

Figure 2. PHEV uses 90% electricity from low power chargers and 10% liquid fuel for long trips a year.

This policy not only allows the use of electricity first, then the gasoline engine to drive the vehicle, but to also add electric energy from the battery/electric motor system to provide acceleration and collect regeneration of energy from deceleration of the vehicle. In this mode the electric energy is cycled in and out of the batteries as needed to maintain the vehicle at the drivers command. This mode is sometimes called the Charge Sustaining (CS) mode. There are many different powertrain configurations to accomplish these three basic modes, All Electric, Regeneration and CS. We will not be discussing the specific configurations except to say that the best configurations are ones that are simple but allows the operation of all the modes at the best possible efficiency. The amount of AER is dependent on the size of the batteries or kWh of energy stored. The commercial vehicles currently being manufactured have enough batteries for AER of 12 miles-20 km, 20 miles-30 km and 40 miles-60 km. Of course, they could be increased to 60 miles-100 km, as was done by the author. If the same battery technology were to be used, then the power available from the battery pack will be proportional to the battery size. Then if all the vehicles are to have the same performance, the gasoline engine power can be so adjusted to make the total power the same as the conventional all gasoline vehicle. Thus, the larger the batteries, the smaller the liquid fueled engine required. This has the effect of increasing the fuel economy or decreasing the fuel consumption without affecting performance. To counterbalance the positive effect on fuel consumption is the added weight of the batteries and the top sustained speed of the vehicle. Notice the top speed may be the same but the sustained speed is lower due to the smaller engine. In addition, the battery costs today are higher than almost all other components in the PHEV. The manufacturers at any time point in the technology development must reach a compromise. So we see the Japanese manufacturers choosing a small battery pack and a larger engine, Ford Choosing a 20-mile AER and a smaller engine and General Motors choosing a 40-mile AER and a small, efficient engine. It is obvious that the overall vehicle cost would be affected by the battery size and powertrain complexity.

The main advantage of a larger battery pack is the displacement of liquid fuel with electricity. Obviously, the larger the battery pack, the less liquid fuel will be used if all vehicles performed equally and were driven similarly with the energy from the battery pack coming from a wall plug or an external source. The key difference between charging PHEV batteries, as opposed to the EV batteries, is the *rate of charge* needed and desired. The PHEV in general, can be charged much slower and be fully effective.

The US Society of Automotive Engineers, SAE, have established standards on charging batteries for safety and other reasons. The standard is referred to by the number J1772 for charging at a rate up to about 10 kW. In addition they have set two levels of charging depending on the power capacity of the electrical outlet. Around the world, for domestic and light duty electricity use, there are essentially two levels of electrical outlets. Less than 2 kW is called Level 1 and less than 10 kW but more than 2 kW, is called Level 2. The low power Level 1 outlets are distributed around the world and essentially in every home, business and industry. Level 2 outlets are used for large electric stoves, electric heaters and clothes dryers, and pumps for swimming pools, etc. By far, there are more Level I plugs in society than Level 2. It is the accessibility and availability of off-the-shelf components that is critically important to charge these PHEVs at the beginning of new technology introduction. So Level 1 charging systems are preferred. The problem has been that more money can be made by a charge station business in Level 2 and above, than in Level 1.

The PHEV is a dual or triple fuel vehicle being able to use electricity from an outside source and liquid or gaseous fuel from one or more sources. This gives the PHEV a big advantage over the EV because it does not have to be electrically capable to carry out a driving mission; there is also gasoline or other energy on board. Thus easing a phenomenon of single-fueled vehicles called Range Anxiety. Gasoline vehicles also experienced this phenomenon before 1920, when there were few gasoline stations and gasoline was mostly distributed in small glass bottles by drug or hardware stores. The fall back in those days was the horse. Today, in most countries around the world, gasoline stations are abundant enough and the range of the gasoline vehicle is far enough to essentially make the liquid fueled vehicle convenient and productive. In fact, some will say that the gasoline vehicle is what led us to our current level of productivity. Thus, as we introduce new vehicle technology it must be as good or better than the current gasoline vehicle in every dimension of measure, and be comparable in cost. The multi-fuel aspect of the PHEV could do just that.

The PHEV will uses electricity to displace liquid fuel if the user plugs the vehicle in when parked. If we look at the typical use of the conventional gasoline vehicle we find from the US National Highway Traffic Safety Administration (NHTSA) that the average driver across the US travels about 30 miles a day and drives his vehicle for about 3 hours. This means he has 21 hours to charge the vehicle. The PHEV will use about 0.3 kWh of energy per mile of travel. So, in 21 hours, using Level 1 plugs at 1.5 kW can provide about 31 kWh of energy, or a range of about 100 miles per day. Thus, Level 1 plugs are capable of supplying 3 times the average mileage required by the normal driver if the vehicle were plugged in every time it is stopped and the batteries are large enough. Hence for PHEVs, the much lower cost Level 1 plugs would more than satisfy today's average driver's need to displace liquid fuel with electricity.

The PHEV then, has little or no need for high-cost level 2 and even less need for much higher cost quick-charge concepts. Of course the amount of liquid fuel displaced is dependent on the battery size. The average US driver travels 30 miles (50 km) round trip a day to-and-from work. So a driver travels 15-or-less miles at a time and then stops for up to 8 hours, thus the battery only has to provide 15 miles of range (25 km) at a time, and be charged twice a day. Such a battery then will be cycled twice a day, and if the car is used for 4 years, the battery will have experienced approximately 2400 cycles, or about the limit of current battery life cycle. On the other hand, if the battery were capable of 30 miles and the person drives 15 miles and charges twice a day, the batteries experience only one cycle per day and the batteries should last 4800 cycles, or about 9 to 10 years. (This is the strategy of the General; Motors Chevrolet VOLT) The down side is that the batteries cost 1/3 more. Thus, the Federal government has provided an incentive for the larger battery, and the State of California provides additional incentive, plus High Occupancy Vehicle (HOV) lane access to allow the driver to drive in the free-flowing HOV lane with only one occupant in the vehicle. This has turned out to be a very popular incentive for people to purchase the PHEV with 30 to 40 miles of AER. This is less than 1/3 the batteries of Pure Electric vehicles with the same capabilities.

Depending on government subsidy is not a sustainable business model but it helps to get the technology into the hands of the general public until the volume drives the battery price down to the point where these PHEVs can be comparable in price as the standard gasoline- or Diesel-fueled vehicle.

So why would the driver of a PHEV plug his vehicle into the electric grid? The base cost of electricity around the world is about US $0.10 to $0.12 per kWh. Gasoline is at approximately $4.00 per gallon in the US (more like $9.00/gallon in Europe), with current gasoline vehicle fuel economy at about 30 mpg, making fuel cost about 13 cents per mile. When driving on electricity, the energy needed is about 0.3 kWh/mile and the cost is about 7 cents per mile, or close to ½ to ¼ the cost of using gasoline. At the current time, most places do not charge for electricity because it would only amount to about 15-to-18 cents an hour. Collecting this is not worth the paperwork for short-term parkers, so it is often free. This provides further incentive to purchase such a vehicle today. On the other hand, with Level 2 charging, it is about 72 cents per hour, making it more difficult to justify free charging. As mentioned above, Level 1 charging is completely adequate for the PHEV.

It should be pointed out that the Japanese PHEVs by Toyota, Honda and Nissan have elected an AER of just 12 miles (20 km). The reason for the smaller battery pack is clear, the average driver in Japan drives a distance of less than 20 km to work, and assuming the vehicle is parked for at least 8 hours at work, then the vehicle can easily accumulate 10 kWh of energy, providing a range capability of 50 km. This is about 5 times what is needed in Japan. The down side is that the engine cannot be downsized as much, even with higher-power batteries because there is not enough energy to account for un-anticipated needs, such as long, steep highway grades where the small battery pack boost will dissipate before the grade needs are met; the vehicle will slow down making it less acceptable by the general public. Never-the-less, the step to PHEVs is headed in the right direction, and as battery costs come down due to higher volume then the battery pack size can be increased and the engine further downsized for better fuel economy in Charge Sustaining or CS mode.

The vehicles constructed by the author have shown that the liquid-fueled engine could be downsized to 1/5 to 1/3 the conventional vehicle and still maintain the performance of the comparable all gasoline vehicle. The battery packs in these vehicles were sized for 60 miles or 100 km AER. The charge-sustained highway fuel economy was close to 3X the conventional car, or the vehicle consumed about 1/3 the fuel of the comparable conventional car at sea level whise sustaining the charge in the traction batteries. A down side of this drastic engine downsize is the performance at high altitude. Thus the engine needs altitude compensation by turbo or super charging the intake air. This is well known technology and there are multiple sources for this. In addition, a turbo system can be designed to regulate the engine intake air to sea level pressure by the use of an electric generator providing a turbo compounding effect at sea level further increasing the engine efficiency and reducing fuel consumption. This works well with the downsized engine because when it is run, it is run at high load providing high exhaust heat energy and the ability to extract energy from the exhaust stream.

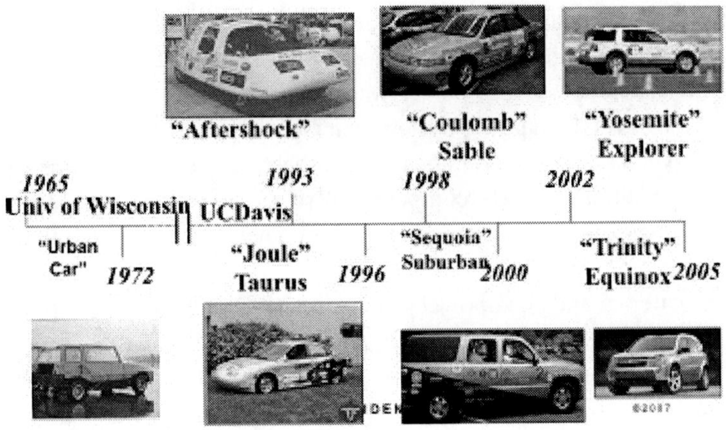

Figure 3. Some production cars and Sport Utility Vehicles (SUVs) modified to PHEV with 60 mile AER by the author.

IMPROVING THE ELECTRIC POWER DISTRIBUTION SYSTEM WITH THE PHEV

This is an area that is not well addressed. If charging, as argued above with Level 1 at less than 2 kW, then 1000 cars distributed in society could consume 2 MW of power. Power plants must regulate their power according to demand because if extra power is generated, it must be dissipated or used in storage somewhere.

PHEVs distributed in society today all have battery systems that are intelligent, so it is a simple matter to allow communication between the battery pack and the power generation system, which have compatible needs. These batteries can be used to take load when there is excess energy available and not take load when the electric power system is overburdened. A simple on-off signal to the charger is all that is needed at this stage. The central power distribution system can cycle the energy to various chargers on line in a sequence that is equitable. For example, if a vehicle signs up for such service maybe the electric rate can be reduced to ½ or so because the electric system receives a valuable benefit. The power companies benefit because they are getting revenue with energy that would have been wasted otherwise due to short term over capacity, and they can use this variable load distributed in society to regulate substations where peak demand may be occurring. If a driver finds he is in the wrong neighborhood to take advantage of this low rate then he can sign up for regular service or move his vehicle's parking location to a more favorable spot.

The net result of this strategy of electric power distribution and management is to utilize the existing electric distribution lines already in place to a higher utility factor. For example, most American homes have 200 amp service at 220 volts, or about 40 kW capability. This capacity is rarely used, and it is calculated that less than 20% of this peak capacity is used in the common American home. This implies that we have installed far greater capacity than is being utilized. By using battery charging in vehicles distributed throughout society we can increase this utilization factor with *no new power-line facilities* by simply allowing the charging system to go off for those times when the local electric grid is highly loaded. For the EV this may not work well since they are one energy source vehicles and already suffer range anxiety, thus they would want high power charging every time they are parked, lowering the utility factor and efficiency of the power lines further. Since PHEVs are on Level 1 low power charge, this Level 1 charging system can become a dream for power-line regulation purposes. The driver of the PHEV may use his secondary energy source, gasoline, a little more periodically. If he is properly compensated so that everyone wins, then such a system can work and it is low-to-no cost to society and the owner or driver of the PHEV.

A further extension of this concept is to design a bidirectional charger that can take power from the batteries and feed power back to the AC lines in proper phase as needed by the power company. This power regulation scheme can be used to compensate for sudden changes in the local electricity demand and the batteries are then used for power line compensation to maintain high efficiency power transfer from the main power plant. Because each battery system is being charged or, in this case, discharged at 1 to 2 kW the regulation of a large load would take many vehicles in a local area. A computer algorithm can be designed to take this distributed resource and allocate it properly to regulate the local grid and greatly improve energy transfer efficiency with no additional resources! The objective would be to improve the grid at the same time providing the customer energy over a period of time such as 4 to 8 hours. People have called this concept "Vehicle-to-Grid," or V2G, energy feedback. Thus, the PHEV at Level 1 charging can be implemented and benefit society with essentially no new power resources because it is getting it's energy from higher utilization of existing resources, thus increasing the efficiency of the electric power system. In addition, the batteries may benefit from the low-level charge-discharge exercise.

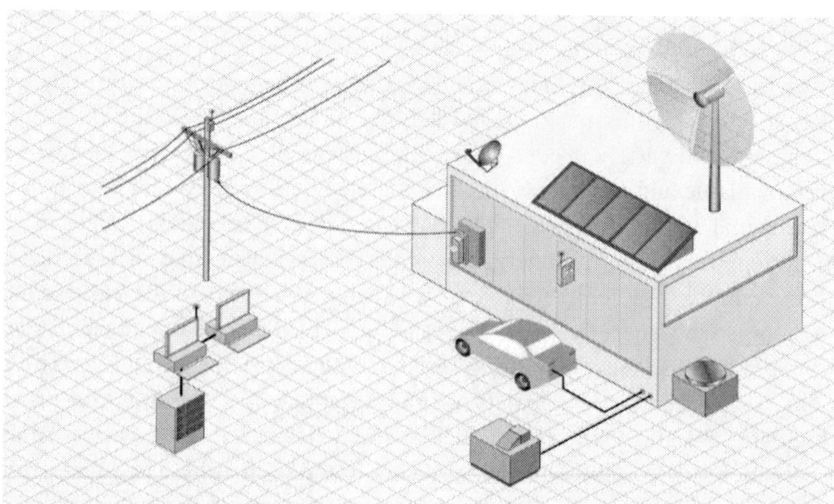

From Southern California Edison.

Figure 4. Total energy management system at work and home with energy stored in PHEV's, most energy can come from renewable sources.

We have shown in this section that the PHEV can be used to greatly improve the Electric Distribution System and the savings can be shared between the customer and the utility, so everyone benefits. In addition, we have indicated that the current Electric grid systems all over the world (even in places like India) have over capacity that is not used. In fact the USDOE Pacific North West Laboratory has calculated that this over capacity in generation and distribution in the US could, under proper management and integration, support 80% of cars on the road today if they were all PHEVs, with no change in the electric power lines.

On the other hand, if we had to go to high power charging with Level 2 and above, for EVs, etc., then new infrastructure would be needed at great expense.

The message of this chapter is that with the PHEV concept, there is no new electrical infrastructure needed. And as far as liquid fuel infrastructure, these vehicles can begin on fossil fuel and transition to biofuel that will be distributed by the same liquid fuel infrastructure we have today. These advantages are lost with the EV concept.

Alternative fuel concepts such as hydrogen, methanol, etc., have been tried in the past and each time they failed after huge amounts of money and time were spent. Why? Simple. The energy distribution system was not there, and to construct a new energy infrastructure means a huge public investment. There is never a good time for such investments so these proposals all die a natural death. Hydrogen is on its last legs now because in addition to the lack of infrastructure and generation from renewable sources, Hydrogen Fuel cell energy "Source-to-Wheels" efficiency has been "brushed under the rug". For example, it is a simple matter to compute the generation of energy from a windmill and trace the electricity path to powering the wheels of a car using a EV or PHEV, and then compare that to taking the same wind energy and forming hydrogen from water then transporting it to a distribution system, and finally using it in a fuel cell at "high efficiency". The current PHEVs and EVs on the road today would be able to go 300%-to-500% farther than the best hydrogen vehicle today at a fraction of the cost. In addition, the batteries in the PHEV or EV can be used to benefit the grid as discussed above, further lowering the system costs.

A Road Map to an All Renewable Energy Society Based on Long-Range PHEVs and EVs

We just mentioned above that PHEV and EV batteries can be used to level the load and better utilize the current grid. In this section let us look at what is needed to transform from a 100% fossil fuel-dependent Society to one that can use 100% renewable energy in the shortest possible time.

It is obvious from the above discussion that nothing will change if we do not address the entire energy sector, both production and consumption/utilization together. We have not discussed renewable energy production and how it can be used in PHEVs and EVs of the future. We did say above that it is a simple matter to use battery energy storage to level the grid and to provide energy back to the grid in times of need. Once this feature is developed, then the technology can be extended to address renewable energy.

All renewable energy systems are discontinuous. For example solar energy only generates for less than 12 hours, and wind blows intermittently, daily and seasonally. Even waterpower is dependent on last year's rainfall, etc. To take these variable sources and integrate them into the electric grid means that we are varying the power output of the grid even more than we are with load. This generally means new power lines and more infrastructure to accommodate the large wind and solar farms located in remote locations. In addition to these large, efficient wind and solar farm installations, we can utilize the batteries and energy storage capability of the PHEV and EV to level the renewable energy sources and relieve the grid from trying to accommodate the added renewable energy by throttling the fossil electric power generation and reducing the fossil or other fuel generation system efficiency due to throttling.

A proposal in a recent paper by Sterling Watson now at MIT, stated that if every new car sold were a Chevrolet Volt and it was sold with a windmill or a set of solar panels, (the car would be sold with "fuel" for the life of the car), and if everyone bought one of these cars or something similar with a renewable energy generator, in a period of about 15 years we can be completely off of fossil fuels, if we supply 10% or so of our energy from liquid biofuel sources. As we mentioned above, this 10% biofuel is already in the gasoline being sold in California by law. Thus, in California we can today begin to use only the ethanol in our fuel and forget about the gasoline with little or no change in any of the energy infrastructure. Then if the energy were generated by local distributed renewable wind and solar and stored in the batteries of the PHEV or EV with G2V (Grid-to-Vehicle) and V2G capability, it can be used by the vehicle, or local homes, or industry. We would suddenly be transitioned to a renewable society with no loss in performance or convenience. Since cars last about 15 years, then, this is the minimum time for this transition. In the paper by Watson and Frank, he they found that the energy for industry would have to come from large wind and solar farms. So simultaneous with the introduction of the PHEVs and EVs, would be the construction of the large energy farms and biofuel production farms. Biofuels are needed for things like aircraft and ships. We currently use oil and jet fuel for these purposes, but biofuels could easily be substituted with no loss in productivity.

Figure 5. Sustainability road map.

The batteries of the PHEV and EV are not worn out after 10 to 15 years of use but they may have degraded in power and energy capacity. None-the-less, they have additional life left for other uses than transportation. These vehicle batteries could be used in homes, business offices and industry to store renewable energy where they can provide a steady load for 24/7 to run buildings and to charge vehicles. Thus, PHEV and EV batteries can be used to provide the total energy store necessary to take energy from both local- and centrally-generated wind and solar facilities, and the energy can be managed for vehicle needs, while also supplying enough energy for all stationary use, as well. The question is where should these generators sit? It is suggested in the paper that they simply sit at the place of residence and at the work place.

Ultimately, such small energy generators will have to be managed by someone other than the home owner or the industrial user. The logical energy managers are simply the existing electric power companies. Many of these companies are called home or stationary energy managers, like Pacific Gas and Electric (PGE). Since energy management is their game, then managing a distributed collection of energy sources could easily be done with today's computer technology. This gives the company much more work, but then management and costs of it's traditional sources of energy from coal, natural gas, water and nuclear would be less or greatly reduced. This allows them to grow and provides additional revenue. The idea is to ultimately reduce coal and natural gas to zero. This gives the existing energy companies a lucrative new business and allows growth for the foreseeable future, as the technology evolves.

CONCLUSION

We have tried to convey a scenario using mostly PHEVs, but also some EVs, where there are no losers and all winners in order to solve the problem of transitioning off of fossil fuels in an acceptable and easily integrated fashion for everyone. This applies to all involved in the energy business, from the individual, to large corporations designing and selling cars and trucks, and the energy industry's oil and gas production and electric energy generation and distribution companies. We are suggesting that all these industries be provided incentives to grow into the new paradigm. The key is that we stop using fossil material for energy and start using fossil material for everlasting products that can be recycled many times *without* being destroyed by combustion and turned into CO_2 or other greenhouse gases.

Our planet is quickly becoming a place no longer habitable by our species. If we do not begin the amelioration of our actions of the past we will essentially extinguish ourselves with essentially no survivors. This scenario is common to all species on earth. The classic high school experiment where a small dish of nourishment is injected with bacteria and allowed to grow and flourish illustrates the potential outcome of the world. In so many days the colony of bacteria grows and grows happily until all resources are consumed then the entire colony suddenly dies. We, as "intelligent" bacteria, have the means to change this scenario because we know a lot about our finite world and we have the technology to affect this change. The question is, can we act fast enough to catch ourselves before the start of the inevitable catastrophic collapse. The bacteria surely couldn't act fast enough and other past living species couldn't act fast enough. The bigger question is, with our knowledge of physics and chemistry do we have the financial and political "will" to work together to stem the tide of exponential growth before it is too late for everyone? We have suggested that it is possible with the technology we currently have in PHEVs, biofuels, and renewable energy from the sun, wind and water. But all these related industries, government, society at large, and the investment community must act together, pushing in one concerted and coordinated direction for everyone's benefit. Then, perhaps, we can avoid the inevitable, sudden collapse for the next 10,000 years or so.

The key is to use the fastest possible way to electrification using renewable, non-greenhouse gas-generated energy, and the road map outlined above. Remember that the end game is not yet visible since we do not currently have the technology with the same energy density and convenience that fossil fuel offers. If we wait until it is developed and distributed worldwide it will be too late. So we must have a road map that will take us from our current social and energy infrastructure and transition seamlessly to the unknown future. The PHEV allows us this beginning. Keep in mind, however, that because we don't know the ultimate end game technology, since it has not yet been invented, that is the PHEV running on biofuels and renewable low-cost electric energy may, in fact, be the end game. It meets the worldwide objective of ZERO CO_2 in the shortest possible time frame. Time to transition to Zero CO_2 is a matter of adoption rate with a minimum time of 15 years simply because that is the average lifetime of vehicles in general. No other technology can claim this seamless and affordable transition from our current Fossil Fueled Society.

REFERENCES

Boschert, Sherry (2006). *Plug-in Hybrids: The Cars that will Recharge America*. New Society Publishers, Gabriola Island, Canada. Chapter 4, A Small Army of Engineers, pp. 68–78. ISBN 978-0-86571-571-4.

Business Council on Climate Change (2013) Electrify Your Business, Moving Forward with Electric Vehicles: A Bay Area Business Guide (http://www.bc3sfbay.org/ev-guide-for-businesses-full-length.html) Retrieved 2013-06-16.

Center for Climate and Energy Solutions (2013) Plug-in Electric Vehicle Dialogue (http://www.c2es.org/initiatives/pev) Retrieved 2013-06-16.

Electric Auto Association (2013) Why Vehicles Need to use Electricity (http://www.eaa-phev.org/wiki/Why_Vehicles_Need_to_use_Electricity) Retrieved 2013-06-16.

Frank, Andrew A. (2007) Plug-in Hybrid Vehicles for a Sustainable Future. *American Scientist* 95 (2), 158 - 165 (http://www.its.ucdavis.edu/?page_id=10063&pub_id=1875)

Gouvernement du Québec (2013) Running on Green Power! Electric vehicles: 2011-2020 Québec Action Plan (http://www.vehiculeselectriques.gouv.qc.ca/english/pdf/action-plan.pdf) Retrieved 2013-06-16.

Humes, Edward (2009) *Eco Barons, The New Heroes of Environmental Activism*, Chapter 11 Andy Frank and the Power of the Plug, pp 221-244. ISBN 978-0-06-135030-6.

McManus, Reed (2013) Rise of the Plug-In Hybrids - Sierra looks at electric vehicles that go the extra mile. *Sierra Magazine* - July/August 2013 issue. (http://www.sierraclub.org/sierra/201307/electric-vehicle-buyers-guide.aspx) Retrieved 2013-06-16.

Peak Moment Television (2008) Episode 107: Plug-in Hybrids Power the Grid. (http://www.wordpress.peakmoment.tv/conversations/?p=142) Retrieved 2013-06-16.

Renewable Northwest Project (2013) Advocating Clean Power Generation for the Next Generation. (http://www.rnp.org/) Retrieved 2013-06-16.

Romm, Joseph and Andrew A. Frank (2006) Hybrid Vehicles Gain Traction. *Scientific American* 294 (4), 72 - 79 (http://www.its.ucdavis.edu/?page_id=10063&pub_id=1876)

Taylor, Kelly (2012) Plug-in Hybrids Are the Future; Not Hydrogen, Not Electric. (http://www.energytribune.com/66493/plug-in-hybrids-are-the-future-not-hydrogen-not-electric-2) Retrieved 2013-06-16.

Watson, Grant and Andrew A. Frank (2012) Electrification of Taxi Cabs in Major Chinese Cities with Range Extended Electric Vehicles. Institute of Transportation Studies, University of California, Davis, Research Report UCD-ITS-RR-12-39 (http://www.its.ucdavis.edu/?page_id=10063&pub_id=1885)

Watson, Sterling and Andrew A. Frank (2012) A Fifteen Year Roadmap toward Complete Energy Sustainability. Institute of Transportation Studies, University of California, Davis, Research Report UCD-ITS-RR-12-35 (http://www.its.ucdavis.edu/?page_id=10063&pub_id=1848)

WGBH-Radio/TV Educational Foundation (2008) Car of the Future. (http://www.pbs.org/wgbh/nova/tech/car-of-the-future.html) Retrieved 2013-06-16.

WGBH-Radio/TV Educational Foundation (2011) Cars That Power the Grid. (http://www.pbs.org/wgbh/nova/tech/cars-power-grid.html) Retrieved 2013-06-16.

In: Grid Electrified Vehicles
Editor: Carla Alexandra Monteiro da Silva

ISBN: 978-1-62808-839-7

Chapter 2

Design Considerations for Vehicle Electronic Control Systems: Technical and Commercial Aspects

Alexander Hanzlik*[*] *and Erwin Kristen
Austrian Institute of Technology, Vienna, Austria

Abstract

This chapter deals with the design of distributed electronic control systems for automotive applications. Due to the ever increasing complexity of automotive electronic control systems and the increasing demands for functional safety for road vehicles, innovative methodologies for system design in this area are needed to be able to bring systems to the market at reasonable costs in reasonable time. We introduce a simulation based design methodology developed at the Austrian Institute of Technology AIT that has been used for the development of an electronic control system for an electric city car. This has been done in the course of the European Union ARTEMIS research project POLLUX, which is related to the design of embedded systems for the next generation of electric vehicles. For illustration of the proposed design methodology, we present an extensive case study, namely the design and performance analysis of a control system for an electric city car. Further, we introduce the notion of a vehicle control network standard as a possible future trend in the development of vehicle control systems. This idea is inspired by the ongoing efforts to establish industry standards for different automotive application domains, like standards for network communication protocols, software architectures, and, more recently, standards for automotive development processes and requirements for validation and confirmation measures to establish functional safety of the vehicle. Finally, we analyze simulation based design approaches for automotive applications with regard to technical aspects, commercial aspects, quality aspects and aspects of functional safety.

Keywords: Simulation Based Design, Automotive Communication Networks, Network Performance Analysis, Functional Safety, Vehicle Network Standard, Electric Vehicles

---

[*] Tech Gate Vienna Donau-City Straße 1, A-1220 Vienna. Email: alexander.hanzlik.fl@ait.ac.at.

INTRODUCTION

Overcoming the problem of steadily increasing system complexity is an engineering challenge today and it can be expected that it will remain to be a challenge in the future. This is especially true for vehicle electronic control systems. The trend in the development of such systems shows that mechanical and hydraulic components are more and more replaced by electronic components. Substantial evidence for this evolution are the emerging brake-by-wire and steer-by-wire technologies [1] that aim on the mechanical de-coupling of driver actions and physical reactions of the vehicle. By-wire technologies are already state-of-the-art in avionics, where control systems have to fulfill stringent requirements to ensure passenger protection. Introducing these technologies in vehicle control systems will not only increase system complexity, but will also increase the demands on functional safety for vehicles. The recently released norm ISO 26262 "Road vehicles - Functional safety" [2] defines, among others, functional safety aspects of the entire development process and provides requirements for validation and confirmation measures. This norm has been released at the end of the year 2011 and it can be expected to become a de-facto standard for the automotive industry.

On the one hand, steadily increasing system complexity and safety requirements raise the need for regulations and standards for the development process of safety-relevant systems. Such standards are already well established in avionics, like the DO-178B [3] released in 1992 that deals with the safety of software used in airborne systems.

On the other hand, development time is costly and the automotive industry operates in a very competitive market. Moreover, it is a market of mass production. Products are released in large quantities, and car manufacturers and suppliers are very ambitious to keep the costs per piece as low as possible.

In this area of tension between system complexity, quality considerations, safety requirements and development time constraints it may become hard to maintain product quality and safety at affordable costs unless there are appropriate methods at hand that support system design processes. Although good engineering skills, practice and know-how will always be the most important factors, it seems reasonable to support engineers with methods and tools that facilitate the design process and that help to avoid costly re-design activities caused by the detection of substantial design weaknesses in later phases of system development.

The challenge in the development process of vehicle control systems is to find an optimal design that fulfills current functional and safety requirements and that is open for future extensions in the course of the implementation of new functionalities. "Optimal" means that the design fulfills all requirements and is as complex as necessary and as simple as possible. Reducing the complexity of control systems facilitates thorough analysis and improves testability which in turn reduces development time and implementation efforts. A clear and mean design will also lead to reduced manufacturing costs of the final hardware implementation which is a meaningful aspect in a market of mass production.

Simulation and Model Based Design [4] are appropriate methods for identification of an optimal control system design because they allow comparing different designs in short time. Usually, the creation of variants of control systems in a simulation environment is significantly less time-consuming than the creation of such variants in hardware. While being able to build up only few alternative designs in hardware implementations, the number of variants that can be created in a simulation environment is higher. This in turn increases the probability that an optimal and cost-effective design can be identified in a simulation environment, because there are more alternatives available whose advantages and disadvantages can be compared and weighted up against each other.

The early validation of system design concepts is a substantial benefit of simulation based design. The assessment and validation of design concepts in an early development phase reduces the probability of revealing substantial design weaknesses in later phases.

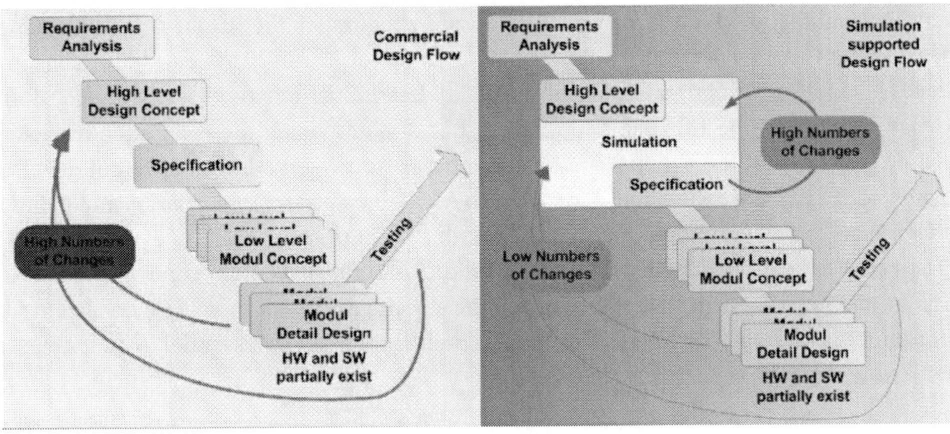

Figure 1. Conventional Design vs. Simulation Based Design.

Figure 1 (right hand side) illustrates the advantages of a simulation based design approach. The design steps *High Level Design Concept* and *Specification* are done in the simulation environment. Additional effort for the simulation requires more time for these design steps, but ideally most of the necessary changes are done in this early design phase where the costs per change are low. In a conventional design process (Figure 1, left hand side), the costs per change are higher because possible requirement violations are detected in the test and integration phase where hardware and software have already been implemented. It is obvious that it is less expensive to implement changes in a simulation environment than in an already implemented hardware and software environment. This fact reduces the costs per change in a simulation based design process, because many changes are done in an early development phase, thus minimizing the demands for costly re-design activities in the test and integration phase.

However, the detection of substantial design weaknesses in a late development phase cannot be completely ruled out by a simulation based design approach: since a simulation model is always an abstraction of reality that focuses on system properties that are considered to be essential, it does not include all details of the final implementation. If one or more essential properties of the system are not considered in the simulation model, the outputs of the simulation experiments are not meaningful. Since the hardware implementation is based on this simulation model, at the latest the validation and test phase using the hardware system will reveal potential design deficiencies and may lead to a re-design of the system. Therefore, the model coverage[1] is an essential factor for the usefulness of the simulation model for the final hardware implementation.

Another factor are the requirements that are the starting point for the design of the system. If these requirements are not complete, clear and consistent, the simulation model may have a poor model coverage, resulting in impracticable simulation results.

An essential precondition for getting the best out of simulation based design is the determinism[2] of the target system. By establishment of well-defined rules of communication and proper definition of range tolerances for timing behavior and values, the system behavior becomes deterministic which in turn facilitates and encourages the use of simulation models for development and analysis of such systems. Simulation results are very significant in the presence of a clear design methodology and a deterministic target system with defined properties that have to be fulfilled and that can be validated using the simulation model. The simulation results are less significant if only best-effort methods on different levels are implemented. In this case the simulation may cover a given subset of scenarios and behaviors that the system shows at runtime, but not all possible scenarios.

The next section starts with an outline of the evolution of vehicle control systems from the past to the present. After that, we focus on the structure of modern vehicle control systems and introduce a system model on the base of which we study essential components and properties of such systems.

VEHICLE ELECTRONIC CONTROL SYSTEMS

In the beginnings of automotive engineering, electronic control systems were mainly based on analogue technique. Mechanic and hydraulic components played a dominant role in the implementation of car functionalities. Consequently, the number of electronic components, in the range of about 100 these days, was small. Wiring of these components was based on point-to-point connections.

Figure 2 shows a typical early automotive electronic control system. In the meantime, the world has become digital. Mechanic and hydraulic parts have more and more been replaced by electronic components. As a consequence, the number of electronic components has dramatically increased and is now in the range of thousands of components in a modern premium car.

The increasing number of electronic components not only dramatically increased system complexity, but also required new wiring methods, like bus systems [5]. Modern automotive control systems are distributed computer systems, divided into different segments that implement different fields of car functions, like the power train, the body electronics or the infotainment system. The structure of a typical modern vehicle control system is shown in Figure 3.

System complexity can be expected to further increase in the future. Evidence for this evolution are the emerging X by wire technologies, which will not only further increase

---

[1] Model coverage in this context is the probability that the behavior of a model of a given system is consistent with the behavior of that system.

[2] Determinism of a system means that the system behavior is predictable by knowing a complete set of preconditions.

system complexity, but also the demands for functional safety and consequently the development effort for electronic control systems.

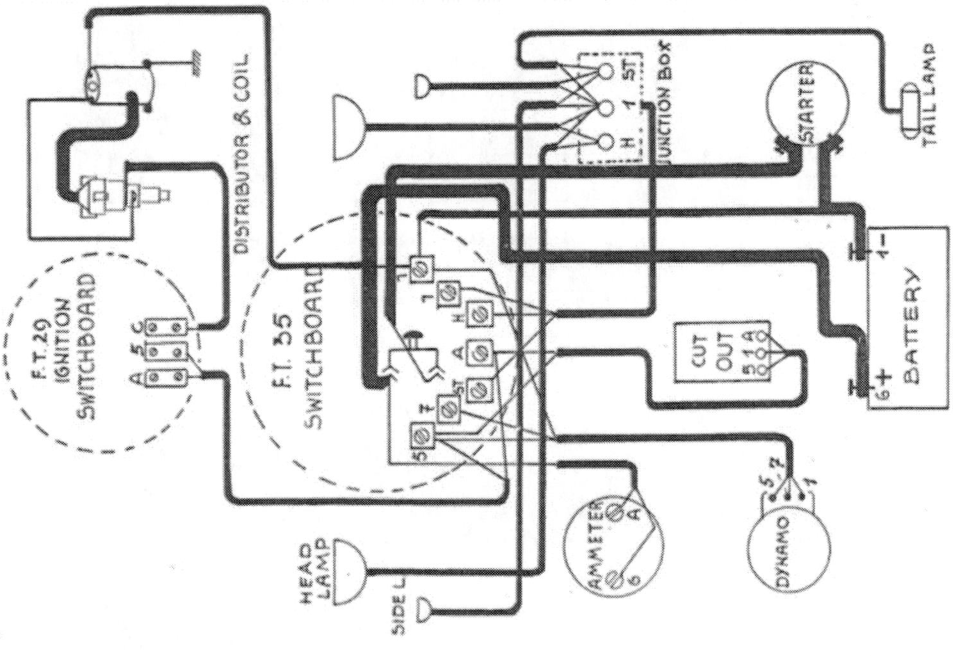

Figure 2. Early Vehicle Control System (1935)[3].

In the following, we present a system model for vehicle control systems. We will use this system model for the further considerations in this chapter. Due to the complexity and variety of the topic, this system model cannot cover every system component in full detail. We will focus on the aspects that are necessary for the considerations in this chapter, especially with regard to the electric city car design and performance analysis case study. A comprehensive approach to the subject of automotive control systems is given in [6].

System Model

A modern vehicle control system is a distributed system [7], built from spatially separated electronic control units (ECUs) interconnected via a shared communication network, like a bus system [5]. ECUs are embedded systems that control one or more of the electrical or mechanical systems of the vehicle. A vehicle electronic control system can be built by repetitive use of the following components:

ECU. An Electronic Control Unit ECU is an embedded system that controls one or more of the electrical systems or subsystems in a vehicle. An ECU receives state signals from sensors and/or other ECUs and issues control signals to actuators and/or other ECUs.

[3] Scanned from the antiquarian book "Werkstatt und Praxis des Auto-Elektrikers. Von Willy Popp. Verlag C.A./WELLER, Berlin, 1935."

Different ECUs are assigned different control tasks, like engine control, damping control or transmission control.

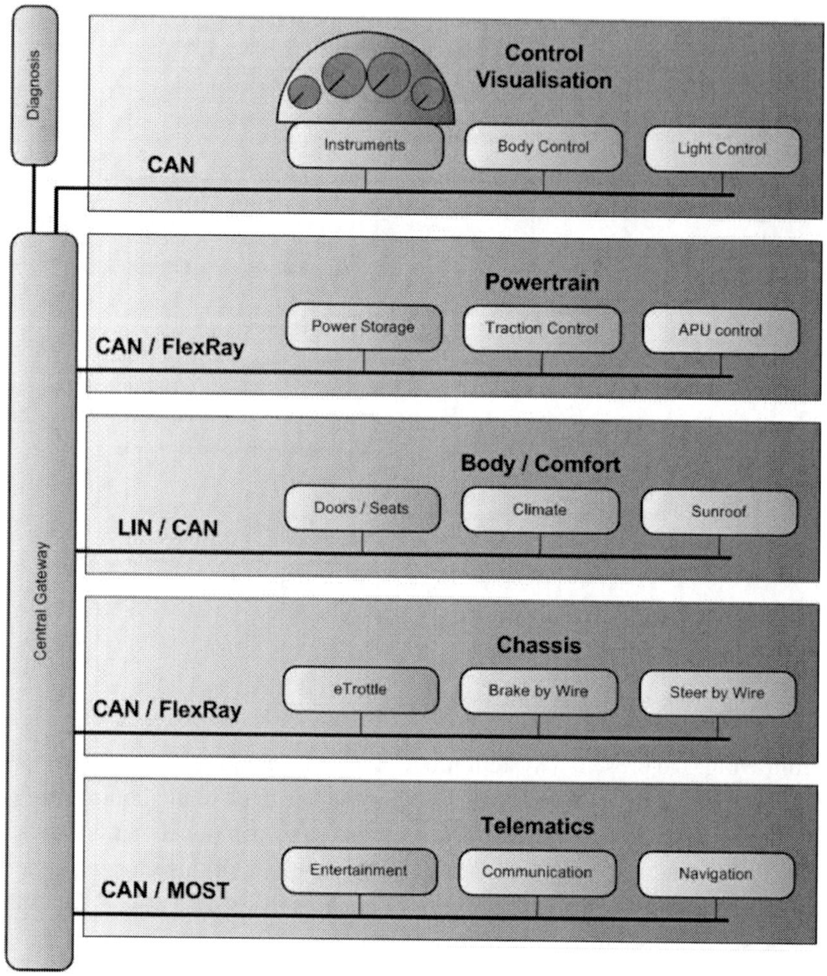

Figure 3. Modern Vehicle Control System.

Communication network. A shared communication network interconnects different ECUs. For communication, dedicated communication protocols are used, like CAN [9] FlexRay [9] LIN [10] or MOST [11] *Segment.* ECUs are grouped into segments, which are sets of ECUs that execute a distributed control task concurrently. The ECUs in a segment are linked via a shared communication network. Typical segments in a vehicle are for instance the power train, the chassis or the body segment.

Gateway. A gateway links different segments and allows for communication between these segments. Further, gateways are used for gathering status information from each segment for diagnosis and maintenance.

Sensor and Actuator. These are the interfaces between the vehicle control system and attached physical devices, like the vehicle brake or the steering wheel. Sensors are used to deliver signals for further computation, like the brake pedal position or the current steering wheel angle. Actuators are used to control mechanical or electrical components, like the brakes or the motor rotational speed.

Figure 3 shows a vehicle control system comprising the five segments Control and Visualization, Powertrain, Chassis, Body/Comfort, Chassis and Telematics. A central gateway is provided for communication between the five segments and the gathering of diagnostic information for vehicle maintenance.

Electronic Control Units

Electronic Control Units ECUs are embedded systems that control one or more of the electrical systems or subsystems in a vehicle. Figure 4 shows a schematic view of a typical ECU structure containing a *host CPU*[4] and a *communication controller.*

Host CPU. The host CPU contains the *host application* and the *controller network interface*. The host application implements the functionality of the ECU. A typical example for a host application is the software for a driving dynamics controller. The controller network interface links the host CPU to the communication controller. Its task is to hide hardware related specific features of the communication controller from the host application which allows to incorporate communication controllers from different vendors without impacts on the host application software. A typical example for a controller network interface is AUTOSAR [12].

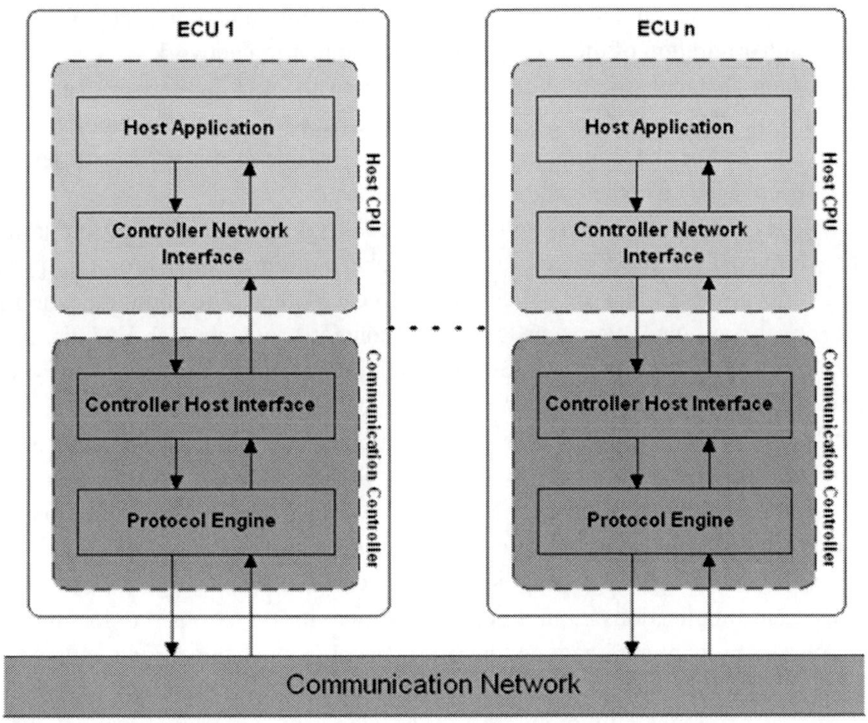

Figure 4. Schematic Vehicle Control System Structure.

[4] CPU: central processing unit.

AUTOSAR[5] is an open and standardized automotive software architecture, jointly developed by automobile manufacturers, suppliers and tool developers. The AUTOSAR consortium provides standard specification that define the interaction between host application and communication controllers via standardized interfaces. To be AUTOSAR compliant, communication controller manufacturers have to deliver the implementation of standardized device drivers accessing the communication controller hardware. For the vendor specific aspects are hidden from the application software components by the standardized AUTOSAR interfaces, an existing AUTOSAR-compliant application software will work with every AUTOSAR-compliant communication controller.

Communication Controller. The communication controller links the host CPU to the communication network. It contains a *controller host interface* and a *protocol engine*. The controller host interface links the host CPU to the communication controller. Its task is to pass commands from the host CPU to the protocol engine and to pass events from the protocol engine to the host CPU. The protocol engine manages the access to the communication network by execution of a dedicated communication protocol.

Communication Network

The control system communication network is a communication resource shared among a set of ECUs in a segment. To coordinate communiation between the different ECUs, communication protocols are used to determine the rules of communication between the the different ECUs. Among others, essential tasks of a communication protocol are message transmission and arbitration of the access to the communication network.

An important characteristic of a communication protocol is the network access strategy. Message collisions (at least two ECUs send a message at the same time) invalidate all messages involved. For the communication network is shared among the different ECUs, message collisions have to be detected or avoided.

Time-triggered communication protocols. In time-triggered communication protocols, all actions are derived from the progression of a global notion of time. For this purpose, all ECUs contain local clocks that are synchronized to each other. The communication protocol contains a global communication schedule in the context of which each ECU is assigned an interval of time during which it is allowed to send a message. For these intervals do not overlap, message collisions do not occur.

This communication medium access strategy implies a deterministic behavior with regard to message transmission durations. Among other benefits, time-triggered communication protocols guarantee a-priori known maximum message transmission durations using a collision-free access to the communication network.

Such a collision-free network access strategy is especially meaningful for safety-relevant applications. Some applications require bound maximum message transmission durations. For instance, in brake-by-wire applications, the brake pedal is mechanically de-coupled from the physical brake. The brake pedal is connected to a sensor that periodically reports the pedal position over the communication network to the brake actuator. It is obvious that the brake pedal position is a safety-relevant signal that requires a predictable and bound message

[5] AUTomotive Open System Architecture. Specification available for download at www.autosar.org.

transmission duration. If the signal gets lost or is delayed due to a message collision, the car (and the passengers) may get into trouble. Therefore, time-triggered communication protocols are usually preferred for safety-relevant applications due to their deterministic behavior.

Important representatives of time-triggered communication protocols are FlexRay [9] and TTP [13]. *Event-triggered communication protocols.* In event-triggered communication protocols, the points in time when different ECUs are allowed to access the communication network are not known a-priori. This implies that it cannot be predicted when an ECU will send a message. Although ECUs may listen on the communication network if it is free before they send messages, message collisions cannot be ruled out because there is always the possibility that at least two ECUs consider the communication network to be free at the same time. This in turn implies the possibility of message collisions that have to be detected and resolved. If a message collision is detected, the involved ECUs try to re-send their messages due to a pre-defined conflict resolution strategy.

This strategy guarantees that eventually all messages will be delivered, but the message transmission duration cannot be exactly predicted for a single message. This property of event-triggered protocols is a potential problem when deployed in safety-relevant applications, especially in the presence of real-time constraints for the transmission durations of safety-relevant signals like the brake pedal position in brake-by-wire applications.

Prominent representatives of event-triggered communication protocols are the Controller Area Network CAN protocol [8] and the Local Interconnect Network LIN protocol [14].

FlexRay

FlexRay is a time-triggered, deterministic and fault-tolerant communication protocol deployed in automotive communication networks. It was developed by the FlexRay consortium founded in 2000. Prominent members of the consortium were, among others, BMW, Daimler Corporation, Motorola, Philips, Bosch, General Motors and Volkswagen. This industry consortium promoted the standardization of a time-triggered fault-tolerant communication system for advanced automotive applications. With the completion and the delivery of the final version of the FlexRay specification the consortium disbanded in 2009. Currently, activities are in progress to integrate the FlexRay standard into the ISO[6] catalogue of norms.

In FlexRay, the communication is completely de-centralized and distributed among a set of ECUs attached to the communication network. Access to the communication network is regulated by a global communication schedule that is stored at each ECU and that assigns intervals of real-time to each ECU. During these intervals, also called *slots*, the different ECUs are allowed to send a message.

Each slot is uniquely assigned to one ECU. A message sent by one ECU is received by all other ECUs attached to the communication network (broadcast). If an ECU has nothing to send, it sends an empty message in its slot. The reception of a message in a slot indicates to the receivers that the sending ECU is operational. The absence of an expected message in a slot indicates that the corresponding ECU has become inoperative. The global communication

[6] International Standardization Organization. www.iso.org.

schedule is iterated repeatedly. One iteration of the communication schedule defines a so-called *cluster cycle*.

Such a coordinated communication mechanism, also referred to as *Time Division Multiple Access TDMA*, requires agreement among the ECUs with regard to a global notion of real-time. For this purpose, ECUs maintain local clocks that are synchronized to each other by periodic execution of a distributed clock synchronization algorithm [15].

TTP

The Time-Triggered Protocol TTP was originally developed at the Vienna University of Technology and is now in the area of responsibility of the TTTech Corporation. TTP evolved in the context of the development of the Time-Triggered Architecture TTA [16] based on the Time-Triggered Model of Computation [17], a design paradigm for time- triggered, deterministic fault-tolerant real-time communication networks. Like FlexRay, TTP belongs to the class of time-triggered communication protocols. The TTP protocol has been released by the Society of Automotive Engineers (SAE) as an open standard (SAE AS6003).

TTP is a predecessor of the FlexRay communication protocol. Many protocol operations of TTP have been implemented in the FlexRay protocol. However, TTP offers additional protocol services not implemented in FlexRay that encourage the use of TTP for safety-relevant applications. Among others, TTP offers a membership and clique avoidance service [18]. The membership service ensures that all ECUs have a consistent view on the state of the communication system, which means that each ECU knows from all other ECUs their operational state (operational or failed). The clique avoidance service shall prevent the communication network from being partitioned into disjoint subsets of ECUs, so-called *cliques*. Cliques are sets of ECUs that consider all ECUs within their clique as operational and all other ECUs as failed.

These additional features make TTP especially suitable for safety-relevant applications beyond the automotive context. Typical fields of application for TTP are aerospace applications like aeronautic propulsion unit control and cabin pressurization control. One of the latest applications of TTP is the communication platform of the power generation and cabin system for the Boeing 787 Dreamliner.

CAN

The Controller Area Network CAN is an event-triggered communication protocol for industrial applications, developed and released by Bosch Corporation in 1987. The protocol was developed for the interconnection of automotive electronic control units (ECUs). CAN is part of the ISO catalogue of norms (ISO 11898). It is extensively used in the automotive world, but is also used as a communication protocol for different industrial applications.

In CAN, the communication is de-centralized. All ECUs are allowed to start sending messages anytime given that the communication network is not occupied by a message in transit. However, the significance of messages is graded according to a static priority scheme which is part of the message conflict resolution strategy. As already mentioned, message collisions cannot be ruled out in event-triggered communication protocols. Therefore, every message contains a priority field that reflects the significance of the message. In case of a message collision, the message with the higher priority wins, which means that the ECU that has sent the message with the highest priority becomes prevalent over all other involved ECUs. The other involved ECUs retreat and try to re-send at a later point in time. If a message conflict happens again among the remaining ECUs, the same mechanism is repeated iteratively until all conflicts are resolved. The priority is an individual characteristic of the ECU. This mechanism implicitly establishes some kind of hierarchy among the ECUs attached to the communication network.

Regardless of the priority of a sender, the send process of messages already in transit is not suspended. That means that if a message is in transit and an ECU is willing to send, that this ECU is delayed with the remaining message transmission duration, even if this ECU has the highest priority assigned.

Depending on the load scenario of the communication network, this network access strategy potentially induces remarkable variations in the transmission duration of single messages in event-triggered communication protocols. This makes these protocols rather inappropriate for safety-relevant applications with stringent timing constraints. Extensions to the CAN protocol solve this problem, e.g. the Time-Triggered CAN [19] approach that allows for real-time control in safety-relevant automotive applications.

LIN

The Local Interconnect Network LIN is an event-triggered communication protocol developed for cost-effective communication between intelligent[7] sensors and actuators in cars. Typical LIN applications are the interconnection of sensors and actuators within a car door (car locking/unlocking), power windows control or car seat adjustment. It is a de-facto standard used for basic and simple automotive applications.

In LIN communication networks the communication is centralized. Communication is triggered by a master that requests data from up to 15 slaves. A communication schedule containing the temporal order of messages required is located at the master. According to this schedule, the master sends commands to and/or requests data from the slaves in the order defined in the communication schedule by sending a message that contains, among others, the address of the slave and the command or data requested. The addressed slaves execute the requested command and answer by sending a message that contains the requested data back to the master.

The centralization of the communication guarantees that no message collisions can occur unless more than one slave is addressed at the same time by the master. This in turn allows for stable and predictable message transmission durations. The obvious weakness of such a communication mechanism is that the master is a single point of failure. If the master fails, the whole application fails.

[7] Intelligent sensors and actuators provide some pre-processing, e.g. an analog-to-digital conversion of the input signal, before transmission.

MOST

The MOST-bus is a serial bus system for the transmission of audio, video, voice and data, using a ring topology[8]. MOST is used for automotive multimedia applications and is based on synchronous data communication. It is used by almost all prominent car manufacturers, like Audi, BMW, Hyundai, Jaguar, Land Rover, Mercedes-Benz, Porsche, Toyota, Volkswagen and Volvo.

A MOST network contains up to 64 MOST devices. MOST supports plug-and-play, thus facilitating adding and removing MOST devices without the need to re-configure the communication network.

Control System Topologies

While the communication network determines the communication between the different ECUs in a network segment, the control system topology determines the communication between the different network segments of the control system. Such a communication is necessary for distributed control tasks that involve different segments. Further, for driver information and diagnostic purposes, information from different network segments has to be retrieved and stored. Two types of communication system topologies are currently used in automotive applications.

Central gateway. In a central gateway control architecture (see Figure 3), the central gateway is the communication center of the vehicle control network. It is connected via separated links to each network segment and to external components, namely the on-board diagnosis unit OBD and the dashboard. The On-Board Diagnostics (OBD) connector gains access to all ECUs for vehicle inspection and maintenance. The dashboard collects all data needed to inform the driver of the status of the vehicle.

The central gateway is a very complex ECU: for each network segment, the central gateway is part of the communication network. That means that the central gateway has to integrate different communication controllers, because different network segments may use different communication networks with different communication protocols. Moreover, the performance of the host CPU of the central gateway is of essential importance, especially in safety-relevant environments with real-time constraints.

Communication is completely centralized: communication across different network segments and between external components and network segments is only possible via the central gateway. The disadvantage of such an approach is apparent: if the central gateway crashes, communication between the control system components is impossible.

Central backbone. In a central backbone architecture (see Figure 5), each network segment system has a dedicated gateway that provides access to the different ECUs in its segment. These gateways are attached to a backbone communication network, together with external components, like the on-board diagnosis unit OBD and the dashboard.

The segment gateways integrate two communication controllers, one for the backbone communication network and one for the segment communication network. If a segment

[8] In ring topologies, the network partners are connected via point-to-point connections such that each communication partner is linked with its two „neighbours“, forming a ring.

gateway fails, only the communication with the associated segment fails. Moreover, the performance of each segment gateway is less critical than that of a central gateway, since the segment gateways only have to manage the communication between the segment ECUs and the backbone communication network.

Communication is completely de-centralized. There is no single point of failure that can cause a complete communication system breakdown in such a topology as long as the backbone network is operational.

Currently, automotive control systems mainly use central gateway topologies.

The next section deals with simulation based control system design with emphasis on essential properties of the communication network. For this purpose, we introduce a dedicated simulation environment for design and performance analysis of electronic control systems, the Data Time Flow Simulator.

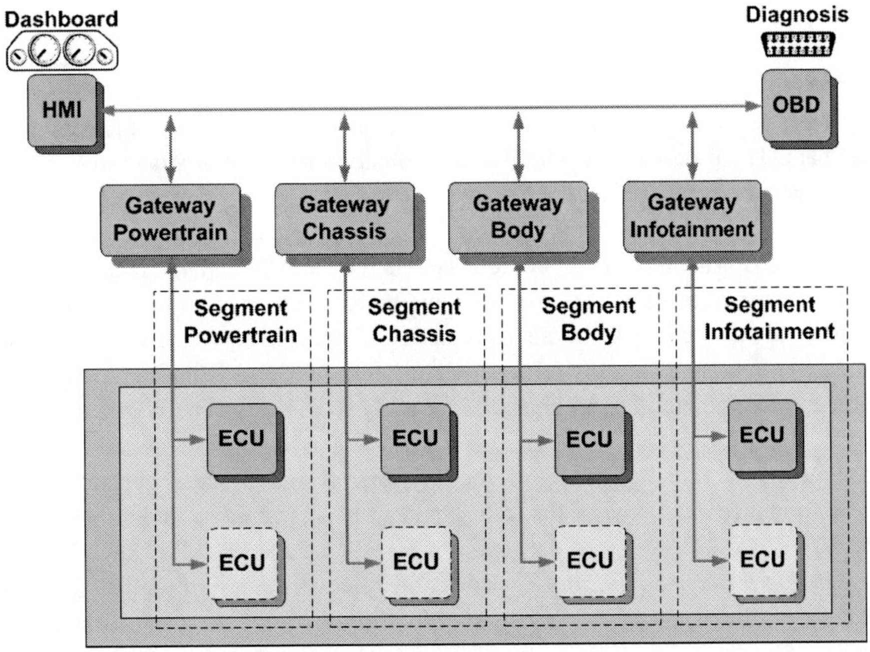

Figure 5. Central Backbone Architecture.

SIMULATION BASED CONTROL SYSTEM DESIGN

From the considerations and from the system model presented in the last section, two essential applications for simulation based system design approaches for vehicle control systems can be identified.

Application Software Design for Electronic Control Units

This topic focuses on the development of application software for electronic control units regarding implementation and validation. It is only marginally discussed in this chapter. An extensive discussion of this topic is given in [20], including a case study of a development process for a driving dynamics controller software component, that illustrates the effective benefit of a simulation based software development approach on the base of a typical practical example.

Communication Network Design

This is the main issue of this chapter. The topic focuses on the development of communication network structures and communication mechanisms regarding design, validation and performance analysis in the presence of safety requirements, legal requirements and industrial norms. The aim of a simulation based network development process is to find an optimal network architecture that fulfills current requirements and that is open for future extensions.

The design of a vehicle communication network architecture by hand is an arduous task and may be an issue of development and test for several months or even years. This considerable effort only pays if the network architecture can be used for different generations of electric cars. The product lifecycle of communication network architectures, consisting of the physical network and the network communication schedule, is expected to be at least a decade or longer. Further, the implementation of new functionalities requires additional network resources which have to be reserved in advance. Finally, since the automotive industry operates in a market of mass production, where the costs per piece are kept as low as possible, there are also cost constraints to consider. The technical challenge is the design of a cost-effective communication network architecture that fulfills existing requirements and that is open for new functionalities without the necessity to re-design the whole communication network.

Consequently, at the beginning of the network design process questions concerning network performance and network flexibility have to be answered to find an optimal network architecture.

Network performance questions. Does the network design fulfill all functional requirements of the application? What are the limits of the network design for stable operation?

Network flexibility questions. Does the network design provide room for future extensions? Does the network design provide reserves for fault scenarios?

These questions can be answered by determination of the following two characteristic properties of the control system:

Network peak load. This characteristic relates to the maximum network load observed in the different segments that constitute the control system. The observed network peak load gives information about the stability reserves and the room for future extensions of the control system.

Control signal latencies. This characteristic relates to the maximum propagation time of control signals. A control signal propagates from a sensor to an actuator, e.g. from the brake pedal sensor to the brake actuator of the vehicle. The observed control signal latencies show if the control system fulfills all timing requirements of the application.

For determination of the essential control system properties, we propose a procedure that consists of three steps:

1. Creation of a simulation model of the control system.
2. Performance analysis of the control system by determination of the essential network properties (peak load, control signal latencies) using the simulation model.
3. Optimization of the control system model to eliminate possible requirements violations and to enhance network performance.

The implementation of this process requires a simulation environment. In the following section, we introduce the DTF Data Time Flow Simulator, a simulation environment for design, validation and performance analysis of electronic control systems.

The Data Time Flow Simulator

The Data Time Flow Simulator DTF [21] is a simulation environment developed at the Austrian Institute of Technology AIT. The vision of the DTF simulator is to improve the design process of distributed electronic control systems by continuous performance analysis and assessment using virtual prototypes of the target system, integrating different physical domains during an overall system simulation. Performance analysis and assessment are based on the monitoring of action chains from sensors to actuators.

The DTF simulator is a discrete-event simulation[9] [22] environment, based on a modular assembly system that provides a set of primitive building blocks, the *elements*. Elements are grouped to form more complex structures, like ECUs. Each element has an input buffer, one output and a propagation delay. When an element receives data in its input buffer, the data is processed and, with the propagation delay defined for this element, the processed value is sent to the output.

The following elements are provided by the DTF simulator:

- **Sensor.** The sensor element is the input interface of the simulator. Its task is to process inputs to the control system.
- **Actuator.** The actuator element is the output interface of the simulator. Its task is to process outputs from the control system.
- **Processor.** The processor element processes data received from other elements, e.g. from sensors. The processor element can be provided with a user defined action routine that defines the data transformation from the input to the output.
- **Memory.** The memory element stores data received from other elements. Its task is to store data and to transmit the data unchanged with a defined propagation delay.
- **Controller.** The task of the controller element is to assemble frames for transmission over a communication network, e.g. a CAN bus.
- **Extractor.** The extractor element disassembles data received from a communication network, e.g. a CAN bus.
- **Line.** The line element is responsible for simulation of the communication network protocol. Typical tasks of the line element are message transmission and arbitration of the communication network.

[9] Discrete-event simulation is based on the modelling of system operation as a chronological sequence of events and the monitoring of the system state at the times when these events occur.

- **Transfer.** The task of the transfer element is to distribute received data unchanged and without delay to its outputs. It is the only element that has more than one output.

For the element input buffer, the following buffer types are available:

- **S_BUFFER.** Stores one value (single buffer).
- **M_BUFFER.** Stores a configurable number of values (multiple buffer).
- **FIFO.** First-in-first-out buffer.
- **RING.** Ring-buffer.

Elements are linked to form so-called *action chains*. An action chain is a directed path from a sensor to an actuator. Figure 6 shows a simple electronic control system consisting of two sensors S1 and S2, two actuators A1 and A2, ECUs A, B C and D and a communication network BUS. The ECUs and the BUS are constituted by elements from the DTF modular assembly system. For example, the ECUs contain a Processor element Proc and a Controller element CC. The control system has four action chains: one from sensor S1 to actuator A1, one from sensor S1 to actuator A2, one from sensor S2 to actuator A1 and finally one from sensor S2 to actuator A2.

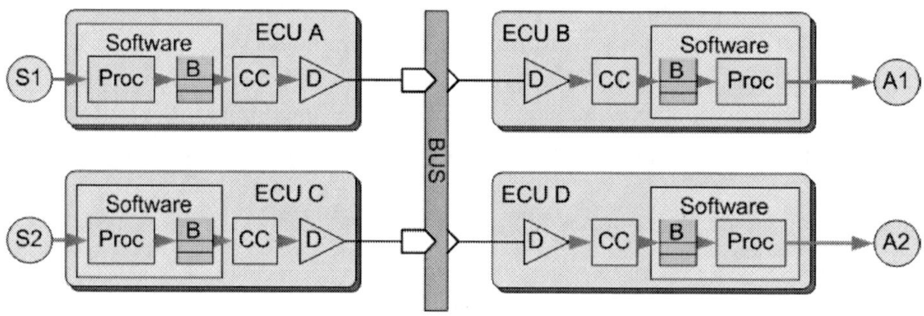

Figure 6. Simple electronic control system.

To analyze the performance of the control system, signals are issued to the sensor elements and signal propagation is observed, both in the domains of value and time, from the sensors over the communication network to the actuators. The analysis of the input buffer fill levels of each element along the action chains over time delivers additional information. If the allowed signal propagation time along an action chain is exceeded, this additional information can be useful to pinpoint one or more elements as the possible reasons.

From the signal propagation time over the communication network important information can be gained with regard to the dynamics and responsiveness of the control system, especially for safety-relevant signals that usually are subjected to real-time constraints, like e.g. the brake pedal position control signal.

Figure 7 shows the workflow for a control system performance analysis using the DTF based on a drive cycle. The drive cycle consists of a set of events that are issued to the sensor elements of the system and that contain typical driving scenarios like the brake pedal positions or the different positions of the steering wheel during simulation time.

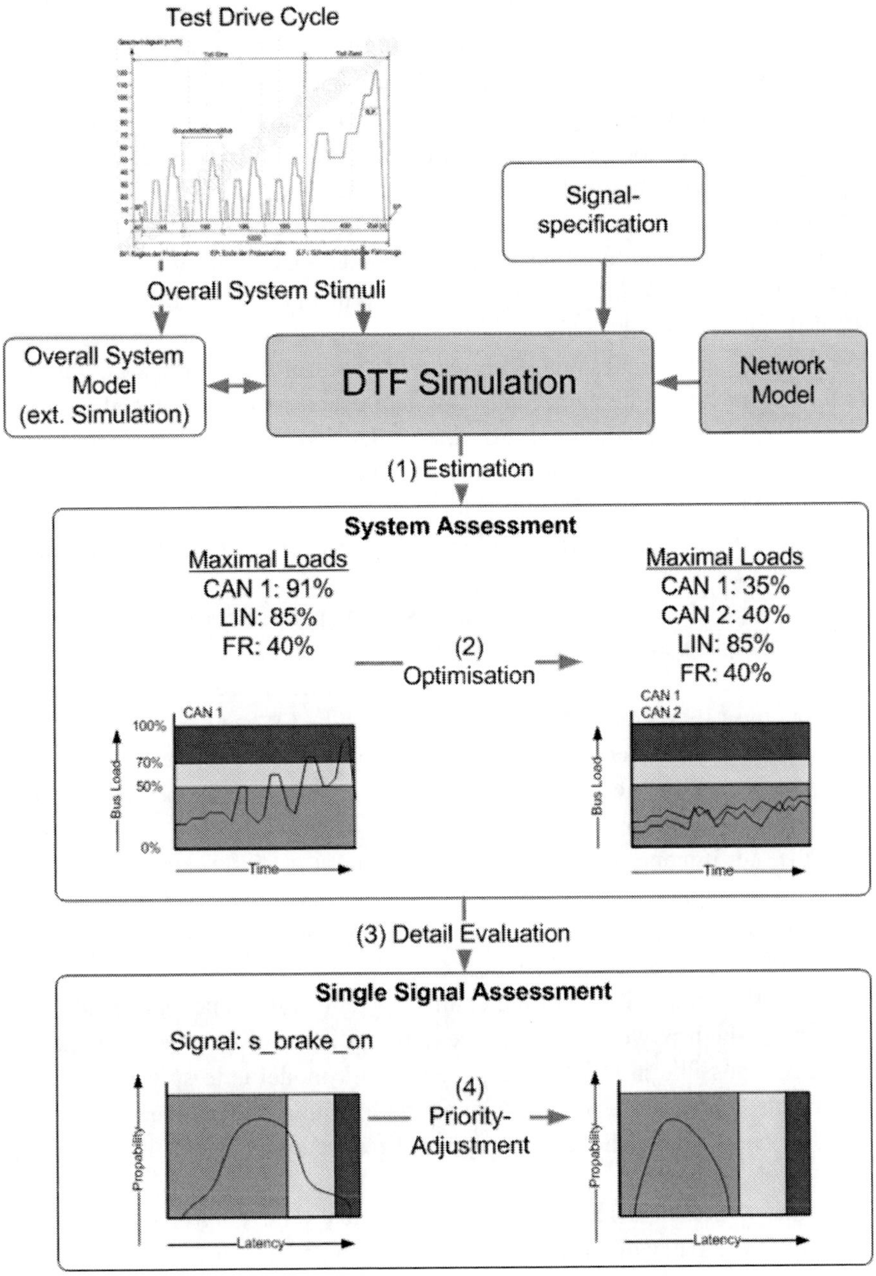

Figure 7. Control System Performance Analysis – Principle of Operation.

Using this set of events, the control system is exposed to load and the system behavior is observed. The assessment of the system performance consists of two phases:

System Assessment phase. In this phase, the maximum network loads of the different segments during the drive cycle are determined. If a given maximum network load is exceeded, the network model is modified. Exampes for possible modifications of the network model are to change the network protocol or to increase network bandwidth.

Signal Assessment phase. In this phase, the control signal latencies are determined. If a given maximum control signal latency is exceeded, the network communication schedule is modified. Examples for possible modifications of the network schedule are the adaptation of message priorities (for event-triggered communication protocols) or the modification of the assignment of communication slots (for time-triggered communication protocols).

This process is iterated until all requirements are fulfilled.

The two network characteristics used for the assessment of the system performance, namely the network load and the control signal latencies, cannot be considered independently of each other. A control system may fulfill all requirements with regard to network load, but may not fulfill all requirements for single control signal latencies. On the other hand, a control system may observe all limits with regard to control signal latencies, but may exceed the limits for the maximum network load. This fact usually requires several iterations of the performance assessment including optimization cycles and aggravates the network design by hand. A simulation based performance analysis facilitates this process, because the necessary modifications in the course of the optimization cycle can be made in the simulation model and the effects can be immediately be observed by execution of the modified simulation model.

Figure 7 shows an example for a control system performance analysis. The *Network Model* contains the physical control system structure. It consists of three segments, namely a CAN1, a LIN and a FR segment. The CAN1 segment uses the CAN protocol for communication, the LIN segment the LIN protocol and the FR segment the FlexRay protocol, respectively. The control system is stimulated using a *Test Drive Cycle* consisting of a set of inputs for the sensors of the network model. The test drive cycle is processed and the control system performance analysis is started.

To see the performance of the control system at a first glance, we introduce an assessment scheme inspired by a traffic light. *Green* means *optimal*, *yellow* means *acceptable* and *red* means *unacceptable*. We use this assessment scheme for the validation of the network load and the control signal latencies.

The System Assessment Phase of the control system shows a network load in the CAN1 segment of 91%, which is well within the red area. To obtain a lower network load in the CAN1 segment, a possible modification of the network model is to split the CAN1 segment into two segments CAN1 and CAN2. After this modification, the network load is 35% in the CAN1 segment and 40% in the CAN2 segment, respectively, which is clearly within the green area for both segments.

In the Signal Assessment Phase, we study the latency for control signal *s_brake_on*. It can be seen from the latency distribution diagram that some latencies of this control signal are in the read area.

To obtain a better signal latency, a possible modification of the network communication schedule is to raise the priority of the message that carries control signal *s_brake_on*. After this modification, the latency of control signal *s_brake_on* is clearly within the green area.

Validation

For validation of the proposed control system performance analysis method and to gain confidence in the correctness of the DTF implementation, we used a Robot Driving Platform

(RDP) developed at the AIT. This validation platform provides the mechanical system to develop control software for the power train as well as the necessary structures for network design. We used the RDP as a reference for development of a simulation model and to compare simulation results and bus measurements from the RDP to optimize the DTF model. The objective of this validation procedure was to show that the proposed development approach is suitable for the development of control systems, on the base of a practical example.

The RDP (see Figure 8, right hand side) is a single-axe driving platform where each wheel is driven by one electromotor, stabilized by a free-running support wheel at the back. If both wheels run with the same speed, the RDP drives straight ahead. To go left, the right wheel has to be accelerated (or the left wheel has to be decelerated or both). In reverse, the same applies for going right. Bigger speed differences between the two wheels result in closer curve radiuses driven by the RDP.

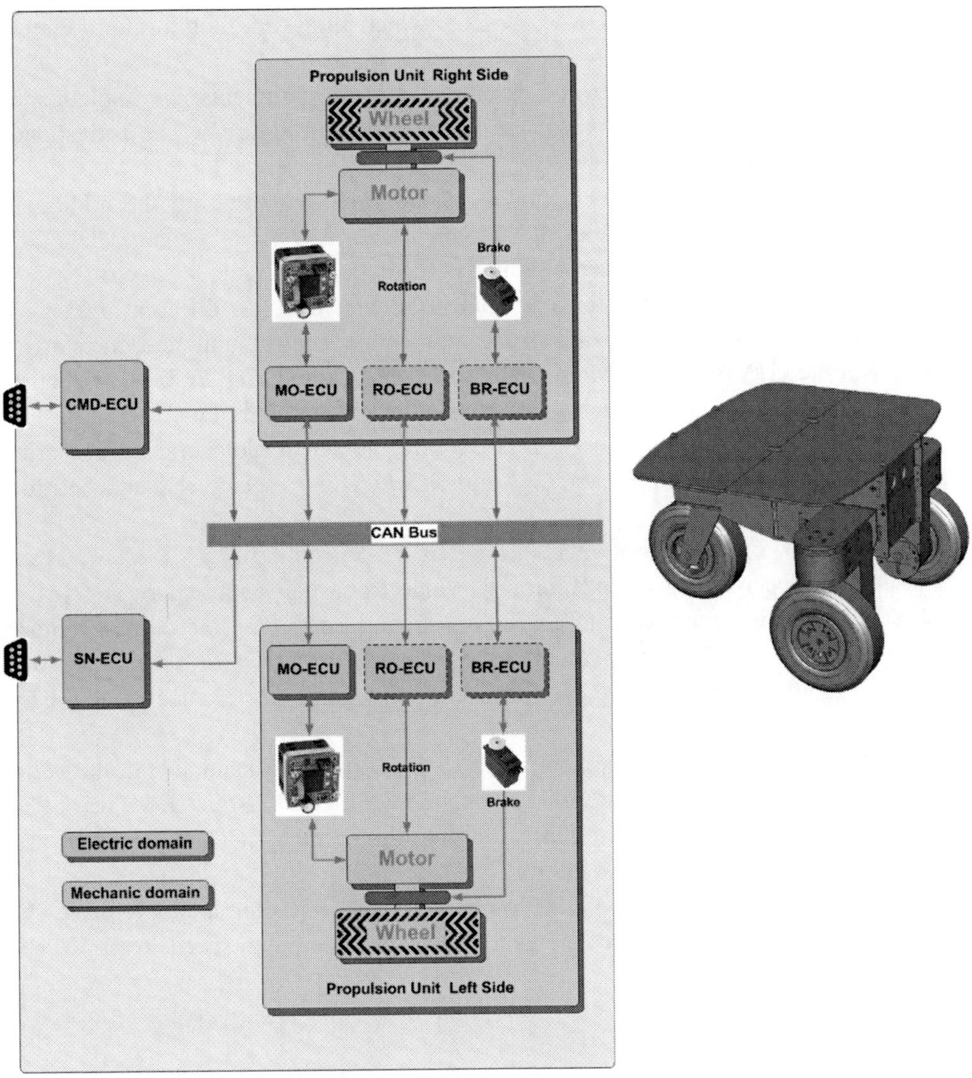

Figure 8. Robot Driving Platform Hardware Structure.

Figure 8 (left hand side) shows the hardware design layout. The RDP is directed via a remote control that issues driving commands to a Command Electronic Control Unit (*CMD-ECU*). The driving commands correspond to the desired rotational speed for each wheel. With a period of 10ms, the *CMD-ECU* samples the values received from the remote control unit, packs them into a CAN frame and sends the frame over the *CAN Bus* to the Motor ECUs (*MO-ECU*) where the frame is stored. Every 20ms, each *MO-ECU* extracts the desired speed for its associated electromotor from the received CAN frame and forwards this value to a control algorithm that calculates a control variable for the motor. Each motor output is attached to one *RO-ECU* that receives the actual value of the rotational speed from the motor. This value is packed into a CAN frame and sent over the *CAN Bus* to the corresponding *MO-ECU* where the frame is stored. After reception of the first actual value from the motor at the *MO-ECU* the control loop is closed. From now on, the control algorithm updates the motor control variable according to the nominal value and the actual value with a sampling rate of 20ms. Also shown in Figure 8 are the brake ECUs *BR-ECUs*, which control an electric brake for each wheel, as well as the sniffer ECU *SN-ECU* which allows for monitoring of the CAN bus traffic.

DTF Simulation Model of the RDP. For the validation experiments, we implemented a simulation model based on the RDP hardware layout shown in Figure 8. The corresponding DTF simulation model is shown in Figure 9.

The main components of the DTF model are the *CAN Bus* (blue, in the middle), the *CMD_ECU* (left-hand side, top) and the Motor ECUs *MO_ECU_1* and *MO_ECU_2*, respectively. The two electromotors can be found at the bottom of Figure 9 (green).

The *CAN Bus* is responsible for communcation between the different ECUs. The *CMD_ECU* receives driving commands from the remote control unit and transmits the commands over the *CAN Bus* to the Motor ECUs *MO_ECU_1* and *MO_ECU_2*, respectively. The Motor ECUs generate the control variables for their attached motors, according to the driving commands received from the *CMD_ECU* and the actual motor values that are fed back over the *CAN Bus* via the *RO_ECU_1* and *RO_ECU_2*, respectively, attached to the motor outputs.

Below the *CMD_ECU*, the two ECUs *MO_ECU_1_SET* and *MO_ECU_2_SET* can be seen. These ECUs receive the current motor set value from their counterparts *MO_ECU_1* and *MO_ECU_2*. Their only task is to send CAN frames containing the current motor set values over the *CAN Bus*. The purpose of these frames is to allow for observation of the current state of motor control over the *CAN Bus* (e.g. via a frame sniffer attached to the CAN bus like in the hardware RDP).

Finally, we provide a disturbance node *Disturb_ECU* (red, left-hand side at the bottom) that allows for intentional disturbances of the CAN network. The *Disturb_ECU* can be used to generate additional traffic on the *CAN Bus*. Further, the *Disturb_ECU* may exhibit various failure modes like e.g. a babbling idiot failure[10].

Drive Cycle Simulation. For the validation experiment, we define a simple drive cycle: from idle state, the RDP gradually accelerates to 50% of the maximum speed, keeps the speed

[10] A babbling idiot is an ECU that disturbs the communication of the other ECUs in a communication network by planless and arbitrary transmission of messages. Bus guardians [23] can be implemented to resolve the occupation of the communication network by babbling idiots.

for a while and then gradually decelerates to idleness. Driving cycle duration is about 40 seconds real-time.

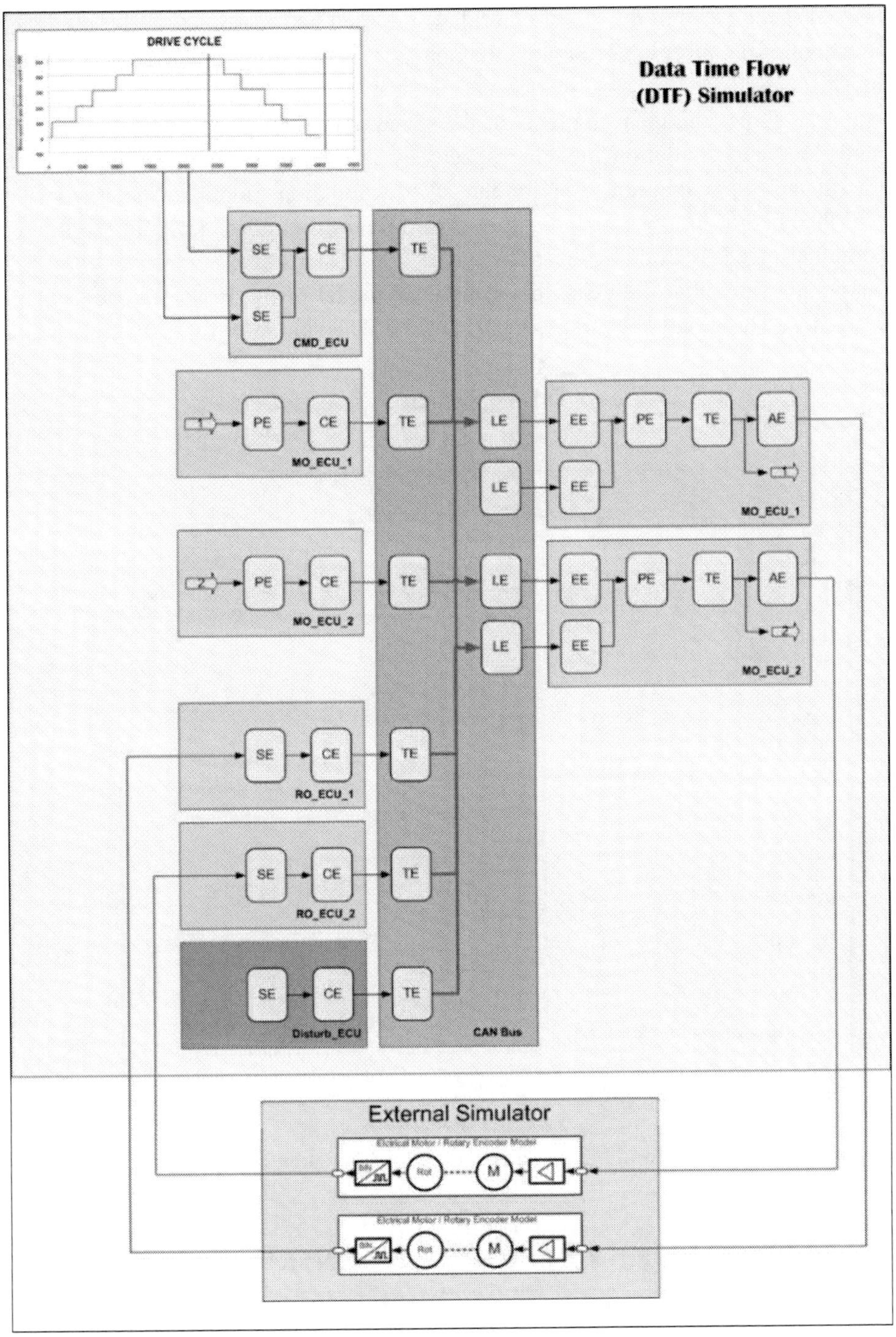

Figure 9. DTF Simulation Model of the RDP.

Each nominal speed change causes the motor to change its rotational speed. During the acceleration phase, an overshoot of the motor speed of about 20% over the nominal speed can be observed. This is caused by the motor characteristics as well as by the delayed processing of the actual speed value in the motor controller. For the same reason, a similar effect, namely an undershoot of the motor speed below the nominal speed during the deceleration phase can be observed.

Figure 10 shows the measurements of the speed during the drive cycle for the hardware RDP and the DTF simulation model, respectively. The curves (blue for the robot driving platform in the upper diagram, red for the DTF Simulation model in the lower diagram) represent the changes in the rotational speeds of the electromotors, according to the changes in the nominal speeds defined by the drive cycle.

The results show that the behavior of the DTF model of the RDP is close to that of the hardware RDP. This result encourages the use of the DTF simulator for further investigations.

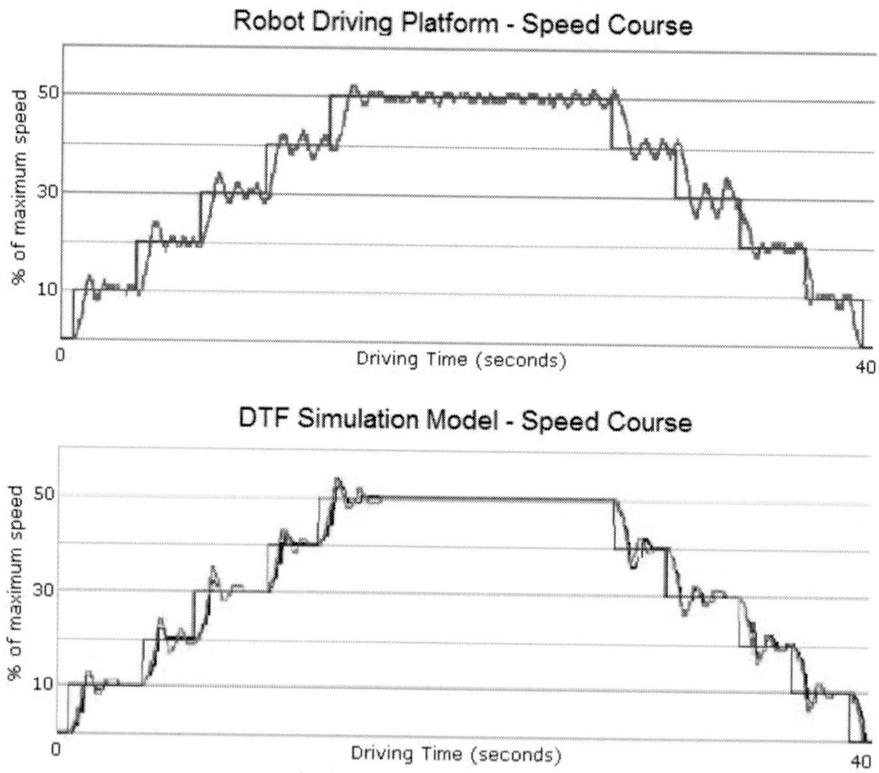

Figure 10. Robot Driving Platform Measurements vs. DTF Simulation Results.

DESIGN OF AN ELECTRONIC CONTROL SYSTEM FOR AN ELECTRIC CITY CAR

We now apply the design methodology presented in the last section on a more extensive application example, namely on the development and performance analysis of an electronic control system for an electric city car. This electronic control system has been developed in the course of the European Union ARTEMIS project POLLUX11 which was related to the development of embedded systems for the next generation of electric vehicles. The consortium was formed by 35 partners from 10 European countries, including prominent partners from the automotive industry, like Centro Ricerche Fiat S.c.p.A. and PSA PEUGEOT CITROEN. In the course of the project, different control system configuration levels for electric vehicles have been designed by the project partners for different vehicle sizes, namely configurations for a microcar, a city car and a family car.

The contribution of the AIT to the POLLUX project consisted of two consecutive steps. Step 1 was the identification of appropriate design methods for vehicle electronic control systems and the use of these methods in the implementation of a simulation environment for system design and performance analysis, the DTF simulator. Step 2 was the use of the DTF simulator for implementation and performance analysis of a concrete electric vehicle configuration level proposed by the project partners, the electric city car.

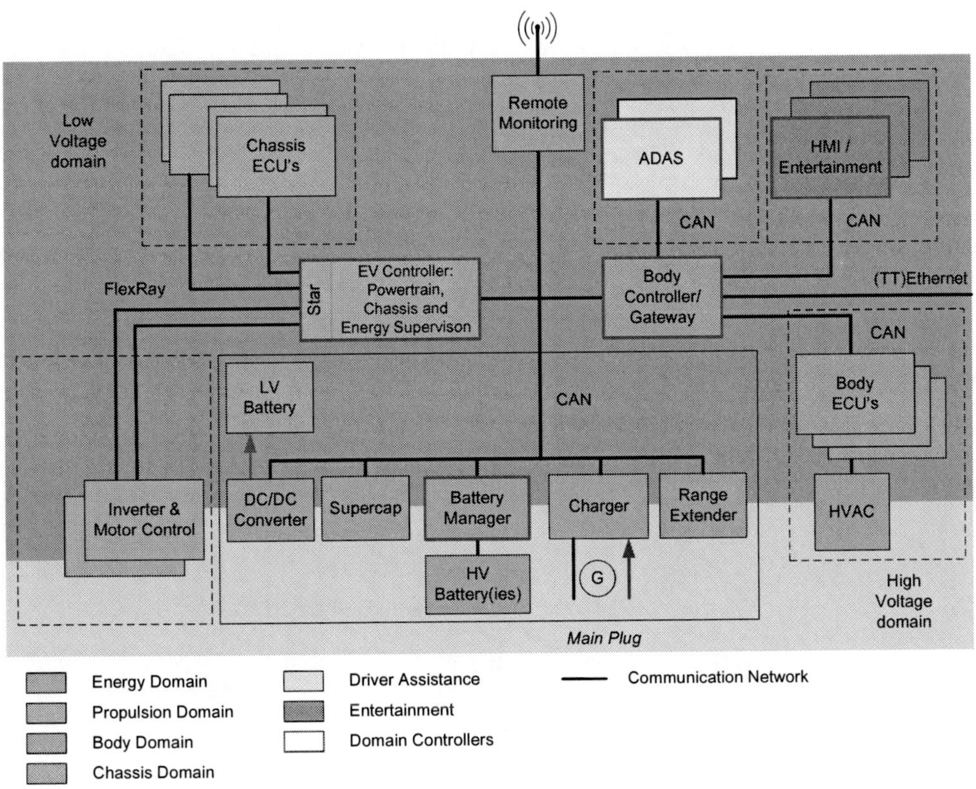

Figure 11. POLLUX Vehicle Network Architecture Proposal "City-Car".

An electric city car is a small-and-light electric vehicle, primarily intended for use in urban areas. However, it can also operate in mixed city-highway environments. Typically, it is a small vehicle for four or five passengers. The vehicle weight is about 1000kg. A fully

[11] Process Oriented Electrical Control Units for Electrical Vehicles Developed on a multi-system real-time embedded platform. POLLUX website: http://www.artemis-pollux.eu/.

charged battery[12] provides an operating range of up to 200km. The maximum speed is 130km, the peak power is 50kW.

Figure 11 shows the network architecture proposal for the electric city car. The proposed architecture consists of different *domains* that control different functional groups of the car. Typical control functions for the different domains are the provision of electric power, charging and power saving for the *Energy Domain*, propulsive power distribution and electronic braking for the *Propulsion Domain*, lighting, door electronic and mirror adjustment for the *Body Domain* and electronic damping and power steering and for the *Chassis Domain*.

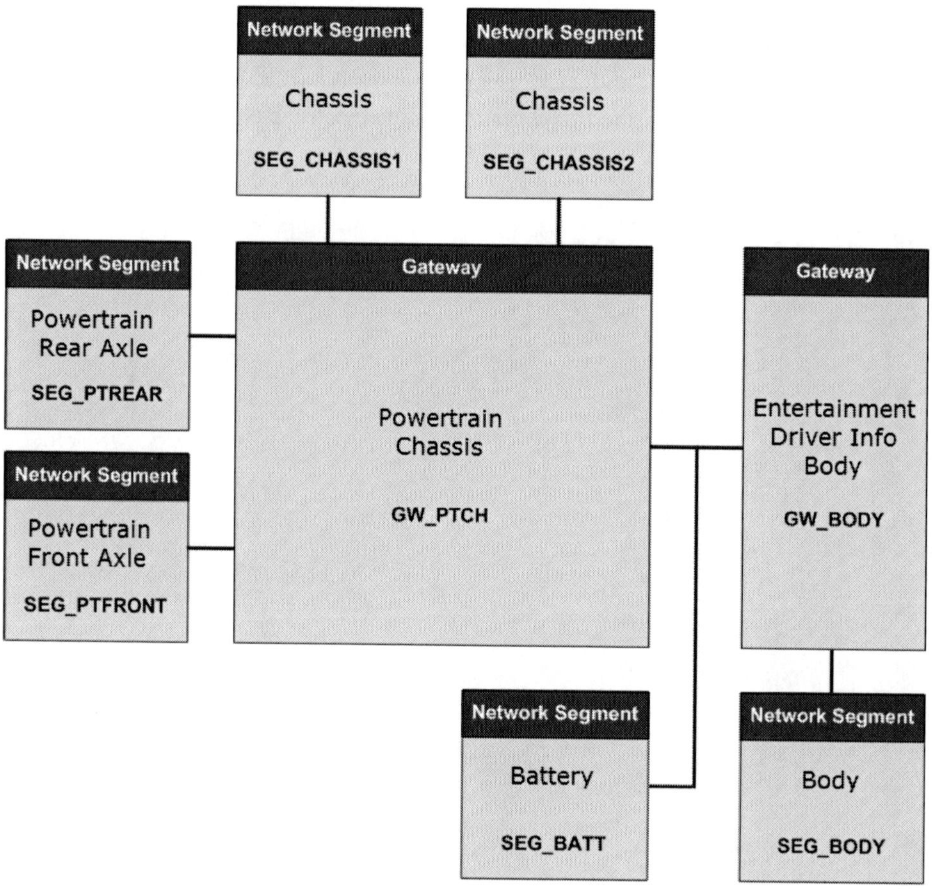

Figure 12. Block Diagram of the DTF Simulation Model.

Figure 12 shows the result of the transformation of the network architecture proposal shown in Figure 11 into a DTF model of the electric city car. The DTF model consists of the following components:

SEG_BODY	Body network segment (*Body Domain*)
SEG_BATT	Battery network segment (*Energy Domain*)
SEG_CHASSIS_1	Chassis network segment 1 (*Chassis Domain*)

[12] Typically, electric city cars use Li-Ion batteries.

SEG_CHASSIS_2	Chassis network segment 2 (*Chassis Domain*)
SEG_PT_FRONT	Front power train network segment (*Propulsion Domain*)
SEG_PT_REAR	Rear power train network segment (*Propulsion Domain*)
GW_BODY	Body / Battery gateway (*Body Controller / Gateway*)
GW_PT_CH	Power train / Chassis gateway (*Powertrain, Chassis and Energy Supervision*)

The topology of the control system corresponds to that of a Central Backbone Architecture (see page 48). The communication between the different network segments is managed by the two gateways *GW_BODY* and *GW_PTCH.* The architecture establishes a close coupling between the domains *Chassis* and *Propulsion* and between the domains *Body* and *Energy*, respectively. Consequently, the *GW_PTCH* is mainly responsible for communication between the network segments *SEG_CHASSIS_1*, *SEG_CHASSIS_2*, *SEG_PT_FRONT* and *SEG_PT_REAR*. The *GW_BODY* is mainly responsible for communication between the network segments *SEG_BODY* and *SEG_BATT.* However, to be able to transfer communication between the *Chassis/Propulsion* and the *Body/Energy* domain clusters, direct communication links between the two gateways *GW_BODY* and *GW_PTCH* are provided.

The *Driver Assistance Domain* and the *Entertainment Domain* have not been considered in the DTF simulation model, since their impact on the overall control system network performance was assumed to be minimal. This assumption is justified by the consideration that that there is a closer coupling between the domains *Body, Energy, Chassis and Propulsion*, resulting in higher communication requirements between these domains than between the *Driver Assistance* and the *Entertainment* domains and the other domains in the control system.

For the performance analysis of the control system, we use the DTF simulation model shown in Figure 13 which is a refinement of the block diagram shown in Figure 12. Figure 13 shows the six network segments, the two gateways, and the connections of the different network segments over the two gateways. Further, it shows two dedicated action paths used for the performance analysis.

Each network segment provides 10 inputs, 10 outputs and consists of a set of *TX_ECU*s, a communication bus[13] and a set of *RX_ECU*s. Each segment input is connected to one *TX_ECU.* The task of *TX_ECU*s is to receive signals, to transform the signals into values, to pack the signal values into message frames[14] and to send the message frames over the communication network to the *RX_ECU*s. The *RX_ECU*s receive the message frames from the communication network, extract the signal value, compute an output value from the signal value and send this value to the output. A network segment input is either connected to a sensor or to a gateway output, a segment output is either connected to an actuator or to a gateway input.

For the performance analysis of the control system, we consider two time-critical action paths:

[13] The propulsion domain network segments *SEG_PT_FRONT* and *SEG_PT_REAR* use the FlexRay communication protocol, all other network segments use the CAN protocol.

[14] Structure and size of the message frames are determined by the communication protocol.

- **Action Path A.** Over this action path, the vehicle speed is communicated, from the speed sensor *S6* in *SEG_PT_FRONT* to the vehicle speed indicator *A8* in *SEG_BODY*. The maximum latency for Action Path A shall be 40ms. This requirement is critical because the signal is used for the cruise control regulator circuit.
- **Action Path B.** Over this action path, the accelerator pedal position is communicated, from the accelerator pedal sensor *S9* in *SEG_CHASSIS_2* to the front left wheel actuator *A8* in *SEG_PT_FRONT*. The maximum latency for Action Path B shall be 30ms. This requirement is critical because the signal is used for the motor speed control regulator circuit.

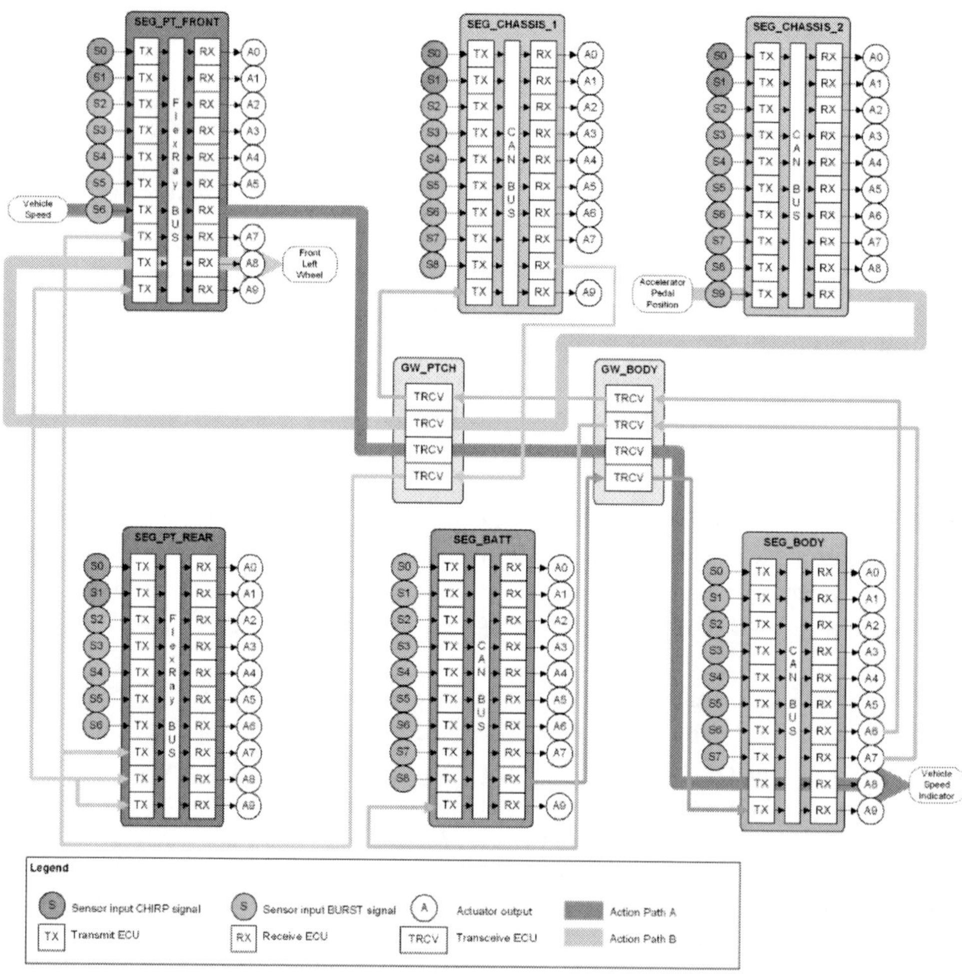

Figure 13. City Car Control System with Action Path A (blue) and Action Path B (green).

The performance analysis of the control system comprises three steps:

- In a **First Assessment**, a simulation experiment using a preliminary set of simulation model parameters is performed.

- Based on the simulation results obtained from the First Assessment, an **Optimization Cycle** takes place in the course of which the set of preliminary simulation model parameters is modified. The aim of this Optimization Cycle is to eliminate possible requirements violations and to improve control system performance with regard to network utilization.
- In a **Second Assessment**, a simulation experiment using the modified set of simulation model parameters obtained from the Optimization Cycle is performed. The simulation results obtained from the Second Assessment are compared to that of the First Assessment. The impact of the model parameters modifications with regard to improvement of the overall control system performance discussed. Further, the potential for future extensions of the control system is evaluated.

Simulation Model Parameters

In the following, the simulation model parameters used for the performance analysis of our city car control system are described. The model parameters comprise the model input stimuli for the sensors, the communication schedule for the transmission of control signals and the network bandwidth settings for the six segments. Further, the conditions that apply for the execution of the simulation experiments are specified. Finally, the requirements that shall be validated regarding communication network load and control signal latencies are defined.

Stimuli. For the stimulation of the sensors[15], we use two different input signal types:

- **Burst**
  The burst signal has a constant period of 20ms and a constant value 10.
- **Chirp**
  The chirp signal has a varying period and a constant value 10. Starting with a period of 10ms, the period is iteratively shortened by 0.1ms until a period of 5ms is reached. From this point in time, the chirp signal behaves like a burst signal with a constant period of 5ms. That means that the chirp signal periods are as follows: 10ms, 9.9ms, 9.8ms, …, 5ms.

Figure 13 shows which input stimuli are assigned to which sensors in the six segments: sensors that process chirp signals are marked red, sensors that proess burst signals are marked grey. The frequencies of the input stimuli have been derived from typical sampling rates used in automotive control engineering applications. Since the network performance analysis focuses on network load and the latencies of time-critical control signals, the signal values are only a side issue. For evaluation of the control network characteristics of interest, only the frequencies of the input stimuli are relevant.

In the different segments, all input signal types are issued at the same time to the sensors. That means, for example, that in the *SEG_CHASSIS_1* segment sensors S0 and S1 are triggered at the same points in time with a chirp signal. The same applies to the sensors S2 –

[15] The collection of input stimuli for all sensors form the drive cycle used for the simulation experiments.

S8 of the *SEG_CHASSIS_1* segment, which are triggered at the same points in time with a burst signal. Since the execution times of all *TX_ECU*s in a network segment are the same, all "chirp" *TX_ECU*s generate message frames at the same points in time. The same applies to all "burst" *TX_ECU*s. The motivation for the simultaneous stimulation of sensor inputs is to exhaust the arbitration mechanisms of the network communication protocol, since the communication network arbitration has an effect on the transmission times of message frames and therefore on the latencies of control signals that are transported by message frames involved in a collision on the communication network. Although it seems to be unlikely that all senders in a communication network are willing to send at exactly the same points in time, the occurrence of this situation, however improbable, cannot be ruled out.

In the segments that use the CAN communication protocol (*SEG_BODY*, *SEG_BATT*, *SEG_CHASSIS_1* and *SEG_CHASSIS_2*), this simultaneous generation of message frames implicitly establishes an ordered sequence of message transmission times. The CAN protocol uses a static priority scheme to resolve transmission conflicts where each sender is assigned a priority. This message priority is included in CAN message frame as the message identifier. In case of a frame collision, the frame with the lowest message identifier is transmitted. The other senders have to wait for the end of the frame transmission and try to send their messages afterwards.

This process is iteratively repeated until all message collisions are resolved. Since in our setup all *TX_ECU*s try to send frames at the same time, the frame transmission times are increasing from the sender with the lowest message identifier (highest priority) to the sender with the highest message identifier (lowest priority). This is the worst-case scenario regarding the frame transmission time for each sender: except the sender with the highest priority that always wins in case of a frame collision, each sender has to wait until all senders with higher message priorites have transmitted their frames. The sender with the lowest priority always encounters the longest frame transmission time.

This effect is extensively discussed in [24]. Among others, it is shown that providing appropriate offsets between the frame send times to avoid collisions remarkably improves the frame transmission times for lower-priority senders. However, to guarantee that these offsets are observed by all senders, agreement between the senders about the frame send times has to be established by additional synchronization mechanisms, like the execution of an agreement protocol [25].

In the segments that use the FlexRay communication protocol (*SEG_PT_FRONT* and *SEG_PT_REAR*), the simultaneous generation of message frames has no impact on the message transmission times and therefore the control signal latencies, because each *TX_ECU* has its own sending slot in the communication schedule. This ensures the avoidance of message frame collisions on the FlexRay communication network and a constant transmission duration for all message frames. Therefore, the control signal latencies are independent from the current network load in the FlexRay segments.

Communication schedule. These simulation model parameters define the assignment of message priorities for the CAN network segments and the assignment of communication slots for the FlexRay network segments. In the different network segments, the sensor input signals are packed into message frames by the *TX_ECU*s. In the CAN network segments, the frames are sent on the CAN bus containing the message identifier corresponding to the assigned message priority. In the FlexRay network segments, the frames are sent on the FlexRay bus in the assigned communication slot. For example, the signal from sensor S0 in segment

SEG_BODY is transmitted on the CAN bus with priority 10, which is the highest message priority in this network segment.

The signal from sensor S1 is transmitted on the FlexRay bus in slot 1, which is the second communication slot in this segment. Table 1 shows the communication schedules for the six network segments. The grey shaded cells in Table 1 indicate that the corresponding segment inputs are connected to gateway outputs. For instance, the input 7, 8 and 9 of the *SEG_PT_REAR* segment are connected to the *GW_PTCH*, all other inputs are connected to sensors.

Table 1. Network Segment Communication Schedules

Segment Input	*SEG_ BODY*	*SEG_ BATT*	*SEG_ CHASSIS_1*	*SEG_ CHASSIS_2*	*SEG_ PT_FRONT*	*SEG_ PT_REAR*
S0	10	10	10	10	0	0
S1	20	20	20	20	1	1
S2	30	30	30	30	2	2
S3	40	40	40	40	3	3
S4	50	50	50	50	4	4
S5	60	60	60	60	5	5
S6	70	70	70	70	6	6
S7/GW	80	80	80	80	7	7
S8/GW	90	90	90	90	8	8
S9/GW	100	100	100	100	9	9

Network Bandwidth Settings. The remaining simulation model parameter defines the network bandwidth settings. Table 2 summarizes the settings for communication protocol and bandwidth used in the six network segments.

Table 2. Network Bandwidth Settings

Segment	*Protocol*	*Bandwidth*
SEG_BODY	CAN	250 kBit/s
SEG_BATT	CAN	125 kBit/s
SEG_CHASSIS_1	CAN	250 kBit/s
SEG_CHASSIS_2	CAN	250 kBit/s
SEG_PT_FRONT	FlexRay	1 Mbit/s
SEG_PT_REAR	FlexRay	1 Mbit/s

Conditions. For the execution of the simulation experiments, the following conditions apply:

- The duration of a simulation experiment is one second real-time (10^6 simulation ticks).
- The payload length of all messages is 8 bytes.
- Each message transports one control signal.
- The FlexRay communication cycle duration is 5ms.

- In FlexRay segments, frames are only sent in the static segment. The static segment has a duration of 2.5ms. Each slot in the static segment has a duration of 0.25ms.
- We assume a fault-free operation of the communication network. That means that all message frames are correctly transmitted, no frame losses occur. However, the results of the performance analysis for fault-free network operation also reveals the system reserves for failure scenarios like frame losses, which would require re-transmission of the lost frames, consequently increasing the network load.

Requirements. The following requirements shall be validated by the performance analysis of the control system:

- In all network segments, the maximum communication network load shall be **70%.**
- The maximum control signal latency for Action Path A shall be **40ms.**
- The maximum control signal latency for Action Path B shall be **30ms.**

First Assessment

The control network simulation using the set of preliminary model parameters shown in Table 1 and Table 2, respectively, delivers the following results:

System Assessment. The System Assessment phase validates the fulfillment of the communication network load requirements in the network segments across Action Path A and Action Path B[16]. The following network loads are observed in the segments *SEG_PT_FRONT*, *SEG_BODY* and *SEG_CHASSIS_2*:

- **SEG_PT_FRONT Segment (FlexRay)**. Despite the low bandwidth of 1Mbit/s[17], the network load of the *SEG_PT_FRONT* segment is well below 20%. This is due to the fact that the FlexRay communication cycle duration is 5ms, while the duration of the static segment is 2.5ms. Frames are only sent in the static segment, the dynamic segment is not used at all. Also, the payload length of 8 bytes is rather small for FlexRay frames[18].
- **SEG_BODY Segment (CAN)**. The network load of the *SEG_BODY* segment is about 25%, using a bandwidth of 250kBits/s. This is about a third of the maximum allowed network load of 70% and gives room for bandwidth reduction in this segment.
- **SEG_CHASSIS_2 Segment (CAN)**.The network load of the *SEG_CHASSIS_2* segment is about 50% and clearly below the allowed network load of 70%.

A first assessment shows that all network load requirements are fulfilled. Further, the network load in the SEG_BODY segment is so low that a reduction of the bandwidth in this

[16] Since we use the same set of input stimuli for all network segments, the observation of the network loads of the segments *SEG_PT_FRONT*, *SEG_BODY* and *SEG_CHASSIS_2* is representative for the other network segments.

[17] The FlexRay protocol support bit rates up to 10Mbit/s.

[18] A FlexRay frame carries up to 254 bytes of payload data.

segment can be taken into consideration. Figure 14 summarizes the results of the first assessment of the network loads.

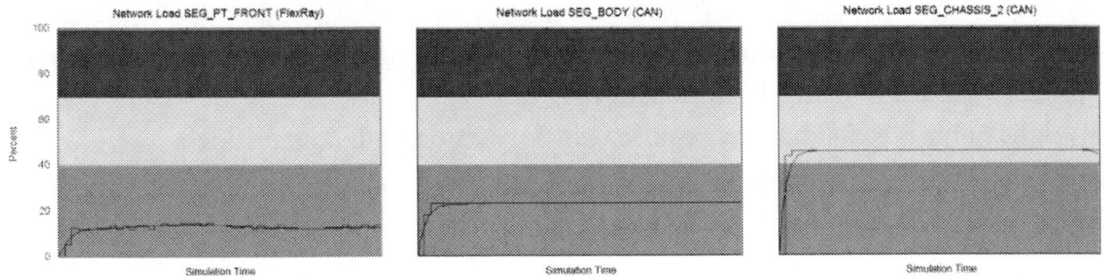

Figure 14. First Assessment – Network Loads.

Signal Assessment. The Signal Assessment phase validates the fulfillment of the control signal latency requirements for Action Path A and Action Path B. The following control signal latencies are observed:

- **Action Path A.** The signal latency is well below 40ms, thus fulfilling the requirement.
- **Action Path B.** The signal latency is exceeding 30ms, thus violating the requirement.

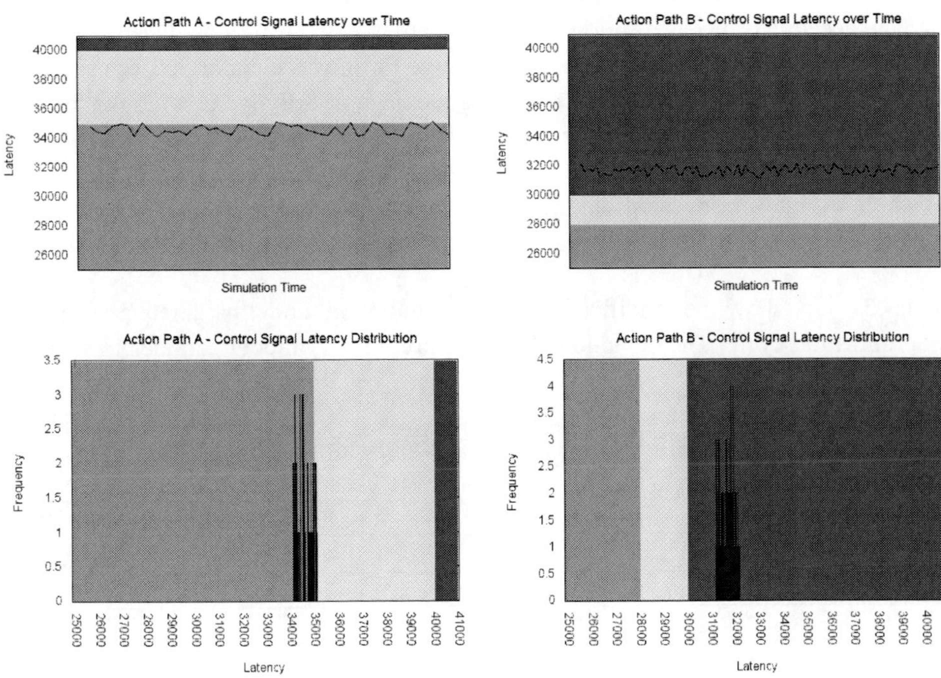

Figure 15. First Assessment - Control Signal Latencies.

A first assessment shows that the control signal latency requirement is fulfilled for Action Path A, but not for Action Path B. Figure 15 shows the latency over simulation time and the latency distribution for the vehicle speed control signal (Action Path A) and the accelerator pedal position control signal (Action Path B), respectively.

Summary. The first assessment delivers the following results:

- All network load requirements are fulfilled. The network load observed in segment *SEG_BODY* segment is particularly low.
- The control signal latency requirement for Action Path A is fulfilled.
- The control signal latency requirement for Action Path B is not fulfilled.

Optimization Cycle

From the observations made in the first assessment, the following optimization actions are derived:

- To improve the bus resource utilization in the *SEG_BODY* segment, the bandwidth is reduced from 250kBit/s to 125kBit/s. This will also improve the electromagnetic compatibility [26] of the *SEG_BODY* segment, since a reduction of the bandwidth also reduces the electromagnetic interference induced by the CAN bus. However, this action will have an impact on the vehicle speed control signal latency due to the longer frame transmission time over the CAN bus of the *SEG_BODY* segment.
- The accelerator pedal position control signal latency has to be reduced. This will be done by changing the priority of the message for the accelerator pedal position in the *SEG_CHASSIS_2* segment. The original message priority is 90, which is the second lowest message priority in the *SEG_CHASSIS_2* segment. By raising the message priority up to 20 (which is the second highest one in the *SEG_CHASSIS_2* segment), the message transmission duration is reduced, resulting in an intended reduction of the overall accelerator pedal position control signal latency to fulfill the requirement.

Tables 3 and 4 show the modifications of the simulation model parameters in the course of the optimization cycle. The deviations from the preliminary model parameters are marked ***boldface underlined***.

Table 3. Modified Network Bandwidth Settings

Segment	*Protocol*	*Bandwidth*
SEG_BODY	CAN	**125 kBit/s**
SEG_BATT	CAN	125 kBit/s
SEG_CHASSIS_1	CAN	250 kBit/s
SEG_CHASSIS_2	CAN	250 kBit/s
SEG_PT_FRONT	FlexRay	1 Mbit/s
SEG_PT_REAR	FlexRay	1 Mbit/s

Table 4. Modified Network Segment Communication Schedules

Segment Input	*SEG_ BODY*	*SEG_ BATT*	*SEG_ CHASSIS_1*	*SEG_ CHASSIS_2*	*SEG_ PT_FRONT*	*SEG_ PT_REAR*
S0	10	10	10	10	0	0
S1	20	20	20	**90**	1	1
S2	30	30	30	30	2	2
S3	40	40	40	40	3	3
S4	50	50	50	50	4	4
S5	60	60	60	60	5	5
S6	70	70	70	70	6	6
S7/GW	80	80	80	80	7	7
S8/GW	90	90	90	**20**	8	8
S9/GW	100	100	100	100	9	9

Second Assessment

A simulation experiment using the modified simulation parameters shown in Tables 3 and 4, respectively, delivers the following results:

System Assessment. We revisit the validation of the communication network load requirements after the modifications of the simulation model parameters in the course of the optimization cycle.

The CAN network bandwidth of the *SEG_BODY* segment has been reduced from 250kBit/s to 125kBit/s. This optimization measure increases the network load from 25% to about 50%. The observed network load is well below the maximum allowed network load of 70%, thus the requirement is now fulfilled, despite the reduction of the original bandwidth. The network loads in the segment *SEG_PT_FRONT* and the *SEG_CHASSIS_2* are not affected by this modification and remain unchanged.

Figure 16 shows the results of the second network load assessment.

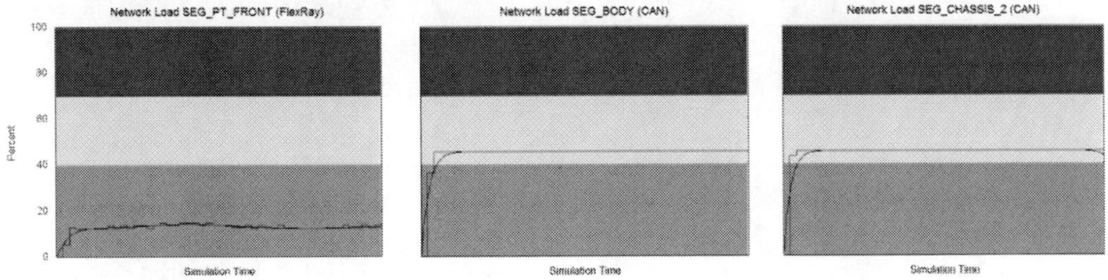

Figure 16. Second Assessment – Network Loads.

Signal Assessment. Now we consider the effects of the optimization measures on the control signal latencies. As already argued in the course of the optimization cycle modifications, a reduction of the CAN network bandwidth in the *SEG_BODY* segment results in an extended latency for the vehicle speed control signal (Action Path A). Considering the control signal latencies for Acion Path A and B after the optimization measures, this assumption proves to be true.

The following control signal latencies are observed:

- **Action Path A.** Compared to the first assessment, the signal latency has increased from 35ms to 39ms. However, this latency is below 40ms, thus the requirement is still fulfilled.
- **Action Path B.** Compared to the first assessment, the signal latency has remarkably improved from 32ms to 27ms. After the message priority adaptation in the course of the optimization cycle, the accelerator pedal position control signal latency now fulfills the requirement.

Figure 17 shows the results of the second control signal latencies. Again, the latencies over simulation time and the latency distributions for the vehicle speed control signal (Action Path A) and the accelerator pedal position control signal (Action Path B), respectively, are shown.

Summary. The second assessment after the optimization measures delivers the following results:

- All network load requirements are fulfilled. A reduction of the CAN network bandwidth in the *SEG_BODY* segment results in a better utilization of the communication network, while preserving the network load requirement.
- The requirement for the vehicle speed control signal latency (Action Path A) is fulfilled.
- The requirement for the accelerator pedal position control signal latency (Action Path B) is fulfilled.

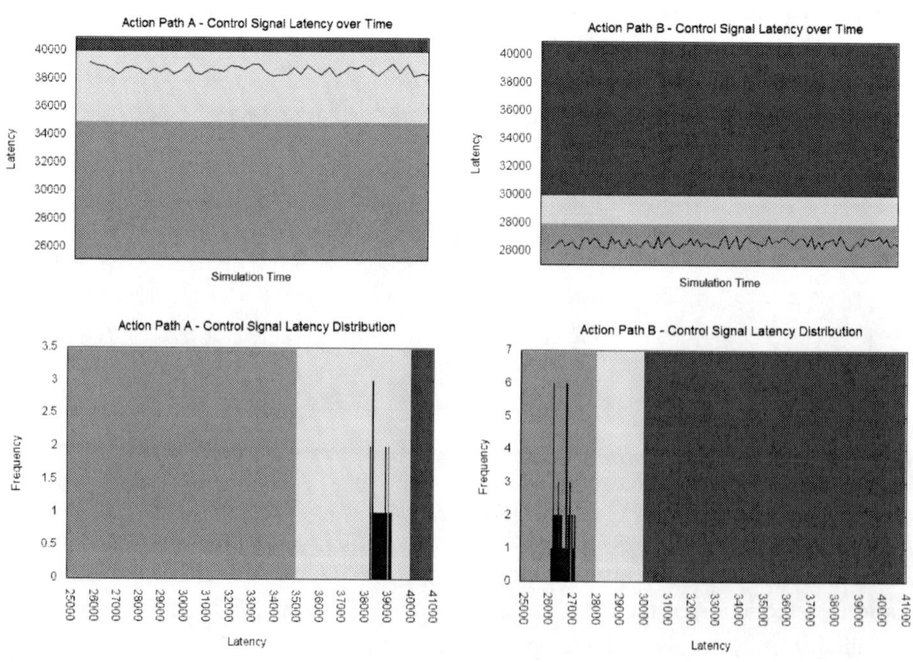

Figure 17. Second Assessment – Control Signal Latencies.

Control System Network Performance Evaluation

This section presented a simulation based control network assessment and optimization workflow based on a practical example, namely the analysis of an electronic control system for an electric city car. In the simulation experiments, we have examined two time-critical control signals. The assessment of the proposed POLLUX vehicle network architecture shows that the architecture has sufficient reserves for extension. The following arguments support that claim:

- **Network bandwidth**. The bandwidths chosen for the six segments are at the lower limit. This is especially true for the FlexRay segments using a bandwidth of 1Mbit/s. Under the given assumptions concerning the input stimuli, even using these considerably low bandwidths the network load in all segments is below 50%. While in the FlexRay segments the network load is constant (as long as only the static segment is used), the network loads observed in the CAN segments provide reserves to preserve the network load requirements in case of frame losses and the need for re-transmission of control signals, without the need to increase the network bandwidth.
- **Network utilization**. The utilization of the available bandwidth is not "optimal". In the proposed configuration, each CAN and each FlexRay frame transports only one control signal. Packing more control signals into one message frame will result in a better utilization of the network resources. Additionally, the chosen payload data length in the FlexRay segments is considerably low (8 bytes), compared to the maximum FlexRay frame payload length of 254 bytes. The dynamic segment is not used at all. So from the point of view of network utilization there is also considerable room for optimization, without the need to increase the network bandwidth.
- **Network load**. The observed network load also provides reserves with regard to the adding of ECUs to the different segments. This is meaningful for the expandability of the control system. While adding components to the control system will increase the network bandwidth requirements and increase the network load, the observed reserves show that there is also room for adding ECUs to the network segments, implementing new car functionalities, without the need to re-design the control system architecture.

This application example shall illustrate the principle of operation and deliver a proof of concept. In a "real" control system, there exist more than two time-critical control signal paths whose requirements have to be validated. Thus, the validation effort will be higher, presumably requiring lots of network modifications and subsequent simulation runs. However, the simulation based design and validation approach allows the detection of requirement violations in early development phases, and shall help to avoid costly re-design activities in later development phases. Considering only three parameters (network bandwidth, control signal latencies and sensor input stimuli frequencies) essential properties of the network architecture can be assessed without the need to consider functional aspects of the system. Therefore, this method is also a step towards a generic approach for vehicle control network design. Providing the methodology for description of vehicle network architectures and the possibility for modification of essential network parameters, this approach shall enable manufacturers and suppliers to develop control systems without the need to share corporate know-how with third parties, like tool developers.

POSSIBLE FUTURE DEVELOPMENTS IN AUTOMOTIVE CONTROL NETWORK DESIGN

This section extends the considerations regarding automotive control network design to reflect about possible future trends. A possible development in this area is the evolution of vehicle control network standards for electric vehicles. The development of such a standard providing recommendations for the design of electronic control networks for electric vehicles seems to be reasonable due to the following reasons:

- Down to the present day, considerable development effort is spent on the design of automotive communication networks. Current automotive communication networks consist of a set of electronic control units communicating over dedicated communication channels. These architectures show a segmentation into different areas of functions, where each function segment covers typical control tasks like engine control, chassis control or infotainment, using standardized communication protocols like CAN, LIN, FlexRay or MOST.
- Typically car manufacturers develop their own vehicle control networks. Although these network architectures usually have a product life cycle of several years and are used in different cars, car manufacturers spend this design effort independently of each other over and over again, whenever new car generations are brought to the market.
- The conventional automotive industry has evolved over more than 60 years, the methodologies and processes for the design of vehicle control networks are well established at the different car manufacturers. Since the automotive industry operates in a very competitive market, it is rather unlikely that manufacturers will leave their well-established design processes that have evolved over many years. However, electromobility is a new and currently evolving market, providing the opportunity for common efforts to establish new methodologies for the design of electronic control systems for electric vehicles. Although approved methods deployed in conventional automotive design have already been used and will further be used for the development of electric vehicles, the next generation of electrical vehicles will require new concepts, especially for the design of the power train and power management functions.

This situation provides the opportunity to revisit conventional control network design processes. Is it really necessary that car manufacturers create and refine their vehicle network designs independently from each other over and over again? Vehicle communication networks based on a standardized generic framework could significantly shorten the design to market time.

But is there a willingness among car manufacturers to work on a common standard for vehicle control networks? Many automotive industry standards have been droven forward by consortia. Representative examples for joint standardization efforts involving different automotive companies are the following:

- The FlexRay protocol for time-triggered network communication was developed by the FlexRay consortium founded in 2000. Prominent members of the consortium were, among others, BMW, Daimler Corporation, Motorola, Philips, Bosch, General Motors and Volkswagen.
- The MOST cooperation, founded in 1998, established the MOST technology as a standard for automotive multimedia networks. The MOST cooperation is open to any company and consists of carmakers, setmakers, system architects and key component suppliers. Currently, the MOST cooperation has more than 70 members.
- The LIN specification was developed by the LIN-consortium, which currently involves almost 100 industry partners.
- The AUTOSAR consortium provides standard specifications for an open and standardized automotive software architecture. It is a worldwide development partnership of car manufacturers, suppliers and other companies from the electronics, semiconductor and software industry.

Scope of the Vehicle Control Network Standard

Given that there is readiness in the automotive community for the development of a vehicle control network standard: what shall be included in such a standard? Reasonable elements are:

Communication Network Protocols. A first step are recommendations for the usage of network protocols according to the characteristics and functions of the different network segments. The Society of Automotive Engineers (SAE) provides a taxonomy for automotive communication systems that consists of three classes, SAE class A, B and C, differing in communication data rates and network arbitration methods, represented by different field bus systems. This taxonomy could be refined and adaptated to the specific characteristics of electric vehicles, providing recommendations for suitable network protocols and bandwidth settings for different network segments.

Communication Network Schedules. In a next step, reommendations for communication network schedules could be defined for the different network segments. These network schedules define the assignment of bandwidth to electronic control units, comprising the assignment of message frame identifiers to ECUs for event-triggered communication protocols like CAN as well as the assignment of communication slots to ECUs for time-triggered communication protocols like FlexRay. Standardized communication schedules would take out the effort of distribution of frame identifiers and communication slots from the design process. Further, all design tools could be aligned to these recommended schedules. This facilitates the automation of the performance and safety analysis of the communication network in the validation and verification phase and supports the implementation of the safety case.

Electronic Control Unit Development and Integration. Once the network protocols and communication schedules are standardized for the different network segments, ECU suppliers could design and optimize their ECUs and the control applications for the specific network segment requirements defined by the standard. This reduces the design and integration effort of electronic control units significantly.

The idea to provide standards for communication protocols and communication networks for a given application domain is nothing new. For instance, the CANopen [27] standard provides specifications for a communication protocol and device profiles for embedded systems used in automation engineering applications. The CANopen standard provides, among others, a set of small, specialized communication protocols for dedicated control and monitoring tasks. Further, standardized device profiles are provided for a set of appliance classes, like I/O modules or sensors and controllers that define functionality and properties of devices. This allows for integration of devices from different vendors, as long these devices correspond to a given CANopen device profile.

CANopen is already well established in Europe, while becoming more and more popular in North America and Asia. Since 1995, the CANopen standard is maintained by the organization CAN in Automation (CIA) and is standardized as a european-wide norm (EN 50325-4)

Why should the establishment of such a standardard not be feasible in the field of communication networks for electric vehicles?

SELECTED FACTORS OF INFLUENCE ON AUTOMOTIVE DEVELOPMENT PROCESSES

To complete the reflections about design methologies for automotive control networks, we finally consider some commercial and technical factors of influence and analyze their impact on development processes. Unsurprisingly, special emphasis is laid on simulation based development approaches. The motivation for these considerations is to give a more comprehensive view on the subject by regarding some representative commercial surroundings and constraints. The following thoughts cover aspects related to cost factors, potential benefits, additional technical issues, potential added value as well as potential competitive advantages that may arise from the deployment of simulation based design methods.

Cost factors

Cost factors increase development costs or development time. Cost factors are assumed to distribute among the development team and over development time. That means that the impact of cost factors on the overall project costs becomes less significant with increasing development team size and increasing project duration.

Acquisition costs for simulation environments. The first step is the acquisition of the simulation environment. High-quality simulation environments have their price. If a license for an expensive simulation environment is not already available, this may be a blocking point.

Initial training for simulation environment. Complex simulation environments require training to qualify the users to achieve correct results within reasonable time.

Adaptation of simulation environment to the development process. The adaptation of simulation environments comprises

- **Integration of simulation environment into existing tool chains.** The interfaces of the simulation environment to other tools already used may have to be adaptated to allow for data exchange.
- **Additional implementation efforts.** The simulation of remote components requires the implementation of simulation models for these components.
- **Application of the simulation environment for the system design phase.** The use of a simulation environment increases the time needed for early development phases. However, this additional effort is expected to pay in later development phases by generating some benefits that compensate for this additional effort.

Potential Benefits

Potential benefits of simulation approaches decrease development costs or development time. Benefits are assumed to multiply with the number of development team members over development time. That means that the positive impact of beneficial factors on the overall project costs becomes more significant with increasing development team size and increasing project duration. Personnel costs are assumed to be the dominating cost factor.

Acceleration of the development process. If a software module is developed from scratch, it makes sense to validate the correctness of the software control flow in a simulation environment and to switch to the hardware environment for development of the hardware related parts. In general, changes induced by design faults are more expensive when hardware and software are already implemented. Changes in a simulation environment are much cheaper. Therefore, simulation based design approaches also help to reduce the average costs per change, thus reducing overall development costs.

Development without hardware components. Hardware components are not yet available when the project starts but are delivered by the manufacturer in a later project phase. Given that the interfaces to the hardware components are specified and available, it may be useful to create simulation models of the components and to develop the host application using the model until the "real" components become available. This avoids "idle" phases in the development process. Of course, the effort for creation of simulation models of the not yet available hardware components has to be taken into consideration.

Improved analysis capabilites. Analysis of the communication with some remote component may be hard because in the hardware environment this component usually is not in the sphere of control of the debugger. Modeling the behavior of a remote component on the base of its interface description and integrating the model into the simulation environment helps to locate and to correct errors.

Reduced costs for test equipment. Dedicated hardware equipment for testing distributed systems is quite expensive. Having a simulation model of the communication network at hand reduces the need for expensive hardware network simulators.

Acceleration of the integration process. A thorough analysis of the control system design in the simulation environment prior to integration helps to diminish the number of errors. Further, the correction of errors may be supported by the improved capabilites for analysis provided by the simulation environment, reducing the overall integration effort.

Further Technical Issues

Further technical issues are influence factors that have an impact on the meaningfulness and significance of simulation results.

Model coverage. A key factor in model based development is the model coverage: How realistic is the simulation model? How good does it reflect the behavior of the physical target system at runtime? A high model coverage is more probably to be obtained in the case of a deterministic target system, compared to a non-deterministic target system, simply due to the fact that the number of possible behaviors and scenarios of a non-deterministic target system increases with the system complexity in an exponential fashion. Consequently, an almost infinite number of test cases would be necessary to cover all possible behaviors of the target system. This is neither feasible nor meaningful.

Determinism of the runtime target system. By establishment of well-defined rules of communication and proper definition of range tolerances for timing behavior and values, the system behavior becomes deterministic which in turn facilitates and encourages the use of simulation models for development and analysis of such systems.

Simulation results are very significant in the presence of a clear design methodology and a deterministic target system with defined properties that have to be fulfilled and that can be validated using the simulation model. The simulation results are less significant if only best-effort methods on different levels are implemented. In this case the simulation may cover a given subset of scenarios and behaviors that the system shows at runtime, but not all possible scenarios.

Potential Added Value

These factors of incluence potentially create added value beyond the already considered potential benefits.

Optimization. Time savings in the development phase can be used for optimization of the product (e.g. an ECU software) such that the resource requirements (processor speed, memory consumption) are minimized. This allows the usage of cheaper components (low-performance processors, less memory chips) which in turn may result in significant cost savings. In a market of mass production, if only a few cent are "squeezed out" of every component, the cost savings are considerable, especially for automotive suppliers that sell standard products in quantity to big automotive manufacturers.

Network design scaling. The creation of a network design in a simulation environment is significantly less time-consuming than the creation of such a design in hardware. While being able to create a few number of alternative designs in hardware implementations, the number of variants that can be created in a simulation environment is much bigger. This in turn increases the probability that an optimal and cost-effective design with regard to hardware requirements can be identified in a simulation environment. Minimum hardware requirements mean lower costs per piece, thus establishing cost-effectiveness due to economy of scale in a market of mass production.

Functional safety. Proper use of simulation environments for control system design supports systematic elimination of design faults, thus achieving design stability. Design stability is a pre-condition for functional safety. Due to the rapidly increasing complexity of automotive control systems, is it necessary to define structures and rules for communication to establish determinism of safety-relevant systems. Clear and mean development methods and alignment of the target architecture to the simulation environment allow for breaking up the complexity of systems and to establish determinism.

Competitive Advantages

Competitive advantages potentially create unique selling propositions for manufacturers and suppliers.

Delivering a proof-of-concept. For a simulation environment can be deployed for cost-effective and time-saving development, it may be used for rapid prototype implementations, e.g. in the course of a tender. If a competitor is able to show a prototype implementation in the course of a presentation at the customer's site, this competitor will probably prevail against all other competitors who sell concepts but have nothing tangible to demonstrate.

Being the first on the market. Simulation based development approaches also help to reduce the time-to-market of innovative solutions. A proper use of simulation during the development process of world-first solutions may result in a decisive advantage with regard to development time in the case of concurrent development of such solutions by different competitors. The winner of this race can make use of the "invented here" advantage.

Avoidance of post-sale efforts. Even failures that occur with a low probability, caused by design faults or bugs in the software of widely used control system components, will eventually occur because large quantities of cars using these components are on the road. It is only a question of time. The occurrence of a component failure may lead, in the course of subsequent analysis, to the detection of a systematic fault.

Systematic faults are in the worst case be caused by network design faults or errors in the software. If a failure can be assigned to a systematic fault in a component, all cars equipped with this component are affected. This leads, in the worst case, to return actions of whole car series initiated by the car manufacturer. Besides exorbitant costs, the negative impact of such return actions on the brand image of manufacturers may be even more significant.

Quality considerations. Quality can be seen as the *improbability of the occurrence of failures.* As stated above, design faults or errors in the software will eventually be detected and may require a visit at the car dealer's garage.

In premium car segments, where the car price does not play a dominant role for customers, quality is of essential importance. Therefore, all efforts are made to avoid the occurrence of failures. A clear and mean control system design, developed, validated and extensively tested with a simulation environment, can help to avoid systematic faults.

Quality is essential for the brand image of a car manufacturer. Further, high quality is a factor for competitive advantage, not only in a premium segment, because a certain level of quality is also a pre-condition for (functional) safety.

Summarily it can be said that *what is cheap is not necessarily of poor quality, but what is expensive is expected to be of high quality.*

Conclusion

This work provides an extensive overview of selected aspects of automotive control system design with the focus on simulation based development methods. Based on the identification of essential characteristics of modern vehicle control systems, a methodology for simulation based control network design was developed. This methodology was used for implementation of the DTF simulator, a simulation environment for design and performance analysis for vehicle electronic control systems.

We presented an extensive application example, in the course of which we used the DTF simulator for the development, performance analysis and optimization of a vehicle control network for an electric city car.

Further, we developed the idea of a vehicle control network standard as a possible continuation of standardization efforts in the automotive industry, especially with regard to the development of next generation electric vehicles. Since such a standard would help to take out considerable research and development efforts out of the design process for vehicle control networks, it may also provide a chance for smaller companies and suppliers to enter the market for electric vehicles.

Finally, we analyzed the impacts of simulation based design approaches on representative commercial surroundings and constraints of automotive development processes, including cost factors, potential benefits, additional technical issues, potential added value and potential competitive advantages.

Acknowledgments

This work has been supported by the European Union ARTEMIS Project POLLUX (JU Grant Agreement number 100205).

References

[1] Juergen R.K. *X-By-Wire Automotive Systems*. SAE International, February 2009.

[2] Hillenbrand M. Funktionale Sicherheit nach ISO 26262 in der Konzeptphase der Entwicklung von Elektrik/Elektronik Architekturen von Fahrzeugen. KIT Scientific Publishing, May 2012.

[3] Hilderman V. and Baghi T. Avionics Certification: A Complete Guide to DO-178 (Software), DO-254 (Hardware). *Avionics Communications*, 2007.

[4] Gabriela Nicolescu G. and Mosterman P. J.: *Model-Based Design for Embedded Systems. Series: Computational Analysis, Synthesis, and Design of Dynamic Systems*. CRC Press, November 24, 2009.

[5] Zimmermann W. and Schmidgall R.: Bussysteme in der Fahrzeugtechnik – Protokolle, Standards und Softwarearchitektur. *Vieweg+Teubner, 4*. Auflage, 2010.

[6] Ulsoy A. G., Peng H. and Cakmakci M.: *Automotive Control Systems*. Cambridge University Press, April 2012.

[7] Tanenbaum A.S., Van Steen M. *Distributed Systems: Principles and Para*digms. Prentice Hall International, September 2001.

[8] Etschberger, K. (Hrsg.): CAN Controller Area Network – Grundlagen, Protokolle, Bausteine, Anwendungen. *Hanser Fachbuchverlag*, Deutschland, 2001.

[9] Paret D. FlexRay and its Applications - Real Time Multiplexed Network. Wiley, 2012.

[10] Grzemba A., Von der Wense H.: LIN-Bus. Franzis, Deutschland, 2005.

[11] Grzemba A.: MOST. *The Automotive Multimedia Network*. From MOST25 to MOST150. Franzis, 2011.

[12] Kandler K.. AUTOSAR - XCP - FlexRay - Design und Implementierung eines AUTOSAR XCP-Transport-Layers fuer FlexRay. VDM Verlag Dr. Mueller, 2010.

[13] Kopetz H. and Grünsteidl G. *TTP-A Protocol for Fault-Tolerant Real-Time Systems*. Computer, Vol. 27, No. 1, pp. 14-23, January 1994.

[14] LIN Consortium. LIN Specification Package Revision 2.1. Specification, LIN Administration, Bernhard Wicki Strasse 3, D-80636 Munich, Germany, 2006.

[15] Hanzlik A. Stability and Performance Analysis of Clock Synchronization in FlexRay. International Review on Computers and Software (IRECOS), 1(2):146–155, September 2006.

[16] Kopetz H. and Bauer G. The Time-Triggered Architecture. *Proceedings of the IEEE*, 91(1):112 – 126, January 2003.

[17] Kopetz H. The Time-Triggered Model of Computation. *Proceedings of the 19th IEEE* Systems Symposium (RTSS98), December 1998.

[18] Bauer G. and Paulitsch M. An Investigation of Membership and Clique Avoidance in TTP/C. *19th IEEE Symposium on Reliable Distributed Systems*, Nuernberg, Germany, October. 2000.

[19] Hartwich F., Mueller B., Fuehrer T. and Hugel R. CAN Network with Time Triggered Communication. *Proceedings of the 7th International CAN Conference*, 2000.

[20] Hanzlik A. Assessment of the Potential for Cost Savings and Quality Improvements by Integration of Simulation Environments into Automotive Development Processes. MBA Master Thesis, PEF Private University of Management, Vienna, Austria, October 2011.

[21] Hanzlik A. and Kristen E.: DTF - A Simulation Environment for Communication Network Architecture Design for the Next Generation of Electric Cars. *Advanced Microsystems for Automotive Applications* 2011, pp. 259-267, 2011.

[22] Banks J., Carson J. and Nelson B.: *Discrete-Event System Simulation*. Prentice Hall, 2000.

[23] Buja, G. Overcoming Babbling-Idiot Failures in CAN Networks: A Simple and Effective Bus Guardian Solution for the FlexCAN Architecture. *IEEE Transactions on Industrial Informatics,* Vol. 3, Issue 3, pp. 225-233, August 2007.

[24] Richter K. and Bartels M. Analyse des Zeitverhaltens von ECUs und Controller-Netzwerken. *ATZ Elektronik Journal* 2/2007, June 2007.

[25] Melliar-Smith, P.M. Broadcast protocols for distributed systems. *IEEE Transactions on Parallel and Distributed Systems,* Vol. 1, Issue 1, pp. 17-25, January 1990.

[26] Noble, I.E. EMC and the automotive industry. *Electronics & Communication Engineering Journal,* Vol. 4, Issue 5, pp. 263-271, October 1992.

[27] Pfeiffer O, Ayre A. and Keydel C. *Embedded Networking with CAN and CANopen.* Copperhill Media Corporation, April 2008.

In: Grid Electrified Vehicles
Editor: Carla Alexandra Monteiro da Silva
ISBN: 978-1-62808-839-7

Chapter 3

ELECTRIC VEHICLES: FROM PRODUCT TO SERVICE

***Adi Wolfson*[1,*], *Dorith Tavor*[1] *and Shlomo Mark*[2]**
[1]Green Processes Center, Sami Shamoon College of Engineering, Beer Sheva, Israel
[2]Negev Monte Carlo Research Center (NMCRC), Sami Shamoon College of Engineering, Beer Sheva, Israel

ABSTRACT

Although the electric vehicle (EV) was actually introduced about a century ago, today's rising energy costs and heightened awareness of environmental issues have prompted a reexamination of that nascent technology. Indeed, the current problems associated with a world dependent on fossil fuels have challenged the automobile industry to design and produce viable, modern EVs that will reduce global greenhouse gas emissions and air pollution associated with the burning of fossil fuels and achieve a balance between nature and economics. But faced with several barriers to its development—especially the ambiguous trust people have in new technologies and the relatively high cost of batteries—the EV market has devised business models that shift the focus from product sales to services. One such service that can offset the initial investment a person makes in an EV is battery leasing. Business models based on leasing offer new opportunities, as supplying a service entails more cooperation between the service provider and the customer than does the one-time delivery of a product. Yet to achieve the ultimate goal of more sustainable transportation, EV service models should also incorporate and structure the life-cycles of both the car and the battery and the way that electricity is produced. Among the variety of EV business models, that of Better Place, whose focus is on supplying the electricity rather than on leasing batteries or cars, includes a wide range of opportunities to increase the service's sustainability. In so doing, the Better Place business model can make an invaluable contribution to global sustainability.

Keywords: Electric vehicles, service, sustainability, sustainable service

* Corresponding author: adiw@sce.ac.il, Tel: 972-8-6475766, Fax: 972-8-6475636

INTRODUCTION

Motor vehicle sales around the world have been rapidly growing over the years, a process that is largely determined by two major global trends: growth in the level of per capita income and the ever expanding human population [1]. An integral part of the lives of people everywhere, cars are increasingly relied on transportation, and vehicles in general play conspicuous roles in the modern economy and in shaping our natural and built environments.

Yet motor vehicles are also major sources of both air pollution (Table 1) and greenhouse gas emissions (Figure 1), particularly in urban areas where vehicle concentrations are highest. Indeed, their negative environmental impacts have reached critical proportions, dictating the need for greater control [2]. Although new technologies and opportunities in the car market have emerged from time to time, the basic technologies and products, e.g., the internal-combustion engine and liquid fuels, have remained the same. However, the pace of technological improvement, influenced by environmental regulatory pressure, continues to grow, and efforts are underway around the world to develop vehicle propulsion technologies that reduce the reliance on and subsequent use of fossil fuels, either by improving fuel economy or by switching to renewable fuels [3, 4].

Table 1. Distribution of air pollutant emissions by source – Canada, 2010 [5][a]

Source	Sulfur oxides	Nitrogen oxides	Volatile organic compounds	Ammonia	Carbon monoxide	Total particulate matter
Industry	65	30.7	40.9	3.3	16.3	66.7
Transportation	6.9	33.0	12.5	4.9	42.8	4.2
Off-road Vehicles	< 0.1	22.2	14.6	0.2	31.9	6.2
Fuel for Electricity and Heating	27.8	13.5	0.3	0.2	1.0	4.8
Home Firewood Burning	0.1	0.5	8.3	0.2	7.8	16.7
Incineration and Miscellaneous	0.2	0.1	5.2	0.4	0.1	1.5
Paints and Solvents	0	0	18.2	0	0	0
Agriculture	0	0	0	90.8	0	0

[a] Percent of national emissions.

One vehicle technology being considered as a viable alternative that will help reduce the environmental impact of automobiles is the electric vehicle (EV) [6, 7]. Among the earliest of automobiles, electric vehicles (EVs), which used batteries for their movement, were already in use during the nineteenth century. But the commercial drilling for and production of petroleum at the end of the nineteenth century promoted automobiles based on the internal combustion engine, which rapidly conquered the private car market in the twentieth century and spelled the demise of the EV as a means of mass transport. However, today's rising fossil-fuel-based energy costs and the heightened awareness of environmental issues have dictated the current rigorous investments by the car industry to design and produce viable EVs. In terms of today's needs, EV viability will be determined not only by a vehicle that fulfills people's transportation requirements, but also by one that will reduce global greenhouse gas emissions and air pollution and promote a balance between nature and economics.

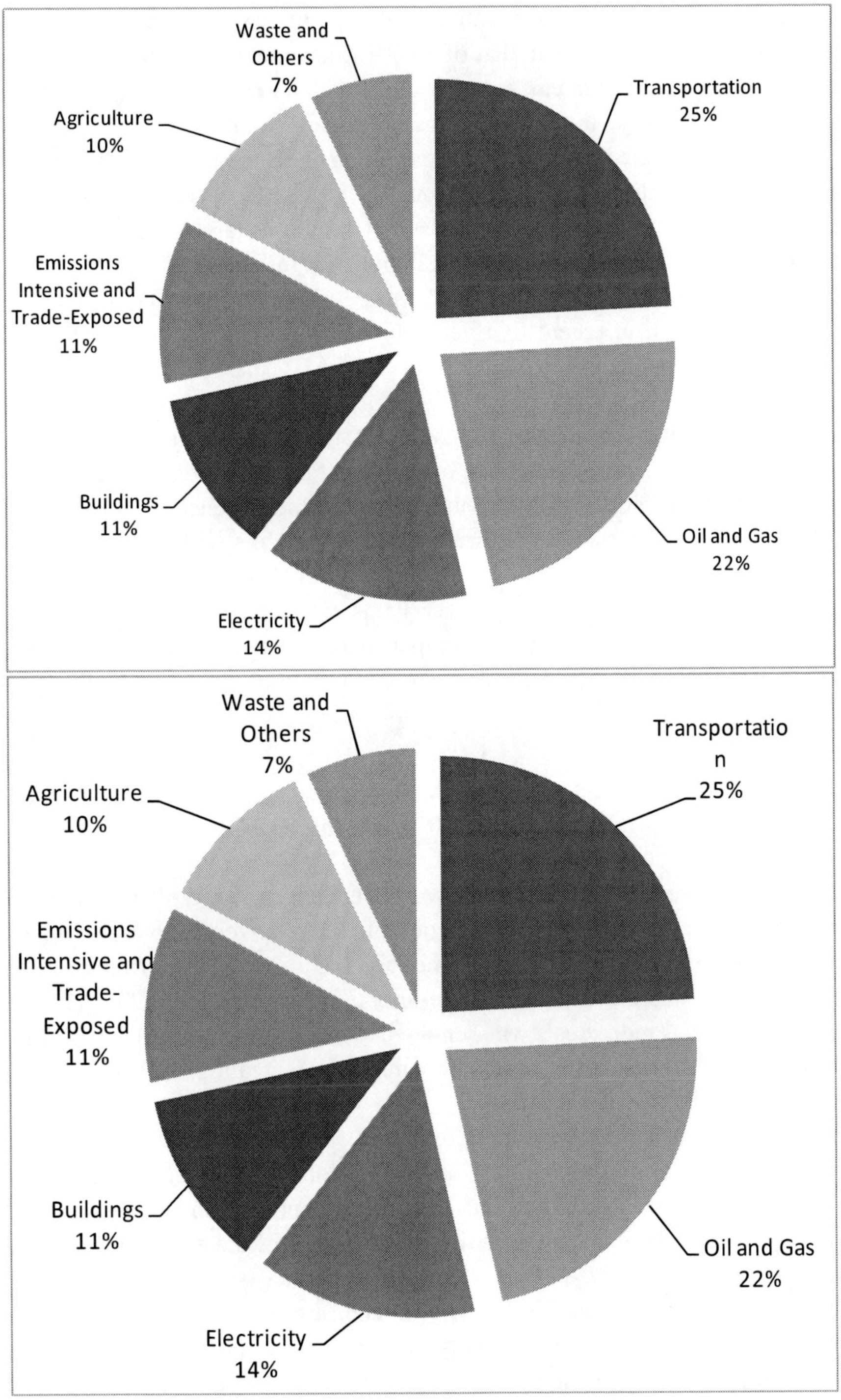

[a] Percent of national emissions.

Figure 1. Distribution of greenhouse gas emissions by economic sector – Canada, 2010 [5][a].

Powered solely by electricity stored in on-board batteries, the EV is charged by plugging into the national grid or, in a limited number of models, swapping the battery [8]. In contrast, a hybrid electric vehicle (HEV) is powered by an on-board engine supplemented by electricity that is recovered via regenerative braking, and a plug-in hybrid electric vehicle (PHEV), whose battery is larger than that of an HEV, is able to recharge via the national grid and operate for short distances in electric-only mode with no option to swap the battery.

The incentive to electrify transportation varies among countries, from the exploitation of wind as a source of green energy for transportation in Denmark to the reduction of the dependence on oil for national security reasons in Israel. Moreover, a number of countries around the world, such as the US and Japan, has set ambitious targets for the future uptake of EVs by supporting national and international policies for tackling climate change while providing financial support to stimulate both the development and market uptake of these types of vehicles. Indeed, in recent five years a wide variety of car manufacturers, such as Chrysler, Chevrolet, Renault, and Volvo, has entered the EV market.

However, the wide adoption of the EV as a transportation solution is fraught with challenges. Current limitations of battery technology still severely limit vehicle range (around 140 km without charging), and the vehicle battery itself sill entails large capital investments. In addition, consumers must be convinced that electric vehicles constitute a robust technology that can fulfill our transportation requirements and reliably replace current technology. Moreover, from the environmental point of view, although a move toward EVs will shift pollution from city centers to electricity power plants, whose production of electricity usually results in reduced emissions of pollution compared to fuel combustion by vehicles, the real environmental impact of EVs will depend on the fuel mix used in the power plants generating the electricity.

From Product to Service

Development of the EV market faces several challenges, not the least of which are the ambiguous trust people have in new technologies and the relatively high capital investment of batteries. In partial answer, the EV business model has shifted its focus from product sales to service provision. By implementing a service-oriented business model, the service provider recruits the consumers as individuals who, on the one hand, share in the risk with the provider but, on the other, are actively invested in and committed to creating the service, i.e. customer (Figure 2).

Service, defined as a transformation of value (intangible product) from the service provider to the client, is a composite of several factors such as customer needs, service supplier abilities, supplies, flexibility, and defined, measurable, and practicable values and methods [9-12]. Today service represents the largest sector of the economy, growing rapidly in size and importance from year to year, while service systems are becoming more comprehensive, complex, and interdisciplinary. In contrast to production and supply systems, based mainly on the manufacture and delivery of a product from the supplier to the consumer, i.e., product in-use, the provision and performance of a service usually require the involvement of both sides, provider and customer, in the delivery of the service, i.e., value in-exchange. As such, embracing the service system mentality presents new opportunities for and different perspectives of sales and consumerism.

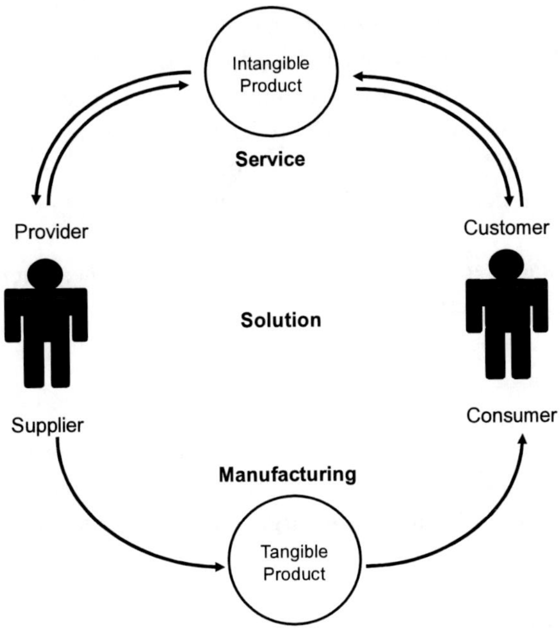

Figure 2. Supplier and customer relationship in service verse production.

In their new paradigm known as "service-dominant (S-D) logic," Vargo and Lusch offered a novel way of thinking about commerce, marketing, and exchange. They claimed that to provide a new foundation for marketing theory that overcomes the limitations of the "good-dominant (G-D) logic" mindset, which arose from the success of the industrial revolution, the focus should be moved from products or goods exchange to services exchange [13, 14]. Moreover, they asserted that within the emerging S-D logic community, service is defined as the application of competences for the benefit of a party and that, in fact, all exchanges between producers and consumers are based on service. With this approach, S-D logic provides companies and organizations with new insights into how to create value, even how to co-create value with their customers.

In general, three EV business models are currently offered in systems around the world (Table 2).

Table 2. Comparison of EV business models

Model	Provider	Main product	Main service
Battery leasing	Battery or car manufacturer, leasing companies	Battery	Leasing
Vehicle leasing	Car manufacturer, leasing companies	Car and battery	Leasing
Charging lease	Assigned company	Charging infrastructure	Charging per km

Battery Leasing Model

A hybrid model—the customer purchases the car yet leases the battery—it entails a substantial cost reduction to the customer, and more importantly, the customer assumes comparatively less of the risk associated with the nascent EV technology. The yearly or monthly fee to rent the batteries covers maintenance, insurance, and battery replacement. Although the flexibility offered by this model reduces the high costs and assumes some of the uncertainty associated with EV technology, this model also allows the provider to improve its technology and maintain its customer base. Indeed, this model offers a balance between customer risk and provider sales that is highly amenable to large scale operations run by existing vehicle manufacturers. In addition, the battery leasing model does not restrict the customer to purchasing a particular car model.

Vehicle Leasing Model

In this model both the car and the battery are leased, thus minimizing the risk to the customer. Although capital investment is also low, the price per month is higher. This model is more likely to be used by companies with large numbers of customers.

Charging Lease Model (Better Place Model)

As a producer of neither cars nor batteries, Better Place [15] offers a service through its charging infrastructures. However, Better Place joined both car and battery manufacturers to create a full package offer based on fixed price per km usage that includes charging stations and prices for electricity, changing batteries, etc.

Sustainable EV Solution

One of the most important challenges facing humanity today is how to design processes, e.g., production and services that satisfy the needs of today's people without sacrificing the ability of future generations to satisfy their own needs, i.e., sustainable processes. Sustainability that integrates economic, social, and environmental elements is defined as the capacity of an ecosystem to bear the stress of processes while maintaining them into the future and preserving enough space for subsequent generations [16, 17]. It can be accomplished by the rational use of resources, the development of efficient and responsible operation modes, and the prevention and/or reduction of harmful effluents and discharges into the environment.

Although for many years the implementation of sustainability was the concern of only the manufacturing and agricultural sectors of the economy, in recent years the service sector, which is steadily growing in size and importance and globally employs more people than the two other sectors of the economy [18], can also promote sustainability. Sustainable service is primary based on the rational use of natural resources and on acting with environmental awareness without negatively affecting either the natural or social environment [19-22]. In addition, a sustainable service facilitates conceptualizing services as shared and dynamic problem-solving endeavors that create value in multiple dimensions, and it takes into account potential long-term effects in a life cycle-oriented perspective on the relationships between consumers, providers, and suppliers. Moreover, a sustainable service goes beyond just fulfilling current customer demands, and shifts the service process boundaries between provider and customer from values that are focused only on personal, local, and current needs and economic profit to values that are broader and more future oriented and that are more socially and environmentally responsible.

We recently proposed a novel perspective and model, according to which we claim that for a service to be sustainable, sustainability should not only be an essential part of each service "core value," but it should also allow and motivate the customer to become a provider of sustainability as a "super value" to the next generation (Figure 3 [21]). In this respect, our approach can be implemented in EV market models, facilitating both their creation and their maintenance.

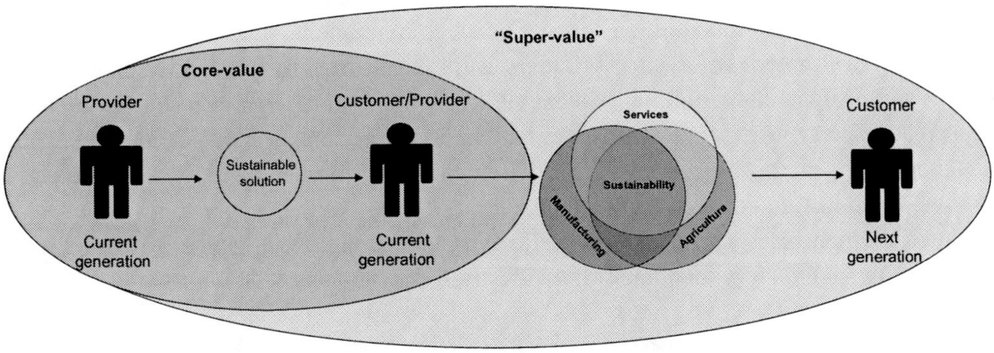

Figure 3. Sustainability as a service system within the framework of service science [14].

To grow the EV market, the emphasis should shift from products to services. But for EVs to constitute a sustainable transportation solution, it is not enough to merely shift the pollution from city centers to power plants. In addition, new modes of operation that consider the sustainability of both the "core" and the "super values" should be adopted. Thus, the design methodologies of EV services should also espouse a holistic perspective in terms of the needs of customers and other stakeholders, not merely meeting customer requirements, but also aspiring to design and implement solutions with new added value. These can include, for example, rethinking the production and supply of electric energy and the protocols for battery switching and treatment. One possible approach can entail a computer controlled charging network that enables energy monitoring and that integrates EVs with public transportation solutions [23].

As previously noted, from a sustainability perspective, perhaps the key advantage of EVs is the opportunity they provide to significantly decarbonize the road transportation sector, and as such, their implementation will reduce global greenhouse gas emissions and ameliorate the effects of climate change. Of course, this is also dependent on the decarbonization not only of grid electricity, but also of both the car and battery manufacturing processes (Figure 4 [23]). Another important sustainability factor to be considered in a shift to EVs is the environmental impact of producing, using and disposing of batteries. To that end, the effect on the environment can be minimized using renewable fuels and clean technologies in the production of electricity, by utilizing smart electricity charging systems, and by developing more efficient batteries and organized second-hand markets to recycle batteries, either for their reuse in cars or to recover their raw materials. Anything but trivial, such goals can be achieved on the route to more sustainable EVs by shifting the EV business models from technologies and production to services.

Core Value

The relatively steep initial investments and the high risk associated with EVs and incurred by the customer are associated mainly with the battery, e.g., its purchase, replacement, or re-sale, and these constitute a significant barrier to the ownership model common in internal combustion engine vehicles. The various business models for EV uptake are therefore based on service models that offer battery lease as the "core value," i.e., main service, with or without the addition of a car or other components to the leasing package. But the principal raw material of every EV battery is lithium, which has a negative environmental impact, and its use may not be sustainable. Thus the main challenges associated with widespread battery use are to decrease battery lithium content and to increase battery effectiveness together with the provision of reliable opportunities to recycle and re-use the batteries. One must go one step further, however, and examine the supply chain of the electric car battery, which faces serious social challenges such as the social rights of workers in the lithium mining and battery manufacturing industries.

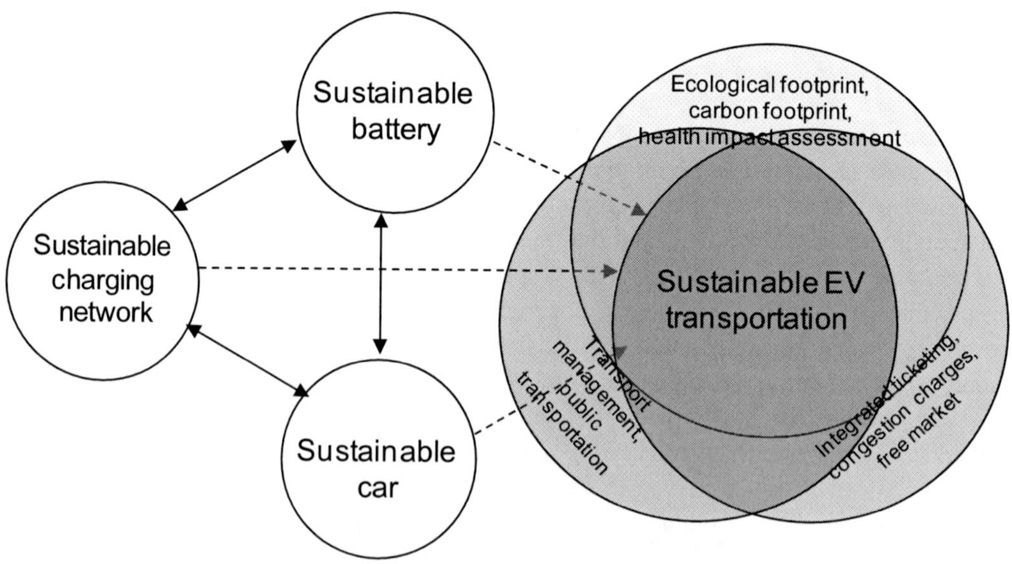

Figure 4. Components of sustainable EVs [23].

However, in terms of sustainability, the most important issue for EVs, which is strongly associated with the "core value" in all EV business models, including that of Better Place, is battery charging, particularly, how the electricity used to charge the batteries is produced. A variety of battery charging options exist, including, in its simplest form, plugging into the electricity grid at home or at the office or using special charging stands in the street. In addition, as one of the major issues regarding EV use is the limited sustained travel distance of which the battery is capable before it has to be recharged, drivers will have to plan for stops to charge their batteries (a time-consuming endeavor). To solve this problem, travelers can plan to switch their batteries at battery exchange stations. Again, because the main difference between EVs and gasoline or diesel cars is the use of electricity as fuel, how the electricity is produced remains the principal issue of EV sustainability, and can also be distinguished between the service that the three models offer.

It is generally clear that to fulfill the environmental promise of EVs and reduce global greenhouse gas emissions and air pollution, the production of electricity must shift toward renewable fuels (Table 2). Both carbon dioxide and air pollution emissions of renewable sources are much lower than those of fossil fuels, but some of the renewable options, like hydroelectric and biomass, entail massive water footprints and high land use, both of which significantly raise the associated environmental costs.

Table 3. Electricity generation with various fuels and their corresponding partial environmental impacts

Energy source	Carbon footprint ($gCO_{2\text{-eq}}$/GWh) [24]	Water footprint (Gal/MWh)[25]	Land use (m^2/GWh)[27]
Coal	888	692	50-400
Oil	733	~300[26]	-
Natural gas	499	172	320
Solar-PV	85	2	250-800
Biomass	45	~500	12,600
Nuclear	29	572	130
Hydroelectric	26	9,000	4,090
Wind	26	<1	1,000-2,100

Because the turbines that generate electricity cannot be easily turned on and off, another important issue central to electricity use and battery charging is the time of day. As most electricity providers prefer to supply a steady stream of electricity throughout the day and night and consumers tend to concentrate their electricity usage in the morning and evening hours, many providers offer discount rates to encourage people to use electricity during off-peak times, e.g., at night. Moreover, from a sustainability point of view, charging the battery at night will not only result in a more efficient use of electricity and a reduction in wasted electricity, it will also prevent the need to increase the level of electricity production during peak demand times to supply the EV customer needs that have been added to the usual electricity demand. Therefore, by exploiting remarkably simple solutions such as changing people's energy use habits, the promise of a more efficient and sustainable transportation solution can be fulfilled.

Of the three above-mentioned models, that of Better Place, with its focus on electricity supply instead of on battery or car leasing as its "core value," seems to offer the most sustainable opportunity. In so doing, Better Place can promote—and even compel governments and electricity suppliers to join—the move to greener electricity using renewable fuels and smarter, more efficient processes. Moreover, Better Place can also control, assist and educate its customers in purchasing greener energy and in adopting more efficient energy use habits as regards the time of electricity consumption. Nevertheless, the two other models that focus on battery leasing can also directly promote more sustainable electricity use by improving the operating efficiency of the battery. But in terms of customer-provider dynamics, the Better Place model includes greater customer cooperation and a value co-creation process, thus increasing the sustainability of the service, but also requires the customers to change their habits.

SUPER VALUE

From a sustainability perspective, other issues related to resource utilization during the production of the infrastructures, including the car, the battery and the charging equipment, and to car use and maintenance schedules, must be considered, as they indirectly affect the environmental footprint of the EV, i.e., its "super value", and of its complementary and supported services. In terms of the EV infrastructure—apart from the use of a battery, for which precautions must be taken regarding radiation, disposal or re-use, and electricity charging and production facilities—EVs are similar to gasoline or diesel cars. The three business models mentioned above use similar infrastructures and resources. However, in formulating their services based on shared maintenance and an offsetting of the responsibility for the technologies (e.g., battery, car, etc.), they are obligated to produce more sustainable products that will incorporate cheap and effective solutions for their maintenance, re-use, and disposal.

Another "super value" the sustainability route can offer is the smart and effective use of energy. This can be achieved, for example, using data tracking tools that compare energy usage patterns against scheduling requirements and optimize the charging systems for performance and cost. In addition, GPS and telematics data can also be coupled with historical data and traffic studies to develop services that map charging infrastructure support and driving routes. In addition, internet or mobile applications can be used to determine the optimum time and place to recharge a battery, in the process choosing between electricity suppliers based on the sustainability of their electricity production.

Another avenue for achieving a more sustainable transportation solution is to integrate EV use with transportation alternatives, such as public transportation systems. Better Place, whose solution could include reserved parking with charging spots or battery exchange facilities adjacent to public transportation stations and the provision of discounts or discount cards for clients who use public transportation, can easily promote a more sustainable solution [23]. Finally, the adoption of EV car sharing models represents another pathway to use the vehicles more sustainably.

CONCLUSION

The penetration of new technologies into any market, especially into heavily entrenched sectors like the automobile market, is accompanied by fear that is tempered by expectations and the possibility of new opportunities for both suppliers and customers. Already used a century ago, EVs have garnered increased attention in recent decades as a transportation solution that can reduce the greenhouse gas emissions and air pollution currently affecting the entire planet. Yet it is clear that to reach this goal, the production of the electricity for charging the EV battery should be based on more renewable fuels. The ambiguous trust in new technologies and the relatively high capital expense of EV batteries have shifted the EV business model focus from product sales to services. Battery leasing, with or without the car included in the leasing package, is one of the incentives that can increase EV uptake, and another is a service based on battery charging or replacement. These solutions demand a shift to a service-oriented business model, which presents a more responsible and sustainable solution as both supplier and customer share and cooperate to promote the service. In addition, inherent to the service model is the co-creation of "core value" and "super value" that further promote the sustainability of the EV as a viable transportation solution. As such, the model advanced by Better Place can make the greatest contribution to reducing the environmental footprint of the EV.

REFERENCES

[1] Gomes, C., (2012). Global economic research, Global Auto Report, December 27, Scotiabank Economics.

[2] Faiz, A., Weaver, C.S., Walsh, M., (1997). Air Pollution from Motor Vehicles: Standards and Technologies for Controlling Emissions, World Bank.

[3] Lave, L., MacLean, H., Hendrickson, C. Lankey, R., (2000). Life-cycle analysis of alternative automobile fuel/propulsion technologies, *Environ. Sci. Technol. 34,* 3598–3605.

[4] MacLean, H. L., Lave, L.B., (2003). Evaluating automobile fuel/propulsion system technologies, *Progress in Energy and Combustion Science* 29, 1–69.

[5] Environment Canada: http://www.ec.gc.ca/indicateurs-indicators/default.asp?lang=en&n=58DE4720-1#sector

[6] Granovskii, M., Dincer, I., Rosen, M.A., (2006). Economic and environmental comparison of conventional, hybrid, electric and hydrogen fuel cell vehicles, *J. of Power Sources* 159, 1186–1193.

[7] Bradley, T.H., Frank, A.A., (2009). Design, demonstrations and sustainability impact assessments for plug-in hybrid electric vehicles, *Renewable and Sustainable Energy Reviews* 13, 115–128.

[8] J. Larminie and J. Lowry, Electric Vehicle Technology Explained, John Wiley and Sons, UK, 2012.

[9] Fitzsimmons, A.J., Fitzsimmons, J.M., (2006). Service Management: Operations, Strategy, and Information Technology, 5th Ed. McGraw-Hill, Boston, USA.

[10] Alter, S., (2008). Service system fundamentals: work system, value chain, and life cycle, *IBM Systems J.* 47, 71-85.

[11] Maglio, P., Srinivasan, S., Kreulen, J., Spohrer. J., (2006). Service Systems, Service Scientists, SSME, and Innovation, *Communications of the ACM* 49, 81-85.

[12] Rust, R., (2004). A call for a wider range of service research, *J. of Service Research* 6, 211-212.

[13] Vargo, S., Lusch, R., (2004). Evolving to a new dominant logic for marketing, *J. of Marketing* 68, 1-17.

[14] Lusch, R., Vargo, S., (2006). The service-dominant logic of marketing: dialog, debate, and directions, ME Sharpe Inc, Armonk, NY.

[15] Better Place: http://www.betterplace.com/

[16] Dresner, S., (2008). The Principles of Sustainability, 2th Ed., EarthScan, UK.

[17] Dryzek, J., (1997). The Politics of the Earth: Environmental Discourses, Oxford University Press, UK.

[18] International Labour Office, (2011). Global Employment Trends 2011, Geneva, Switzerland.

[19] Halme, M., Hrauda, G., Jasch, C., Kortman, J., Jonuschat, H., Scharp, M., Velte, D., Trindad, P. (2007). *Sustainable Consumer Services: Business Solutions for Household Market,* EarthScan Ltd, UK.

[20] Galbraith, J.R., (2002). Organizing to deliver solutions, *Organizational Dynamics* 31, 194-207.

[21] Wolfson, A., Tavor, D., Mark, S., Schermann, M., Krcmar, H., (2010). S^3-Sustainability and services science: novel perspective and challenge, *Service Science* 2, 216-224.

[22] Wolfson, A., Tavor, D., Mark, S., (2011). Sustainable services: the natural mimicry approach, *JSSM* 4, 125-131.

[23] Wolfson, A., Tavor, D., Mark, S., (2011). Better Place: A case study of the reciprocal relations between sustainability and service, *Service Science* 3, 172-181.

[24] Cameco Corporation (2010). Comparison of lifecycle greenhouse gas emissions of various electricity generation sources, Lifecycle GHG Emissions.

[25] Wilson, W., Leipzig, T., Griffiths-Sattenspiel, B., (2012). Burning our rivers: The water footprint of electricity, River Network, Portland, Oregon.

[26] Wu, M., Peng, M.J., (2010). Developing a tool to estimate water use in electric power generation in the United States, Center for transportation research, energy systems divisions, Argonne National Laboratory.

[27] Fthenakis, V., Kim, H.C., (2009). Land use and electricity generation: A life-cycle analysis, *Renewable and Sustainable Energy Reviews* 1465–1474.

In: Grid Electrified Vehicles
Editor: Carla Alexandra Monteiro da Silva

ISBN: 978-1-62808-839-7

Chapter 4

HOW DRIVER CHOICES INFLUENCE CHARGING INFRASTRUCTURE EFFECTIVENESS IN THE UNITED STATES

Scott Peterson *

Carnegie Electricity Industry Center, Pittsburgh, PA, US

ABSTRACT

Federal plug-in electric vehicle (PEV) policies in the United States (US) currently include vehicle purchase subsidies linked to PEV battery capacity and subsidies for installing charging stations. This chapter will explore how consumer charging behavior can affect gasoline consumption over the life of a plug-in hybrid electric vehicle. The total fuel savings are calculated as a function of dwell time and infrastructure capability across a range of plug-in hybrid vehicles (PHEVs). I show that PHEVs with a range of 15 to 20 miles derive the most benefit in terms of gasoline savings. When the value of owner's time is accounted for instead of simply fuel savings PHEVs with a 20 mile range have the highest value. The total value to an owner for plugging in is small and non-monetary reasons will likely drive decisions.

Keywords: Charging, plug-in hybrid electric vehicle, infrastructure, dwell time, gasoline displacement, income

INTRODUCTION

Plug-in hybrid electric vehicles (PHEVs) have been on the market in the US since the end of 2010 when General Motors introduced a PHEV called the Volt. At the same time Nissan introduced a battery electric vehicle (BEV) called the Leaf, and since then other automakers have also begun to offer PHEV and BEV models -- collectively referred to here as plug-in electric vehicles (PEVs). These types of vehicles use grid electricity to displace gasoline. As

* Email: sxotty@gmail.com

of January 2013 three major manufacturers of vehicles in the US have at least one PHEV for sale (GM, Ford, and Toyota). These PHEVs may partially be in response to new corporate average fuel economy standards (CAFE) recently finalized in August 2012. These rules published by the National Highway Traffic Safety Administration (NHTSA) push CAFE standards higher to 40.3 mpg by 2021 and 48.7 mpg by 2025 [1]. In addition to CAFE standards the Environmental Protection Agency (EPA) released greenhouse gas (GHG) standards of 163 g/mile by 2025 that will match (these values correspond with 54.5 mpg, but the EPA expects other improvements such as a decrease in coolant leakage to reduce GHG emissions) [1]. The EPA standards include a number of incentives to encourage the use of "game changing" technologies [1]. These incentives begin with "off cycle" credits for model year (MY) 2012-2016 vehicles that implement specific technologies. Alternative fuel vehicles such as PHEVs are permitted a multiplier for model years 2017-2021 in the GHG calculations [1]. The rule also treats emissions from electric travel as 0 g/mile in model year 2022-2025 [1]. Thereafter upstream fuel emissions will be taken into account [1].

PEVs are currently more expensive than conventional vehicles (CVs). To help overcome obstacles to PEV adoption, policymakers have provided purchase incentives based on battery size and incentives for installation of charging infrastructure. The American Recovery and Reinvestment Act of 2009 provides a tax credit of $2,500 per plug-in hybrid electric vehicle sold (requires at least 4 kWh battery capacity) and an additional $417 for each additional kWh of battery capacity in excess of 4 kWh (capped at $7,500 for vehicles with a gross vehicle weight less than 14,000 pounds) [2]. This subsidy for a specific manufacturer's vehicles declines to 50% then 25% in a phaseout period, which begins in the second calendar quarter after that manufacturer has sold 200,000 vehicles and lasts four calendar quarters [2].

The US Department of Energy (DOE) also granted $37 million for installing 4,600 charge points in specific markets around the nation (over $8,000 per charge point) [3] and granted $99.8 million to fund the EVProject, which is installing 14,000 level 2 (208-240V) chargers and a variety of other infrastructure and monitoring equipment [4]. BEVs require a large number of charge points and advancements in technology to act as primary vehicles. It is possible that the charge points could benefit PHEVs as well.

It is questionable whether total fuel use will decline because of efficiency requirements under CAFE standards. However, the gasoline use by an individual who chooses to purchase a PHEV instead of a CV or hybrid electric vehicle (HEV) will likely be diminished depending on how the vehicle is used. Furthermore calculations for CAFE standards are based on EPA efficiency ratings of vehicles. The behavior of drivers of PHEVs could significantly alter the actual efficiency ratings of vehicles. What sets a PHEV apart from a BEV or HEV is an internal combustion engine (ICE) and a substantial battery pack. This allows them to take long trips using gasoline, while short trips can be powered mostly or entirely using electricity. Because of this, a PHEV designed for 40 miles of electric range will weigh and cost significantly more than a HEV with comparable interior space and performance.

Depending on how PHEVs are used more or less gasoline will be consumed. PHEV drivers will not need to change their habits at all, but it is hoped they will plug the vehicles in otherwise there would be an increase in fuel consumption compared to HEVs due to increased weight. Recent publications have suggested PHEVs with small all-electric range (AER) can save gasoline most cost effectively [5] and that charging them during the day will lead to greater reductions in gasoline consumption than higher AER PHEVs [6, 7]. A broad range of possible AERs were not considered and the time it would take to plug a vehicle in was also ignored. This chapter will examine how these factors could influence charging behavior and affect gasoline consumption.

METHODS

The basic method for calculating fuel use in this chapter is taken from [8] with slight changes described below. The necessary time for a consumer to be parked in order to choose to charge their vehicle is varied from five minutes to four hours. A variety of possible values are attributed to the time taken to plug a vehicle in.

Fuel Use Model

The fuel use model uses two main data sources. The first is the National Household Travel Survey 2009 (NHTS) which includes travel information from diaries of over 150,000 households [9]. The NHTS day trip file lists trips taken by each household on a randomly-assigned day. The second data source is the Department of Energy's Greenhouse Gases, Regulated Emissions, and Energy Use in Transportation (GREET) model, version1.8d [10]. Units in this report will often be in US customary units because this aligns with the GREET model. The GREET model assesses energy use in transportation and includes simulations for passenger cars, SUVs, and light trucks with over 80 fuel systems and technologies. I use GREET 1.8d estimates of year 2015 vehicle efficiency in charge depleting (CD) and charge sustaining mode (CS). It spans a set of PHEVs with battery capacity values producing all-electric range (AER) values from 5-60 miles. It is possible to combine these efficiency numbers with vehicle travel patterns from the NHTS day trip file allowing the prediction of estimated gasoline and electricity consumption from a set of PHEVs of a given AER. A basic overview of model is shown in Figure 1. The GREET model assumes that PHEVs with AER less than or equal to 25 miles utilize a split powertrain and blended control strategy, while PHEVs with AER in excess of 25 miles utilize a serial hybrid design, which is far less efficient in charge sustaining mode (the Chevy Volt with AER of 35 miles does not use a serial design [11]). This assumption causes the model to predict lower efficiency for vehicles with AER over 25 miles (as shown later in Figure 3). To avoid the serial powertrain assumption, I consider only AER of 25 miles or less in this chapter.

The NHTS data were processed following [8] with minor changes as described below. The daily vehicle travel data were extracted and weighted according to the vehicle weights assigned in the sample. To estimate the timing of fuel savings, vehicles are partitioned by age and vehicle class because their travel varies significantly along both dimensions (Figure 2, Figure 3). They are further partitioned into trips taken on weekdays and weekends and then the results weighted by the number of weekend (104) and weekdays (261) in the year (see equation set 1). The trip chain and the resulting total distance traveled in CD and CS mode for each vehicle under each charging strategy scenario is averaged over the vehicles within each class and age and within each weekend or weekday group in the NHTS data set. This approach represents the driving patterns of an aggregate US vehicle for age, and I do not examine heterogeneity for different vehicles that have different driving patterns within each class-age group [12-, 13, 14, 15].

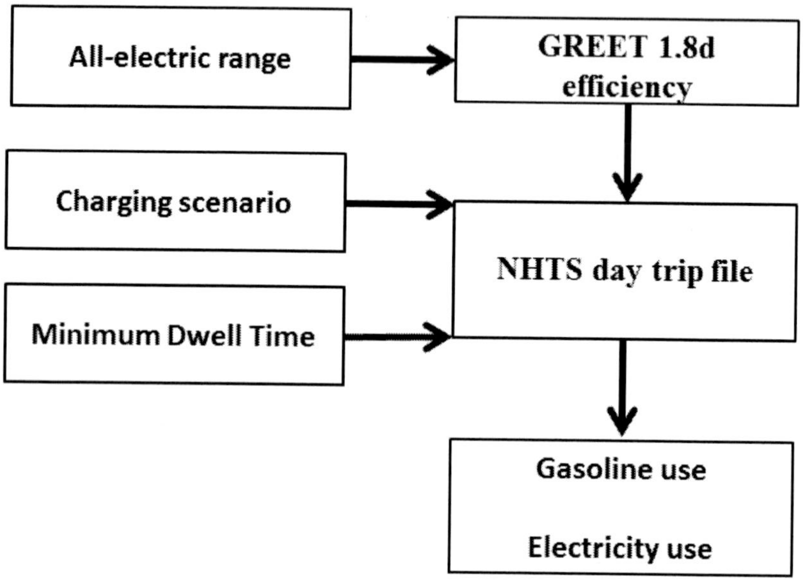

Figure 1. Basic fuel use model overview.

Changes in efficiency and AER as vehicles age were ignored. PHEV battery capacity will decline with age, which would shorten AER; however, vehicles are typically designed to use only a portion of the battery's available energy, in part to delay reduction in AER perceived by the consumer. Whether PHEVs will encounter more or less engine wear than CVs depends on how PHEVs use their engines. If the engine speed is partially decoupled from the wheel speed using the electric motor to compensate, this could prolong engine life by enabling the engine to run mostly at steady state, but if the engine starts and stops often and revs to improve the perceived responsiveness then engine wear could increase. Unlike in [8] I consider only cars in this chapter. PHEVs of other types are not yet on the market in the US, but the overall results should still be applicable even if they vary slightly in use patterns.

The energy consumption of each vehicle in the sample is calculated assuming vehicles began each day fully charged and charge completely after the last trip of the day. Throughout this chapter unless otherwise noted I will assume that owners will plug their vehicle in at least in the evening after they are finished driving. Several charging scenarios were included to determine how much additional gasoline could be displaced with electricity by also charging vehicles at work or also at every location where the vehicle parked for at least 30 minutes. Table 2 describes these charging scenarios.

Specifically, the total electricity use f^{ELEC} and gasoline use f^{GAS} are calculated for each vehicle age a (years), vehicle AER β (miles), and charging scenario γ. For each case, I compute average distances traveled in CD and CS modes separately for weekday (WD) vs. weekend (WE) data, and I use these averages to compute annual and total distances, as shown in Equation set 1, where L is the vehicle life (12 years in base case), j indexes the vehicle driving profiles taken from the NHTS day trip file, $J_{a,\text{WE}}$ is the set of NHTS car profiles of age a surveyed on a weekend, $J_{a,\mathrm{WD}}$ refers to those surveyed on a weekday, $|J_{a,\mathrm{WE}}|$ is the number of NHTS cars of age a surveyed on a weekend, $|J_{a,\mathrm{WD}}|$ is the number of cars of age a surveyed on a weekday, and $\eta^{\text{CD-E}}$, $\eta^{\text{CD-G}}$, and $\eta^{\text{CS-G}}$ are the vehicle's electrical efficiency in CD mode (mi/kWh), gasoline efficiency in CD mode (mi/gal, for blended operation), and gasoline efficiency in CS mode (mi/gal) as estimated by GREET (shown in Table 1). The CV and HEV are assumed to have efficiencies of 27 and 38 miles per gallon respectively. These GREET values are estimates for efficiency of future vehicles as indicated by year. Note that the PHEVs run blended operation in CD-mode, meaning that they consume both gasoline and electricity.

$$f_a^{\text{ELEC}}(\beta,\gamma)=104\,\frac{\sum_{jJ_{a,\text{WE}}} d_j^{\text{CD}}(\beta,\gamma)}{|J_{a,\text{WE}}|\eta_c^{\text{CD-E}}}+261\,\frac{\sum_{jJ_{a,\text{WD}}} d_j^{\text{CD}}(\beta,\gamma)}{|J_{a,\text{WD}}|\eta_c^{\text{CD-E}}}$$

$$f_a^{GAS}(\beta,\gamma)=104\left(\frac{\sum_{jJ_{a,\text{WE}}} d_j^{\text{CD}}(\beta,\gamma)}{|J_{a,\text{WE}}|\eta_c^{\text{CD-G}}}+\frac{\sum_{jJ_{a,\text{WE}}} d_j^{\text{CS}}(\beta,\gamma)}{|J_{a,\text{WE}}|\eta_c^{\text{CS-G}}}\right)+261\left(\frac{\sum_{jJ_{a,\text{WD}}} d_j^{\text{CD}}(\beta,\gamma)}{|J_{a,\text{WD}}|\eta_c^{\text{CD-E}}}+\frac{\sum_{jJ_{a,\text{WD}}} d_j^{\text{CS}}(\beta,\gamma)}{|J_{a,\text{WD}}|\eta_c^{\text{CS-E}}}\right)$$

$$f^{\text{ELEC}}(\beta,\gamma,L)=\sum_{a=1}^{L} f_a^{\text{ELEC}}(\beta,\gamma)$$

$$f^{\text{GAS}}(\beta,\gamma,L)=\sum_{a=1}^{L} f_a^{\text{GAS}}(\beta,\gamma)$$

Equation set 1

Table 1. Vehicle efficiencies for 2015 vehicles from GREET 1.8d. $\eta^{\text{CD-E}}$ in mi/kWh, - $\eta^{\text{CD-G}}$ in mi/gallon, and $\eta^{\text{CS-G}}$ in mi/gallon

	$\eta^{\text{CD-E}}$ **(mi/kWh)**	$\eta^{\text{CD-G}}$ **mi/gallon**	$\eta^{\text{CS-G}}$ **mi/gallon**
PHEV 5	5.2	74	43
PHEV 10	5.2	74	43
PHEV 15	5.3	78	43
PHEV 20	5.3	82	42
PHEV 25	5.3	82	42

The functions d^{CD}_j and d^{CS}_j use the NHTS data to compute the distance that a vehicle with AER of β under charging scenario γ traveling on vehicle day trip profile j would travel in CD mode and CS mode, respectively. I examine $\beta\in\{5,10,15,20,25\}$ miles and $\gamma\in\{$*HomeEve, HomeAll, WorkHomeEve, WorkHomeAll, AllStops*$\}$, where the charging scenarios are described in Table 2.

The procedure for computing d^{CD} and d^{CS} for each vehicle in the set of a given age, AER, charging scenario, and weekday or weekend starts by assuming each vehicle begins the day fully charged. The vehicle is tracked through all reported trips and it is assumed it operates first in CD mode, where it consumes both electricity and gasoline, switches to CS mode once the battery drops to its target state of charge (SOC) (40% of battery energy remaining according to GREET), and fully recharges after the last trip of the day. I use trip distances and times specified in the NHTS dataset and assume a constant efficiency per vehicle miles travelled (VMT) from GREET ($\eta^{\text{CD-E}}$, $\eta^{\text{CD-G}}$, and $\eta^{\text{CS-G}}$), ignoring differences in efficiency for different drive cycles [16]. For each trip if a vehicle's battery is above the target SOC then the total battery energy required to complete a trip is calculated. If the battery has enough energy, the SOC is decremented by the energy requirements of the trip. If the SOC is too low to complete the trip in CD mode then the SOC is decremented to the target SOC and the portion of the trip not travelled in CD mode is travelled in CS mode. If the vehicle battery is at the target SOC at the beginning of a trip the entire trip is travelled in CS mode. When a vehicle parks, the time between trips is calculated. If it is greater than or equal to the minimum dwell time required to charge (5-240 minutes) then the vehicle can charge if the designated charging scenario allows charging at the location where the vehicle is parked. The charging rates are assumed to be 1.4 and 7.7 kW for home and nondomestic (work and public) charging, respectively. Higher rates have small effects on gasoline and electricity consumption.

Table 2. Charging scenarios

Charging Scenario	Brief Description
HomeEve	Vehicle charges after arriving home on last trip of the day
HomeAll	Vehicle charges anytime it is parked at home for at least 5-240 minutes
WorkHomeEve	Vehicle charges when it first arrives at work and is parked for at least 5-240 minutes and at home after last trip of the day
WorkHomeAll	Vehicle charges anytime it is parked at either home or work for at least 5-240 minutes
AllStops	Vehicle charges anytime it is parked anywhere for at least 5-240 minutes

GREET also accounts for reduction in real-world efficiency compared to test cycle efficiency, where the AER is rated. This means that simulated AER may be shorter than rated AER (but this is primarily pronounced for the serial hybrid configuration that is not included in this analysis). Vehicles that are not driven on the survey day have a d^{CD} and d^{CS} of zero but are included so that the average total mileage found when simulating the trips (found by adding d^{CD} and d^{CS} of all vehicles in a set and dividing by the total number in that set) and fuel use estimates properly reflect all cars in NHTS. The resulting daily consumption is multiplied by either 104 for weekends or 261 for weekdays and summed to convert to annual consumption for a given age, AER, and charging scenario. The NHTS file does not specify if travel was on a holiday, nor are vacation days specified, so such days are included in computing average weekday or weekend travel. Calculating fuel use by CVs and HEVs is accomplished in the same manner, but total mileage is used instead of tracking d^{CD} and d^{CS} separately, since these vehicles do not use multiple energy sources.

The results of total distance travelled annually by all vehicles in a set of given age divided by the number of vehicles (both those driven and not driven on the day surveyed) in that set is shown in Figure 2.

To simulate the life of a vehicle it was assumed that it was driven in a manner consistent with reported NHTS data for a vehicle of its age and class. Thus, for the base case of a 12 year vehicle lifetime I assume a car drives roughly 14,000 miles in the first year, 13,000 in the next year and so on for each year until it reaches year 12 (Figure 2). Calculation of fuel consumption treats each vehicle age separately so any changes in travel patterns as vehicles age are accounted for.

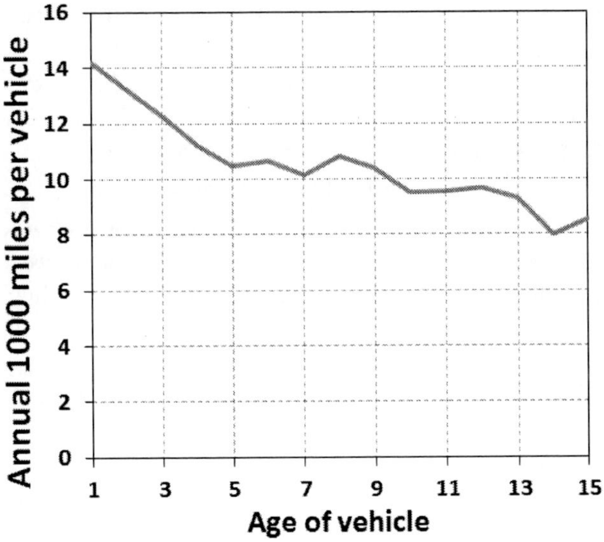

Figure 2. Change in annual VMT with vehicle age as found from NHTS data.

Gasoline consumption for PHEV cars, with AERs from 5 to 60 miles, is shown in Figure 3. Gasoline consumption, over the AER from 5 to 20 miles in *HomeEve* charging scenario, seems to decline almost linearly. This is due to most driven vehicles driving far enough to make use of their battery packs. Ignoring the jump in gasoline consumption created by the change to serial hybrids a more pronounced curve can be seen as the vehicle AER increases further than many vehicles travel in a day.

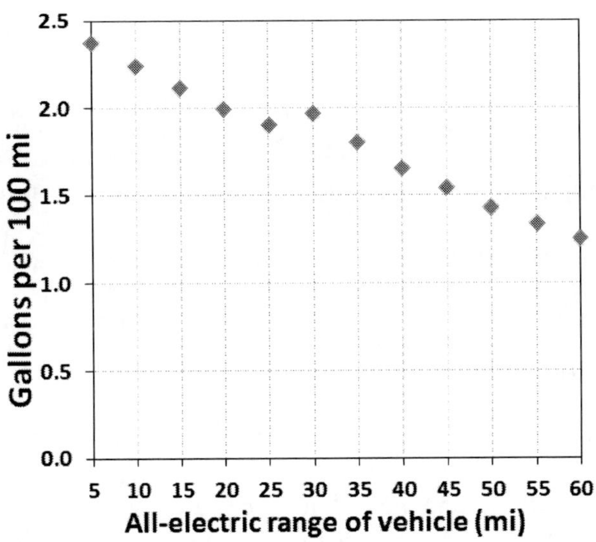

Figure 3. Gasoline consumption over a 12 year life for cars of various AER using GREET 1.8d 2015 efficiency estimates and *HomeEve* charging scenarios. AER corresponds to GREET AER. Gasoline use includes both CD and CS travel. The increase at AER of 30 is due to the assumption in GREET that PHEVs with AER above 25 have a series drivetrain rather than a split drivetrain.

Dwell Time, Charging, and Reduced Gasoline Consumption

Common sense suggests charging more often would result in a lower amount of gasoline consumed, but the amount of charging that consumers desire or are willing to do might change in response to a number of different factors. The availability of charging infrastructure, charge rate, time necessary to begin charging, cost of gasoline, and cost of electricity are all possible factors in the decision. Sensitivity analysis was conducted varying the dwell time requirement for each charging scenario, to determine the number of additional charges beyond once in the evening, that owners would conduct if they chose to plug their vehicles in when parked somewhere between 5 and 240 minutes on weekdays and weekends. The results are shown below (Figure 4). It is apparent that workplace charging is far less sensitive to dwell time requirements than home charging. This makes sense as most vehicles parking at workplaces tend to spend a longer time continuously parked than a vehicle stopping to pick up groceries or running other errands.

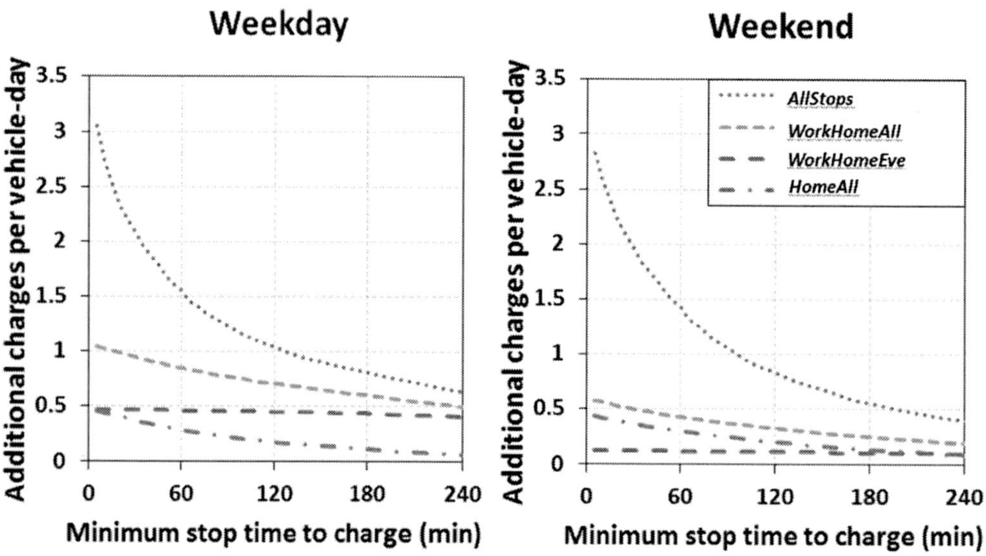

Figure 4. The number of additional charges per vehicle-day for vehicles driven; beyond one in the evening after trips are finished. The number of charges for all scenarios, except work charging, increase significantly in response to decreases in required minimum stop time for a consumer to plug their vehicle in. Note that the number of charges is the same across plug-in hybrids with different all-electric ranges, but fuel savings is not.

When looking at the number of additional charges, each charging scenario can be thought of as additive. For example, *WorkHomeAll* is the sum of *HomeAll* and *WorkHomeEve* and any additional charges at work. To find the additional charges conducted at work a person could subtract the two known values. The same is true for *AllStops* charging. It is important to note that while the amount of charges per vehicle day is increasing rapidly as dwell time requirements decrease the same is not necessarily true for gasoline savings given that each shorter stop has less charging time.

When the gasoline savings over the lifetime of a vehicle is calculated it is apparent that there are diminishing gasoline savings as minimum charge time decreases. The results are shown below for a PHEV with AER of 15 miles (Figure 5).

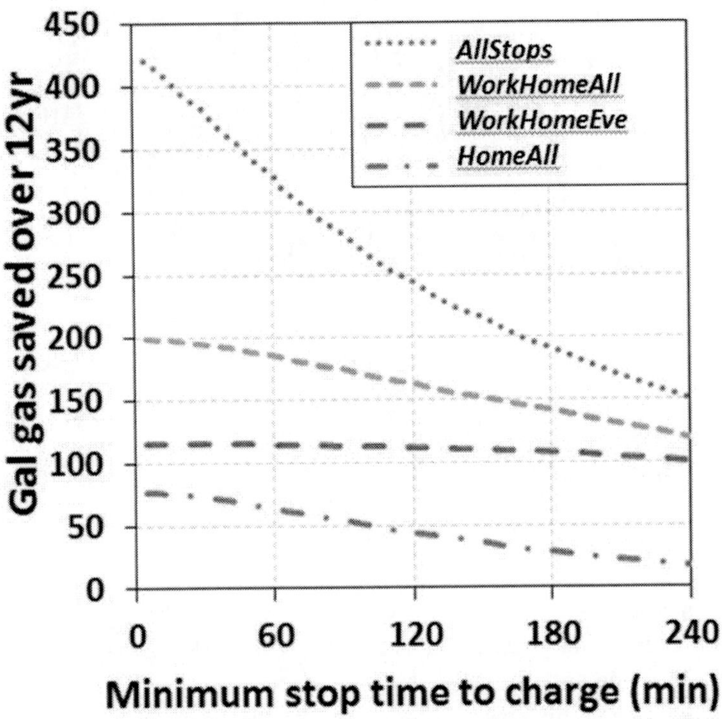

Figure 5. The additional volume of gasoline saved over lifetime of vehicle by charging scenarios other than *HomeEve* with various dwell time requirements for a PHEV with AER of 15 miles. All charging scenarios considered are shown.

Work charging appears to offer consistent gasoline savings, regardless of minimum stop time to charge, between five minute and two hours. Gasoline saved when charging at home or at all non-domestic locations is much more sensitive to the dwell time requirement for plugging in. The change in fuel savings is less sensitive than the change in number of charges.

The amount of gasoline savings also varies in response to changes in vehicle AER as shown below (Figure 6). This figure shows the gasoline saved as a function of AER and minimum stop time to charge in the *HomeAll* case. It is clear that at the shortest stop time an AER 15 PHEV gets the most gasoline savings from charging more often (this is why Figure 5 shows results for a PHEV with AER 15). As minimum parking times to charge increase the most gasoline saved shifts toward PHEVs with higher AERs. In no case does charging the PHEV with smaller AER save as much gasoline. It is likely because the AER is so small that it is depleted on most trips. This is precisely why PHEVs with small AERs are deemed the most cost effective. They utilize their entire battery on almost any trip. The other charging scenarios are not shown, but see a similar pattern except PHEVs with an AER of 20 miles have the maximum gasoline savings from additional charging though it is very close to the results for PHEVs with AER of 15.

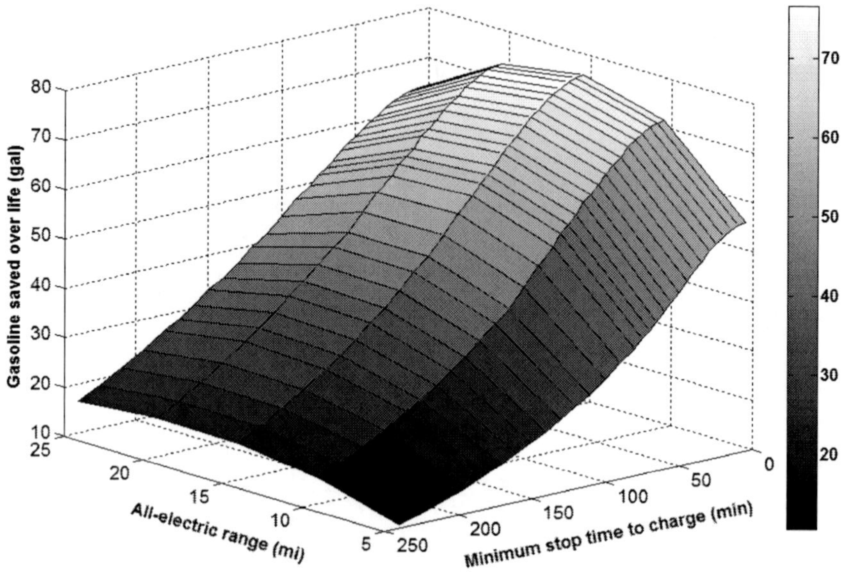

Figure 6. The gasoline savings compared to *HomeEve* nightly charge over 12 year life of vehicle given *HomeAll* charging and various all-electric ranges and minimum stop times to charge.

Vehicles with larger AER began to see smaller gasoline savings because they did not actually need the interim charge as much as PHEVs with AER of 15-20 miles. This means it is likely that as AER increases past 25 miles (outside the scope of the GREET model for power split designs) then the gasoline savings associated with additional charging will continue to decrease on average.

When the cost-effectiveness of non-domestic charging infrastructure was investigated and sensitivity was conducted on charge rate, a similar result was found (Figure 7) [8]. Vehicles with the shortest AER could charge rapidly (compared to battery capacity) on low rate chargers. Vehicles with AER 5 could often charge completely with a 30 minute minimum charge time as considered in [8]. As AER increases the difference in charge rate becomes more useful because it takes longer to charge a large depleted pack than a small depleted pack. However, as pack size continues to increase though there is a good chance that the battery will not be depleted between possible interim charges in the day. In fact, when PHEVs with AER of 30-60 miles were considered, it was found that the difference began to decrease.

If it is necessary to run a new circuit for charging then the marginal cost of a higher rate charger is negligible if panel upgrades are not necessary, but otherwise additional expenditure for home charging infrastructure to support higher rate charging seems largely superfluous. Similar results were found for the other non-domestic charging scenarios except a PHEV with AER of 20 received the peak benefit of higher rate charging and overall higher rate charging showed a greater benefit for non-domestic charging. Because most non-domestic charging will require some sort of installation it makes more sense to use higher rate chargers and that was the assumption throughout this chapter.

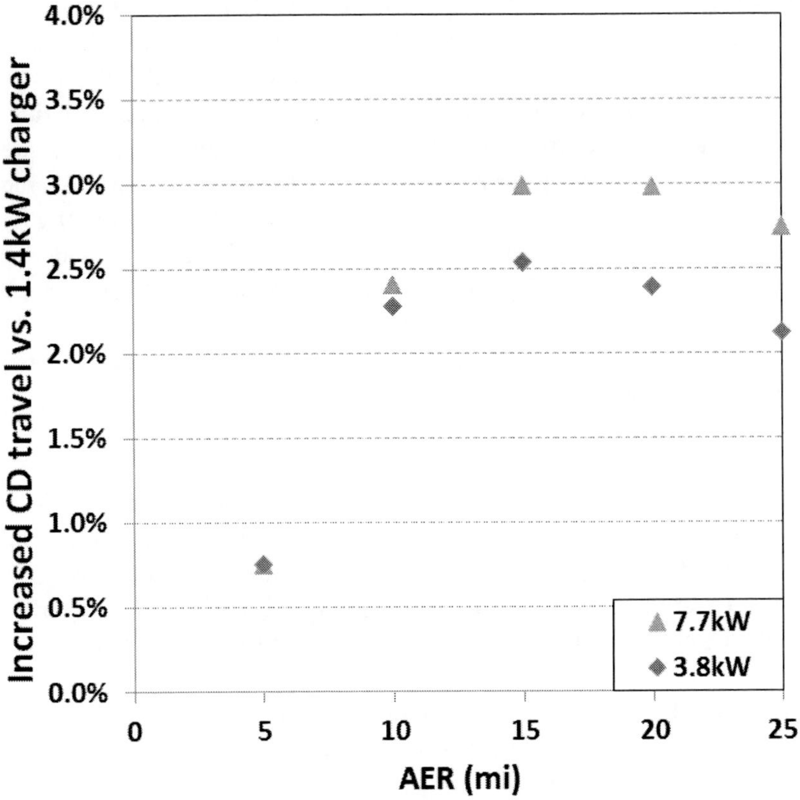

Figure 7. The increase in charge depleting mode travel as a response to home charging rates in the *HomeAll* charging scenario with a 30 minute minimum for charging.

RESULTS

Buyers of HEVs in the US have a higher than average income [17]. The same has so far been true for buyers of PHEVs. The average income of Prius buyers was $100,000 and Volt buyers $150,000 [17]. Other publications have stated the average household income of a purchaser of the Volt was $175,000 [18]. It is possible to use these numbers to make estimates about the decisions owners will make in terms of charging. To investigate the cost of plugging in I conducted analysis using separate income cases. In all cases I assumed that it took an owner ten seconds to plug their vehicle in at home and thirty seconds at non-domestic chargers. This difference is based on the assumption that domestic charging could be set up for maximum ease of use for a given vehicle (for example hanging a cord from the garage ceiling) and non-domestic charging may have to serve more than one vehicle type and require a transaction. These are just example times that seem reasonable.

The first case assumes an income of $175,000 and an after tax income of $140,000 [19]. The next case assumes a household income of $100,000 and an after tax income of $75,000 [20]. The final case assumes an average income of $47,000 or $36,000 after taxes [21]. I assumed 2,000 hours of work a year. This value is likely high for those in the upper incomes considered, but higher annual hours of work causes a lower value of time and therefore improves the likelihood that plugging in will make financial sense. The value of an hours work is $70, $36, and $18 respectively for each case considered. I consider fuel prices from the Energy Information Agency's Annual Energy Outlook (AEO) 2012 that lists retail prices including taxes [22]. I consider both the high oil price and reference scenarios for energy prices. Because the decision to charge is made at that moment no net present value calculations are needed at each decision point. Instead the current difference between electricity and gasoline prices can be used. The average cost for a gallon of gasoline in the reference case over the 12 year lifetime of the vehicles is $3.70 in the reference scenario and $5.11 in the high oil price scenario. Electricity prices are stable at $0.12 per kWh in each scenario. If charge point operators for nondomestic charging charge a higher rate for electricity the value of plugging in will decline from that shown here, if they charge a lower rate (or are free) then the value will increase.

The first case considered is the $70 per hour case. The *HomeAll* case is shown in Figure 8 and the process described below for a PHEV with an AER of 5 miles. The number of additional charges per day is shown in Figure 4 for weekdays and weekends. For the *HomeAll* case and a 60 minute dwell time requirement and additional 0.29 charges occur on weekdays and 0.30 on weekends. Over a 12 year life of a vehicle there are 1248 weekends on which 60% of vehicles are driven and 3132 weekdays on which 67% of vehicles are driven. This is a total of 830 additional times plugging a vehicle in over the life of the vehicle (as compared to 2800 if only plugging in after the last trip of the day). Using a 10 second time to plug in at home this works out to 7.9 hours spent plugging a vehicle in and at $70 per hour this is a $160 cost to the owner. The same owner saves 42 gallons of gasoline and consumes an extra 810 kWh of electricity by electing to charge more often, saving $120 in fuel costs. This results in a loss of $40 by plugging in more often, when the PHEV has an AER of only 5 miles.

For owners making $70 an hour PHEVs with AERs ranging from 10 more than 25 miles likely make sense to charge in the high oil price scenario (Figure 8). The solid line on the followingse figures represents the breakeven point where the average vehicle driven by such an owner would offset enough petroleum to make interim charging worthwhile. In other cases the cost of an owner's time to plug a vehicle in offsets savings from displacing gasoline with electricity in their PHEV.

In these cases an owner would be better off burning additional gasoline in their PHEV instead of taking the time to plug-in. This shows that PHEVs with an AER of 20 have the highest value from additional charging and that the maximum occurs at 75 minutes. So for an average driver earning more than $70 per hour it makes sense to plug in if they are parking for more than an hour at home. It would not cost them money to plug in for shorter times on average, but they would not have as high a value for those short duration charges.

For owners making $36 an hour, PHEVs with all AERs likely make sense to charge in the high oil price scenario (Figure 9). All values in this case are above zero so for the average owner making $36 an hour it would almost always make sense to plug in when parked if gasoline prices are $5.10 a gallon and it only takes ten seconds to plug in. Once again PHEVs with an AER of 20 have the highest value from additional charging and that the maximum occurs at 35 minutes.

So, for an average driver earning $36 per hour, it likely makes the most sense to plug their vehicle in, when parking for more than thirty minutes.

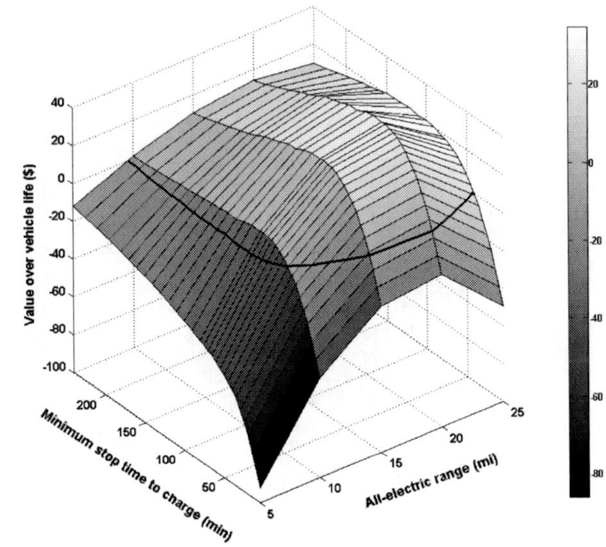

Figure 8. This plot shows the value of *HomeAll* charging scenario given different AERs and minimum stop times to charge. Assuming PHEV owner makes $70/hour, it takes 10 seconds to plug vehicle in, and fuel prices are from the 2012 Annual Energy Outlook high oil price scenario over the life of the vehicle. Solid line indicates breakeven value.

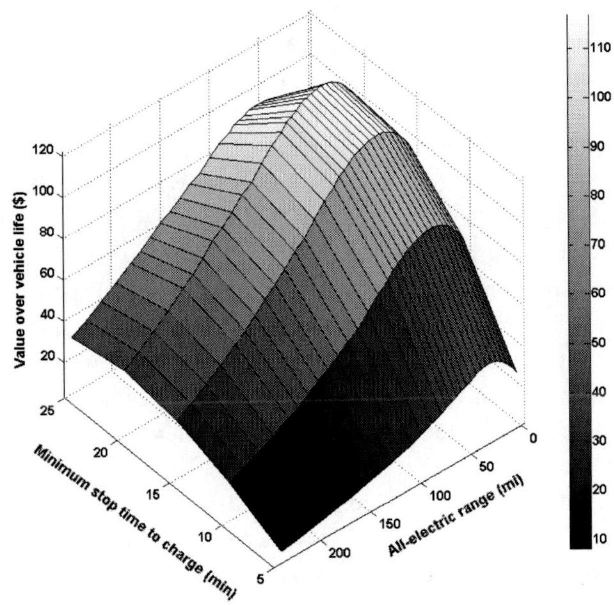

Figure 9. This plot shows the value of *HomeAll* charging scenario given different AERs and minimum stop times to charge. Assuming PHEV owner makes $36/hour, it takes 10 seconds to plug vehicle in, and fuel prices are from the 2012 Annual Energy Outlook high oil price scenario over the life of the vehicle. All values are above zero.

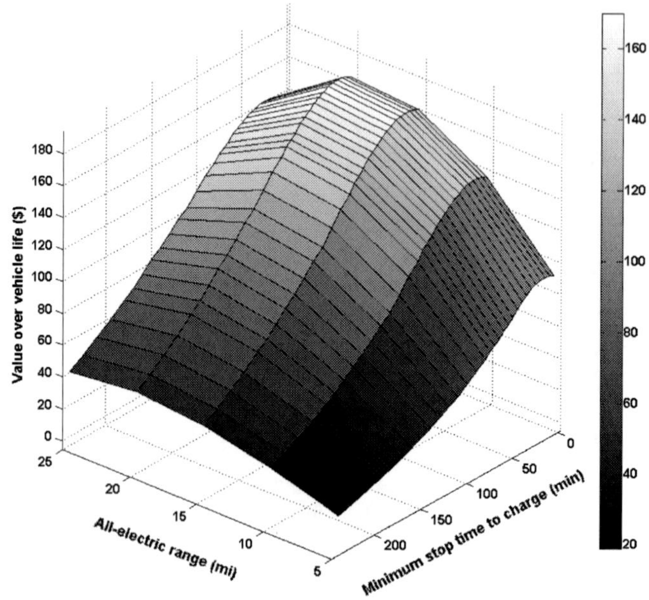

Figure 10. This plot shows the value of *HomeAll* charging scenario given different AERs and minimum stop times to charge. Assuming PHEV owner makes $18/hour, it takes 10 seconds to plug vehicle in, and fuel prices are from the 2012 Annual Energy Outlook high oil price scenario over the life of the vehicle. All values are above zero.

For owners making $18 an hour, PHEVs with all AERs likely make sense to charge in the high oil price scenario (Figure 10). All values in this case are above zero so for the average owner making $18 an hour it would almost always make sense to plug in when parked if gasoline prices are $5.10 a gallon and it only takes ten seconds to plug in. Once again PHEVs with an AER of 20 have the highest value from additional charging and that the maximum occurs at 20 minutes. So, for an average driver earning $18 per hour, it likely makes the most sense to plug their vehicle in, when parking for more than twenty minutes. It is unlikely that individuals earning $18 an hour will be able to afford PHEVs in the near future. High oil prices and consistent electricity prices, combined with a lower value of time shift the optimal time to plug in to a lower bound in this case. It is therefore likely that very few people would want to plug their vehicle in when parked as short as twenty minutes.

For owners making $70 an hour in the reference case scenario it never makes sense to charge any of the PHEVs considered over the times considered (Figure 11). It is however still worthwhile to plug the PHEVs in at night, one time after the last trip of the day. The maximum value is at 240 minutes associated with a AER 20 PHEV though it is still negative.

For owners making $36 an hour, PHEVs with AERs ranging from 10 to greater than 25 miles likely make sense to charge in the reference case price scenario (Figure 12). The solid line on these figures represents the breakeven point where the average vehicle driven by such an owner would offset enough petroleum to make charging worthwhile. This shows that PHEVs with an AER of 20 have the highest value from additional charging and that the maximum occurs at 60 minutes. So for an average driver earning $36 per hour it makes sense to plug in if they are parking for more than an hour at home. It would not cost them money to plug in for shorter times on average, but some more individuals would find the charges of less value so the overall average declines for those short duration charges.

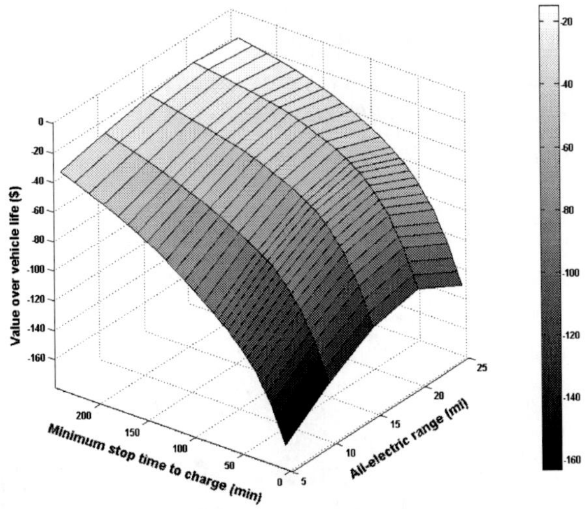

Figure 11. This plot shows the value of *HomeAll* charging scenario given different AERs and minimum stop times to charge. Assuming PHEV owner makes $70/hour, it takes 10 seconds to plug vehicle in, and fuel prices are from the 2012 Annual Energy Outlook reference case scenario over the life of the vehicle. All values considered are negative.

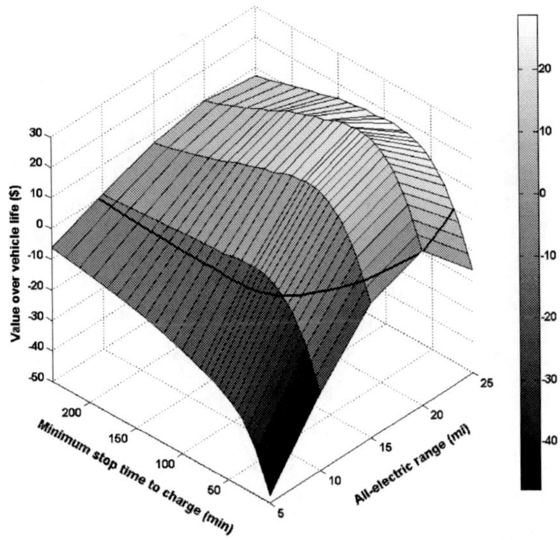

Figure 12. This plot shows the value of *HomeAll* charging scenario given different AERs and minimum stop times to charge. Assuming PHEV owner makes $36/hour, it takes 10 seconds to plug vehicle in, and fuel prices are from the 2012 Annual Energy Outlook reference case scenario over the life of the vehicle. Solid line indicates breakeven value.

The average income of purchasers of PHEVs exceeds $100,000; however it is reasonable to think that the average income of PHEV buyers will decline to this level as technology developsmore entrants to the marketplace appear.

Work place charging is fairly insensitive to the dwell time requirement for charging as shown above (Figure 5) so results are not included. For the *WorkHomeEve* and owners making $36 per hour, It was found that PHEVs with AERs from 15 to greater than 25 made sense to charge, in the reference case energy price scenario. There is very little response to dwell time requirements. In the *WorkHomeAll* scenario, charging at home dominates the results and this is shown above (Figure 8 - Figure 12).

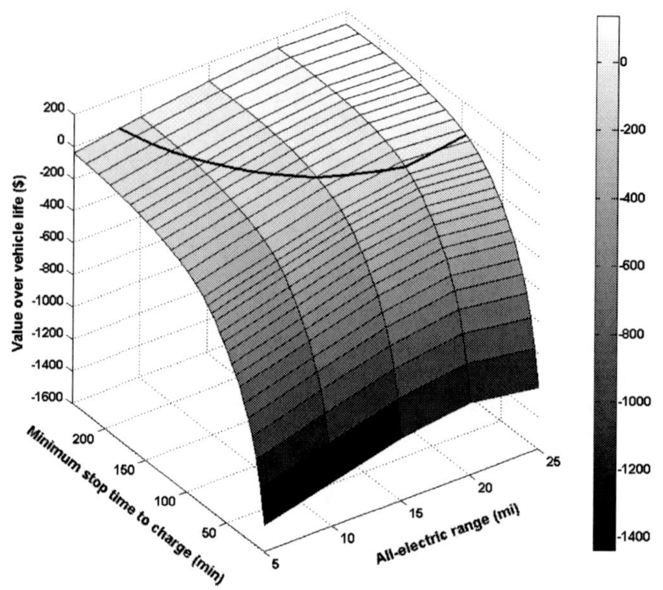

Figure 13. This plot shows the value of *AllStops* charging scenarios given different AERs and minimum stop times to charge. Assuming PHEV owner makes $36/hour, it takes 10 seconds to plug vehicle in at home, it takes 30 seconds to plug in at nondomestic charging points, and fuel prices are equal to the 2012 Annual Energy Outlook reference case price scenario over the life of the vehicle. Solid black line indicates break even transition.

The *AllStops* scenario shows a far greater drop off in value than the *HomeAll* scenario. For owners making $36 an hour, PHEVs with AERs ranging from 10 to slightly greater than 25 miles likely make sense to charge in the reference case price scenario (Figure 13). The solid line on these figures represents the breakeven point where the average vehicle driven by such an owner would offset enough petroleum to make charging worthwhile. This shows that PHEVs with an AER of 20 have the highest value from additional charging and that the maximum occurs at 240 minutes. Because this is the bound of the analysis conducted additional cases were run for 270, 300, 360, and 420 minutes. The results showed that 240 was still the maximum point, but because the results are not continuous it is possible that the true maximum is somewhere between 225 and 270 minutes. The minimums approach a cost of $1,500 over the life of the vehicle.

LIMITATIONS

The results from this analysis have a number of limitations that must be considered. The most obvious flaw in presenting the cost benefit analysis of the decision to plug a vehicle in or not is that most purchasers of these vehicles are not necessarily purchasing them for cost savings. For example, HEVs were not initially cost competitive with CVs, yet individuals adopted them for other reasons, including symbolism and image [23, 24]. Perhaps individuals purchase such vehicles because they feel it is in the collective good of society, much like sorting recyclables [25, 26]. It is likely that purchasers of PHEVs initially will show similar considerations. It is possible they enjoy driving using electricity and will therefore take the time to plug in even if it does not make sense from a financial point of view.

Other more utilitarian reasons to purchase PHEVs include preferred treatment such as high occupancy vehicle lanes, or parking spots. Anecdotal evidence can be found of individuals purchasing low AER PHEVs who plan never to plug them in. Using the percent of vehicles driven on a weekday from NHTS data over a 12 year vehicle lifetime and the value of time it is possible to make an estimate of the potential value of high occupancy vehicle (HOV) lane access. If an individual can save 30 minutes a day by using the HOV lane this has a value of \$41,000 or \$80,000 for the \$36 and \$70 per hour cases. The net present value given a 20% discount rate is \$15,000 and \$30,000. That benefit dwarfs any potential gasoline savings and approaches the cost for an entire vehicle. Most programs have an end date though so in the long run these sorts of programs will not distort the market toward low AER PHEVs, or support the sale of higher AER PHEVs. Preferred parking spots likewise could save significant time if they are closer to the destination. At the beginning of January 2012 a new California law came into effect requiring PEVs to be plugged in when parking in charger spots [27]. The law authorized towing of vehicles not plugged in. This was done to combat the tendency of those that did not need to charge to use spots parking there because they were convenient and to update a preexisting law that only allowed EVs and not PHEVs to park in charging spots and did not require them to charge.

These results are based on aggregate data. It is certain that there are individuals who would find charging at work worthwhile in monetary terms for example even of high income groups. These results can be thought of as determining how often the average driver of a given income level would find it worthwhile to charge. However, it is possible that widespread adoption of PHEVs would change driving patterns. Drivers might purchase PHEVs that match their expected driving habits, or they might bin their trips to the AER of the PHEV they purchased. Until there is more widespread adopting of PHEVs it will be difficult to determine how they might affect aggregate driver behavior. For example, showing that drivers of PHEVs drive longer distances does not show that this was in response to lower fuel costs. It is quite possible that drivers with longer commutes are more interested in purchasing an efficient vehicle. A more intelligent algorithm might attempt to predict future travel and determine whether plugging in at a given time makes sense, however it is not clear how precisely drivers predict their days travel ahead of time.

The lack of scenarios where a PHEV with AER 5 makes sense to plug in also might illustrate why no automakers have introduced PHEVs with such a low AER. The finding that PHEVs with AER of 20 miles show the maximum cost benefit from plugging in is robust across a variety of assumptions. This chapter does not attempt to analyze the benefits of charging infrastructure specifically. The results here in terms of the benefits to a consumer plugging in roughly mirror the results from a cost benefit analysis of infrastructure [8]. It shows that plugging in for interim charges at home makes the most sense, followed by work charging, and all stops charging.

Electricity costs taken from Energy Information Agency (EIA) predictions for residential rates might not reflect the price of electricity at work or other non-domestic locations. Currently there is a variety of pricing strategies for charging stations ranging from free, subscriptions based, to hourly rates. If owners have to pay significantly more for electricity than at home it is unlikely PHEV drivers would bother with charging unless the preferred parking space (or absence of paying a metered parking alternative) makes it attractive. In the $70 per hour case with EIA's reference energy prices it is never worth the time plugging the vehicle in for interim charges. If electricity is free then PHEVs with AER 15-25 make sense to plug in at work. Other non-domestic stops with a dwell time of more than 120 minutes also save money by charging.

I consider only gasoline savings and the time associated with plugging a vehicle in. Other benefits include changes in negative externalities from life cycle air emissions, ground emissions, and noise, and as well other as positive externalities associated with knowledge spillover in technology advancement [5]. Differences in these negative externalities among vehicles and charging scenarios are relatively small on average. Advancement of charger technology is more likely to occur with larger battery packs because smaller packs do not need high rate charging (only onboard thermal management) which also aligns with the results of this chapter.

It is possible that plugging in more often would decrease the number of stops necessary to fill a vehicle with gasoline. Initially estimates of these changes were included; however current PHEVs and HEVs often have smaller gas tanks than their comparable conventional vehicle counterparts. It was found that drivers of conventional vehicles would spend 36 hours filling them if it is assumed that they refuel every 11 gallons and spent 5 minutes. An HEV driver refueling every 9 gallons would spend 32 hours. The PHEV drivers who refuel every 8 gallons would spend 23-30 hours refueling depending on AER, but they would spend an additional 8 hours plugging their vehicle in at night. This key difference means that an owner of a PHEV5 would spend more total time than either an HEV or CV driver plugging and refueling. It is important to note that for the *HomeEve* scenario the time spent plugging a PHEV in is independent of AER. So as the AER increases the relative importance of time spent plugging in at night declines. Indeed all but the PHEV25 spent more time than the HEV refueling and plugging in for the evening. However, the cost of time spent refueling is highly uncertain due to possible variation in fuel tank size, changes in how often consumers refuel (perhaps they would be willing to let the fuel tank come closer to empty if they were driving on electricity anyway), and a variety of other factors so it was not included in this chapter.

Charging a vehicle more often has the potential to change battery life. Work looking at lithium nickel cobalt aluminum oxide (NCA) chemistry found that PHEVs with AER of 10 exhibited decreased battery life with opportunity charging (charging whenever they are parked), but PHEVs with AER of 40 actually showed an increased lifetime in response to opportunity charging because the battery was kept in its preferred state of charge window more of the time [28]. The PHEV with an AER of 20 that was found to give the most value to consumers for interim charges would likely see something in between or little change in battery lifetime. Other battery chemistries appear more robust across different states of charge and might only see a decrease in lifetime proportional to increased use in charge depleting mode [29].

Discussion

The goal of PHEVs is to enable drivers to displace petroleum products with electricity. Consumer charging behavior and vehicle choice could significantly affect the total savings. Although it has previously been found that installation of charging infrastructure is not cost effective for PHEVs, some amount on infrastructure will likely be installed for EVs and having PHEVs utilize the infrastructure could be beneficial [8]. This chapter shows that for average drivers PHEVs with an AER of 20 miles have the maximum benefit from charging. In the high oil price and $36 per hour income scenario the maximum value for *AllStops* charging was $430 when charging only when parked for over 150 minutes continuously.

If average consumers can afford the vehicle and the effective value of time falls to $18 per hour then the value reaches a maximum of $700 at 60 minutes of dwell time. Simply decreasing the price of PHEVs will therefore not significantly increase the likelihood of people charging their vehicles throughout the day. If electricity for charging away from home were free (as it currently is in some locations) then this value would be roughly $1,300 over 12 years. It seems quite unlikely, that in the near future, the average purchaser of a PHEV with AER 20 will make $40,000 a year. Even if that were the case and oil prices continued to rise to $5.10 per gallon the total benefit ($1,300) from this increased charging is still quite small. This benefit spread is a very small amount over a 12 year period and suggests that it is unlikely to play a key role in a person's decision making.

The gasoline savings enabled by charging at all stops is only 10% of the total gasoline consumption of a CV. Transitioning to PHEVs saves between 40 and 53% of the gasoline consumption of a CV, but a large part is the transition to a hybrid architecture that no longer wastes braking energy. Compared to HEVs the transition to PHEVs saves only 18 to 34% of the gasoline used by an HEV, however additional charging saves a slightly higher percentage because HEVs use less fuel. A 10% reduction in fuel consumption is worthwhile, but it is likely that decisions to charge will be based on advantages related to parking, electricity rates, and proenvironmental behavior.

Making electricity free at charging points will likely lead to greater use than would be merited simply by the savings a driver would experience. This is likely due to framing it as free even though there is a cost associated with spending the time to plug in. It is also likely that those who choose to drive a PHEV primarily for the idea of doing something good for the planet will plug in their vehicles more often. If the time to begin charging is removed it would make about the same difference as free electricity. In the spring of 2013 a company is attempting promising to deliver wireless charge points for Nissan Leafs and Chevy Volts. They say the wireless chargers would be 90% efficient and cost $2,500 [30]. Given a $36 per hour time value home charging with the wireless charger would save $160-$230 for interim charges and an additional $400 for evening charges. This means time savings over the life of the vehicle do not offer payback for a wireless charger, and it is unclear whether wireless technology will be adapted for public charging points and how much the cost might decrease. However it is clear that this would remove one limitation for charging infrastructure use. If future vehicles were produced to support wireless charging, a more elegant and cost effective system could also be developed that combined autonomous parking as well. At the moment there are a variety of wired charging protocols for PEVs so it could take a significant amount of time before any wireless standard was developed. Further consolidation would be beneficial for charging infrastructure utilization. While it is possible to have adapters to convert between some of the charging protocols it makes charging into aa more time consuming process greater hassle.

Parking locations that are more desirable combined with laws forcing vehicles to be plugged in when occupying them could increase charging, but it is also possible that these laws could backfire. If a single charging station can serve multiple parking spots then after one vehicle was finished another might plug in, but laws requiring a vehicle to be plugged in to be allowed in the parking space would make sharing of charge points impossible. In many of the cases the maximum average benefit occurs when parked around an hour and it is quite possible that an owner might not come and unplug their vehicle and move it when charging is finished.

Conclusion

Plug-in hybrid electric vehicles can reduce their gasoline consumption by charging throughout the day. It was found that cars with all-electric ranges between fifteen and twenty miles saved the most gasoline from charging between trips. The total savings depends on how often owners are willing to plug in their vehicles to charge. Accounting for the time spent plugging vehicles in and the fuel savings, cars with all-electric ranges of twenty miles gave the most value to owners who conducted interim charges. In likely scenarios Plug-in hybrids with a five mile all-electric range (the smallest considered) were never worthwhile to plug in for additional charging. The total value depends on the owner's income level, but was relatively small (less than $640 over twelve years) compared to other factors that will likely affect owner behavior such as preferred parking, or free electricity at charging points. Efforts to encourage interim charging may meet with public opposition if they are seen as preferential treatment for well off individuals. Charging at home between trips offers the most value and requires no additional infrastructure investment.

Nomenclature

AEO	Annual Energy Outlook
AER	all electric range
AllStops	Vehicle charges anytime it is parked anywhere for at least 5-240 minutes
BEV	Battery electric vehicle
CAFE	corporate average fuel economy
CD	charge depleting
CS	charge sustaining

CV	conventional vehicle
DOE	Department of Energy
EIA	Energy Information Agency
EPA	Environmental Protection Agency
GHG	greenhouse gas
GREET	Greenhouse Gases, Regulated Emissions, and Energy Use in Transportation
HEV	hybrid electric vehicle
HomeAll	Vehicle charges anytime it is parked at home for at least 5-240 minutes
HomeEve	Vehicle charges after arriving home on last trip of the day
ICE	internal combustion engine
MY	model year
NCA	lithium nickel cobalt aluminum oxide battery
NHTSA	National Highway Traffic Safety Administration
NHTS	National Household Travel Survey
PEV	Plug-in electric vehicles
PHEV	Plug-in hybrid electric vehicle
SOC	state of charge
US	United States of America
WE	weekend
WD	weekday
WorkHomeEve	Vehicle charges when it first arrives at work and is parked for at least 5-240 minutes and at home after last trip of the day
WorkHomeAll	Vehicle charges anytime it is parked at either home or work for at least 5-240 minutes
VMT	Vehicle miles travelled
a	vehicle age years
d^{CD}_j	distance travelled in charge depleting mode
d^{CS}_j	distance travelled in charge sustaining mode
f^{ELEC}	function tabulating electricity usage by PHEVs
f^{GAS}	function tabulating gasoline usage by PHEVs
j	index of vehicle driving profiles taken from the NHTS day trip file
$J_{a,\text{WE}}$	the set of NHTS car profiles of age a surveyed on a weekend
$J_{a,\text{WD}}$	the set of NHTS car profiles of age a surveyed on a weekday
$\lvert J_{a,\text{WE}}\rvert$	the number of NHTS cars of age a surveyed on a weekend
$\lvert J_{a,\text{WD}}\rvert$	the number of cars of age a surveyed on a weekday
L	vehicle life (12 years)
β	vehicle AER miles
γ	charging scenario
$\eta^{CD\text{-}E}$	electrical efficiency in charge depleting mode
$\eta^{CD\text{-}G}$	gasoline efficiency in charge depleting mode
$\eta^{CS\text{-}G}$	gasoline efficiency in charge sustaining mode

REFERENCES

[1] US Environmental Protection Agency, and US National Highway Traffic Safety Administration. 2012. *2017 and Later Model Year Light-Duty Vehicle Greenhouse Gas Emissions and Corporate Average Fuel Economy Standards*. Washington, DC: US Environmental Protection Agency. http://www.nhtsa.;gov/staticfiles/rulemaking/pdf/cafe/2017-25_CAFE_Final_Rule.pdf.

[2] The American Recovery and Reinvestment Act of 2009, 2009. PublicLaw.111–5, Congress. http://frwebgate.access.gpo.gov/cgi-bin/getdoc.cgi?dbname=111_cong_bills&docid=f:h1enr.pdf (accessed November 26, 2010).

[3] S. Dogget. "Coulomb gets $37 million from energy dept. to deploy 4,600 ev-charging stations." http://blogs.edmunds.com/greencaradvisor/2010/06/coulomb-gets-37-million-from-energy-dept-to-deploy-4600-ev-charging-stations.html (accessed February 20, 2011).

[4] *Advanced Vehicle Testing Activity*. Washington, DC: US Department of Energy. http://www1.eere.energy.gov/vehiclesandfuels/avta/light_duty/m/arra_vehicle_rpts.htm l (accessed December 14, 2011)

[5] Michalek, Jeremy J., Mikhail Chester, Paulina Jaramillo, Constantine Samaras, Ching-Shin Norman Shiau, and Lester B. Lave. 2011. "Valuation of Plug-in Vehicle Life-cycle Air Emissions and Oil Displacement Benefits" (September 26).

[6] Markel, T, K Smith, and A. A. Pesaran. 2009. "Improving Pertoleum Displacement Potential of PHEVs Using Enhanced Charging Scenarios." In Stavanger, Norway. http://www.nrel.gov/vehiclesandfuels/energystorage/pdfs/45730.pdf.

[7] Dong, Jing, and Zhenhong Lin. 2012. "Within-day Recharge of Plug-in Hybrid Electric Vehicles: Energy Impact of Public Charging Infrastructure." *Transportation Research Part D: Transport and Environment* 17 (5) (July): 405–412. doi:10.1016/j.trd.2012.04.003.

[8] [1]. Peterson, Scott B., and Jeremy J. Michalek. 2013. "Cost-effectiveness of Plug-in Hybrid Electric Vehicle Battery Capacity and Charging Infrastructure Investment for Reducing US Gasoline Consumption." *Energy Policy* 52 (January): 429–438. doi:10.1016/j.enpol.2012.09.059.

[9] National Household Travel Survey, US Department of Transportation, Washington DC, 2010.

[10] The Greenhouse Gasses, Regulated Emissions, and Energy Use in Transportation (GREET) Model, version1.8d. US Department of Energy: Washington DC, 2010

[11] Markus, Frank. 2010. "Unbolting the Chevy Volt to See How It Ticks", October 12. http://www.motortrend.com/features/editorial/1010_unbolting_the_chevy_volt_to_see_how_it_ticks/.

[12] Raykin, Leon, Matthew J. Roorda, and Heather L. MacLean. 2012. "Impacts of Driving Patterns on Tank-to-wheel Energy Use of Plug-in Hybrid Electric Vehicles." *Transportation Research Part D: Transport and Environment* 17 (3) (May): 243–250. doi:10.1016/j.trd.2011.12.002.

[13] Shiau, Ching-Shin Norman, Constantine Samaras, Richard Hauffe, and Jeremy J. Michalek. 2009. "Impact of Battery Weight and Charging Patterns on the Economic

and Environmental Benefits of Plug-in Hybrid Vehicles." *Energy Policy* 37 (7) (July): 2653–2663. doi:10.1016/j.enpol.2009.02.040.

[14] J. Neubauer, A. Brooker, E. Wood, Journal of Power Sources 209 (2012) 269–277.

[15] E.C. Traut, C.T. Hendrickson, Y.L. Klampfl, J.J. Michalek, Optimal design and allocation of electrified vehicles and dedicated charging infrastructure for minimum life cycle greenhouse gas emissions and cost . Energy Policy (2012). To appear.

[16] R. Patil, B. Adornato, Z. Filipi, Impact of Naturalistic Driving Patterns on PHEV Performance and System Design, SAE International, Warrendale, PA, 2009.

[17] Taylor, Alex. 2011. "Chevy Volt Vs. Nissan Leaf: Who's Winning?" *Fortune*, September 26.

[18] "Why The Chevy Volt Is Attracting Wealthy Buyers." 2013. *Aol Autos*. Accessed February 22. http://autos.aol.com/article/why-the-chevy-volt-is-attracting-wealthy-buyers/.

[19] The Congressional Budget Office. 2008. *Historical Effective Tax Rates, 1979 To 2005: Supplement with Additional Data on Sources of Income and High-Income Households*. Washington, DC: The Congressional Budget Office. http://www.cbo.gov/sites/default/files/cbofiles/ftpdocs/98xx/doc9884/12-23-effectivetaxrates_letter.pdf.

[20] The Congressional Budget Office. 2012. Illustrative Examples of Effective Marginal Tax Rates Faced by Married and Single Taxpayers: Supplemental Material for Effective Marginal Tax Rates for Low- and Moderate-Income Workers. Washington, DC: The Congressional Budget Office. http://www.cbo.gov/publication/43722.

[21] Taxing Wages 2011. 2012. Paris: Organization for Economic Co-operation and Development. http://www.oecd-ilibrary.org/content/serial/20725124.

[22] Annual Energy Outlook 2012, US Energy Information Administration, Washington DC, 2012.

[23] L.B. Lave, H.L. MacLean, Transportation Research Part D: Transport and Environment 7 (2002) 155–162.

[24] R.R. Heffner, K.S. Kurani, T.S. Turrentine, Transportation Research Part D: Transport and Environment 12 (2007) 396.

[25] Bruvoll, Annegrete, Bente Halvorsen, and Karine Nyborg. 2002. "Households' Recycling Efforts." *Resources, Conservation and Recycling* 36 (4) (November): 337–354. doi:10.1016/S0921-3449(02)00055-1.

[26] Berglund, Christer. 2006. "The Assessment of Households' Recycling Costs: The Role of Personal Motives." *Ecological Economics* 56 (4) (April 1): 560–569. doi:10.1016/j.ecolecon.2005.03.005.

[27] Butler. 2012. *Assembly Bill No. 475*. http://www.leginfo.ca.gov/pub/11-12/bill/asm/ab_0451-0500/ab_475_bill_20110907_chaptered.pdf.

[28] Smith, Kandler, Matthew Earleywine, Eric Wood, and Ahmad Pesaran. 2011. "Comparison of Battery Life Across Real-World Automotive Drive-Cycles" presented at the 7th Lithium Battery Power Conference, November 7, Las Vegas, NV. http://www.nrel.gov/docs/fy12osti/53470.pdf.

[29] Peterson, Scott B., Jay Apt, and J.F. Whitacre. 2010. "Lithium-ion Battery Cell Degradation Resulting from Realistic Vehicle and Vehicle-to-grid Utilization" (April 15): 2385–2392.

[30] Motavalli, Jim. 2013. "A Wireless Charging Solution for the Leaf and Volt." *Wheels Blog*. Accessed February 28. http://wheels.blogs.nytimes.com/2012/10/30/a-wireless-charging-solution-for-the-leaf-and-volt/.

In: Grid Electrified Vehicles
Editor: Carla Alexandra Monteiro da Silva

ISBN: 978-1-62808-839-7

Chapter 5

VARIATION IN CHARGING OF PRIVATELY-HELD PEVS: IMPLICATIONS FOR ANALYSIS, MARKETS, AND POLICY

Jamie Davies and Kenneth S. Kurani*

Plug-in Hybrid and Electric Vehicle Research Center, Institute of Transportation Studies, University of California, Davis, CA, US

ABSTRACT

As the markets for plug-in electric vehicles (PEVs) and the deployment of electricity infrastructure to charge them are in an initial, dynamic launch phase, there is an absence of stable data on PEV purchase and charging behavior. How then are social, economic, and environmental effects of PEVs being estimated? How are plans for PEV and electric vehicle service equipment (EVSE) production made? How are the effective means to manage that behavior anticipated? In the absence of data, analysts make assumptions. These are often simple assumptions, or perhaps simplifying assumptions. In this chapter we compare PEV charging assumptions to real world measures and assess the implications of changing these assumptions for analysis, markets and policy.

Keywords: Plug-in Electric Vehicle (PEV), charging impacts, utility factor

INTRODUCTION

Plug-in hybrid and electric vehicles (PHEVs and EVs, collectively referred to as plug-in vehicles (PEVs)) are now being sold in countries around the world. Most large international automobile manufacturers and many regional ones have joined this market launch or are about to join. Concomitant developments in charging infrastructure for these plug-in vehicles are necessarily localized. PEV charging infrastructure development is localized, especially in parts of the world such as the United States of America (USA) where the electrical grid is

* Corresponding Author address: 1605 Tilia, Suite #100, Davis, CA 95616; Email: jdavies@ucdavis.edu

operated through the coordination of thousands of different jurisdictions and entities ranging in size from a single municipal electric utility to large grid interties spanning multiple states and crossing national boundaries. Within this context ranging from a global market for vehicles to local electricity distribution, buyers of PEVs and the electricity to power them are the nexus that determines actual vs. estimated demand as well as social, economic, and environmental impacts. Assuming for a moment the purchase of PEVs, the nexus is literally the act of connecting PEVs to the electric grid, i.e., charging.

In an effort to characterize the social, economic, and environmental impacts of PEVs, analysts have devised numerous what-if scenarios. Although these analyses vary in context, they generally employ similar travel data from existing internal combustion engine vehicles and assumptions about consumer PEV purchase, driving, and charging behavior. As these analyses aim to simulate lifestyles, vehicle technologies, energy prices, markets, and charging networks that do not yet exist, the scenarios are ultimately limited by the creativity and imagination of researchers. Of all their assumptions, the ones we address here are those for charging behavior, especially assumptions about how many PEVs charge per day and how many times per day. For example, what are we to make of an assumption that all PEVs are charged once per day? More subtly, how do we interpret an *average* charging frequency of 1.0 charging event per day? And if an average daily frequency is assumed, what is the distribution around that mean? What does it matter if we use an assumed average for all PEV charging, or simulate charging by sampling from a distribution?

In this chapter we first briefly describe common assumptions about PEV charging, including commonly assumed data sources. Then we describe some early data on PEV charging. The implications of deviation of these data from the common simplifying assumptions include the question of whether the analytical community would better serve markets and policy by making more complex assumptions when little is known about essential variables.

Once-Per-Day Charging and Its Variations

Perhaps the simplest assumption one can make about PEV charging is one of the most often made: each and every PEV is charged once per day, everyday [1-5]. A more restrictive version of the assumption is that all such charging events begin when or after a PEV is returned to the driver's home. In such analyses, the PEV may be assumed to begin to be charged immediately [6, 7] or at some point later in the evening [6-8]. According to such assumptions, over the course of a week, the number of charge events per vehicle per day would be as shown in Figure 1. Anticipating the discussion to come, the line in the figure is not an average; there is no distribution around it.

To further foreshadow the discussion below, the EV Project funded by the USA Department of Energy reported that in the third quarter of 2012 the mean number of charges per day of the PEVs in the project across 14 USA cities in which the project is installing charging infrastructure was 1.1 charge events per day for the Nissan Leaf [9] and 1.4 for the Chevrolet Volt [10]—on days the vehicles were driven. Across the cities, the mean for Leafs varied from 1.0 in Tuscon, AZ to 1.3 in Washington, DC and for the Volt, from 1.3 in San Diego, CA to 1.6 in Oregon.

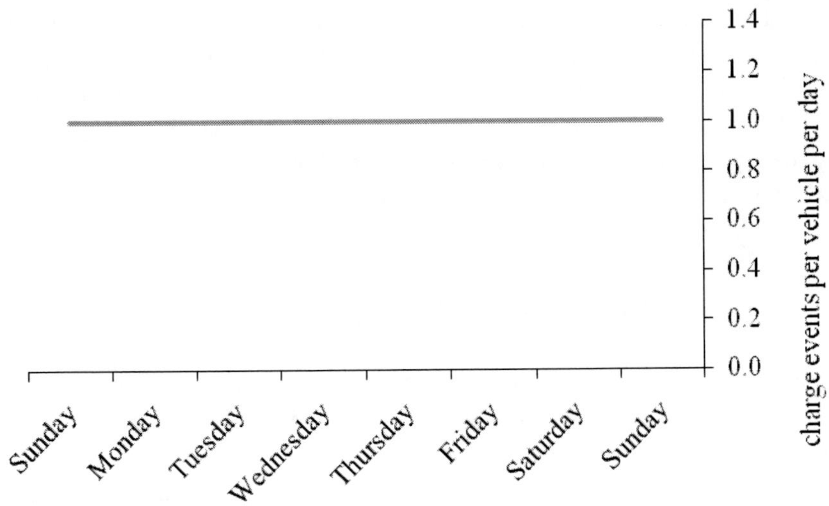

Figure 1. One PEV charge event per vehicle per day over the course of one week.

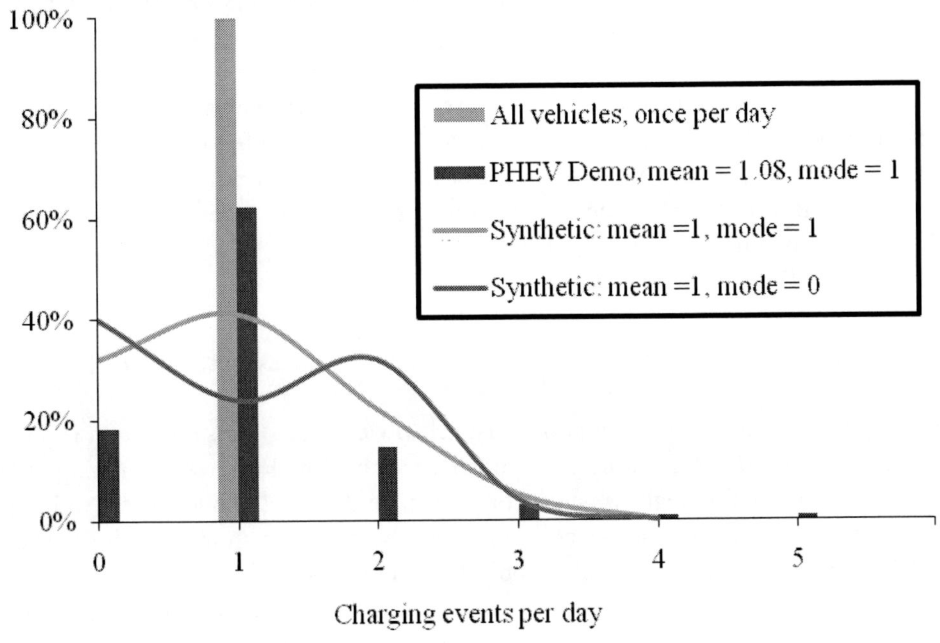

Figure 2. Examples of daily charging frequency distributions, mean =1.

We might reasonably ask whether an average 1.1 per day is sufficiently close to 1.0 to allow the once-per-day assumption to stand. However, a mean number of charging events per day near 1.0 is not the same as an assumption that all PEVs are charged once on every driving day. To illustrate the possible differences, four distributions of daily charging frequency are shown in Figure 2. Each distribution has an average 1.0 charging event per day per vehicle. The synthetic distributions merely illustrate two of all the possible shapes that charging distributions with means of 1.0 could take. The fourth distribution shown is from observed consumer PHEV charging behavior described below in the observed PEV charging section of this chapter. Further, recognizing that as practical matter of scale, i.e., a maximum number of charges per day of 3.0 is likely to be in the high 90th percentiles, an average of 1.4 charge events per day cited for the Chevrolet Volt is not as close to 1.0 as it might seem. Related to correlation between vehicle capability, travel behavior, and charging, the further question is raised by the reported difference between the means for the Volt and the Leaf: why is the car that does not have to be charged being charged, on average, more often than the car that does?

The Generalization As Average

If a numeric average focuses our attention on some centroid of a distribution, statements of generalizations also tend to focus our attention on specific cases. From an interview with a representative of the EV Project we have this statement:

"While we still expect the majority of charging to take place at home, we're excited to see that in the past three months [3rd quarter, 2012], public charging has increased by 25 percent. This shows that *drivers are becoming more and more comfortable with stretching the range of their vehicles and utilizing the public infrastructure available* [emphasis added]." [11]

During the third quarter of 2012, we spoke with PEV drivers in one of the EV Project's initial cities. While some drivers say they are utilizing the increasing public charging infrastructure, others have taken the position that if they are unable to complete their day's trips in their EV with the charge it has when they leave home, they don't leave home in their EV. The generalization obscures the distribution of responses to (even the growing) away-from-home charging infrastructure.

What Can We Know about Variability in Charging?

A critique of any charging assumption must acknowledge the limits on what it is possible to know about PEV charging during the dynamic launch of both the vehicles and charging infrastructure. In general though, we argue that what we can't yet know appears to limit the variability that can presently be observed, i.e., as the conditions we describe below are relaxed, the variability that it is possible to observe will increase.

Some PEV driving and charging data are available from short term vehicle trials in which households drive—but do not have to buy or lease—PEVs for periods of days or weeks, e.g., [12]. Other data comes from households who have purchased or leased PEVs and have to-date driven and charged them for periods ranging from a few to several months. Regarding access to charging infrastructure, the data we will discuss includes only households who can charge a PEV at home. The data are also from a period when away-from-home charging was free.

What are missing are data from households who have purchased or leased more than one PEV from a wide variety of design and performance capabilities and charged them within anything like the competing or complimentary versions of charging infrastructures imagined by, for example, Project Better Place's battery swap stations [13], Tesla's "superchargers" located to facilitate longer trips between regions [14], or any other models of away-from-home networks. Further, the makes, models, and body styles of vehicles that are offered as PEVs are limited at present to generally smaller sedans and crossover vehicles. The limited vehicle offerings may shape the types of travel, e.g., trip purposes and distances that can be presently observed. The variability of PEV driving range—whether the total driving range per charge for EVs or the electric-equivalent driving range for PHEVs—is increasing and one manufacturer has introduced an EV for which driving range is a buyer-selectable option.

OBSERVED PEV CHARGING

We illustrate variability in PEV charging, starting with data from a household PHEV trial. Seventy households in northern California were provided with a Toyota Prius PHEV-conversion to drive for periods varying between four and six weeks. Details are available in [12]; notably, in addition to second-by-second driving data and charging data, each household completed entry and exit questionnaires and three in-home interviews. We do assume that the structure of weekdays/weekends imposes enough regularity to sum all 67 weeks into a single week despite the fact the actual individual weeks occurred over the period from summer 2008 and to summer 2010.

In Figure 3 we overlay the average number of daily charging events from the last week of 67 of these PHEV demonstration households on the every-vehicle, once-per-day assumption from Figure 1. These daily averages show a mid-week maxima (1.30) and a weekend day minima (0.84). Again, one might ask, is the deviation from once per day so severe as to warrant complaint? The problem is we are still arguing about averages with averages. In general, but especially during such a dynamic phase as we are in regarding PEV sales and infrastructure deployment, the quality of any average is assessed by how well it summarizes a distribution—and by the prior question, is any summary the most informative and useful measure?

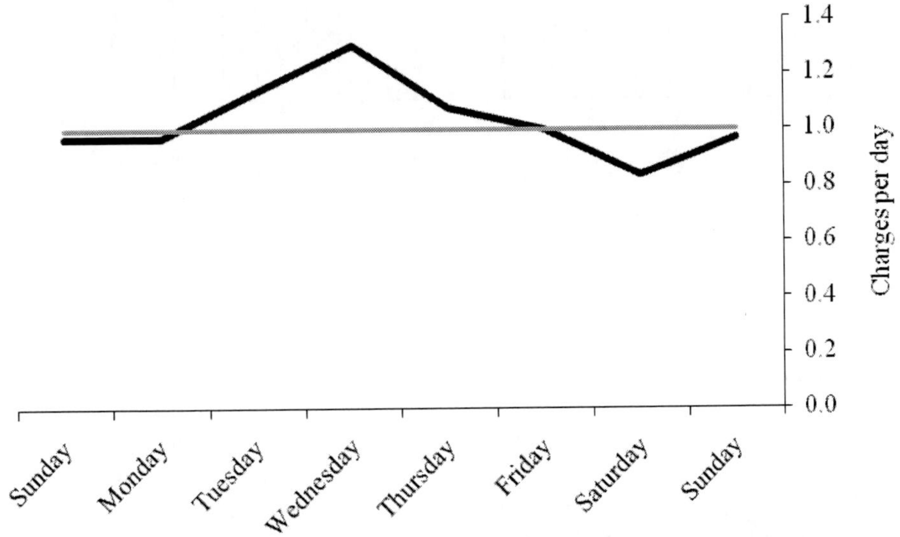

Figure 3. Mean number of charges per day (67 one-week data sets from the last week of a one-month household PHEV trial).

To begin to unpack the summary view of charging in Figure 3, we overlay the daily distribution of number of charging events by household in Figure 4. At the level of disaggregation in Figure 4, we can begin to query the data about why the daily mean values vary and whether any "central-measure" of daily charging behavior is suitable. For example, the highest weekday average (Wednesday) occurs primarily because it is the day the fewest households don't charge the PHEV at all and Saturday's low average is because it is the day the most don't charge the PHEV at all. From interviews with all the households we know the reason that so many vehicles are not charged on Saturday is the greater likeliness that trips involving an overnight stay away from home are made on weekends. Underneath this general cause, we hear from these households that the reasons for not charging the PHEV on Saturday are both that the PHEV is the vehicle taken on these overnight trips and the PHEV is the vehicle left home for these trips. In the first instance, the PHEV is taken away from its primary or sole charging location. In the second, the PHEV would have been plugged in upon its return to home on Friday evening (and thus that charge event was recorded as occurring on Friday) and no one was home to drive and charge the vehicle on Saturday. In the first instance, a growing infrastructure of charging opportunities would lead to increased charging opportunities for PEVs taken for overnight trips. In the second instance, the PEVs would continue to not be charged on Saturday.

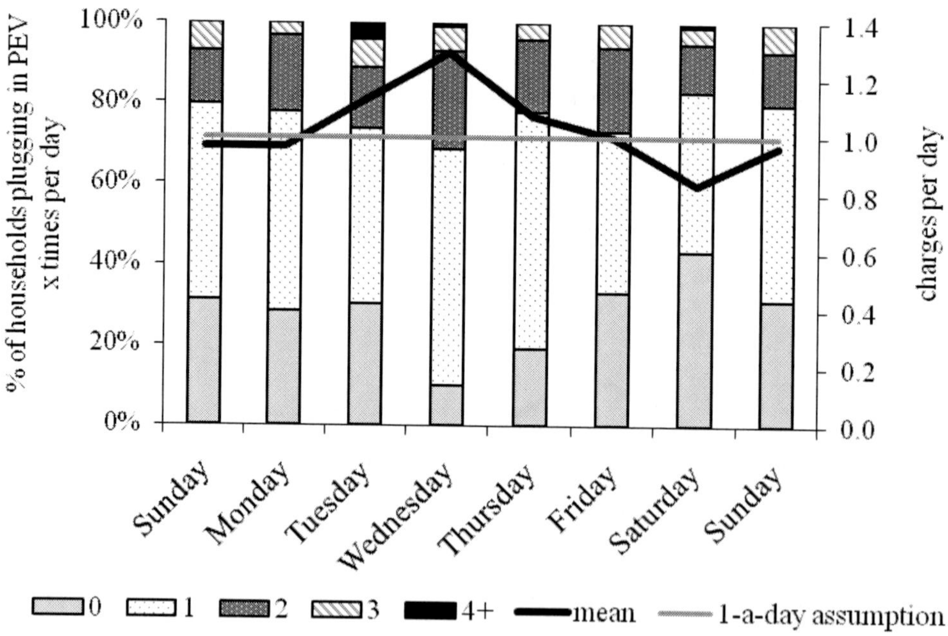

Figure 4. Distribution of number of charges per day (67 one-week data sets from the last week of a one-month household PHEV trial).

Just as averages obscure the distributions of observed charging behaviors, so do generalizations about which group of consumers would be interested in purchasing a PEV. For example, EVs are often described as good cars for "commuters," i.e., people with (an assumed) regular in time and distance workday travel pattern. But from these PHEV demonstration households, the answer to who is charging multiple times per weekday is both commuters and non-commuters. A few of the households that used the PHEV for commuting found an electrical outlet in their workplace parking lot. They regularly plugged in the vehicle upon arriving at work in the morning, and almost as regularly, plugged it in again upon returning to work after any lunchtime errands. Returning home at the end of the day, they would plug in again, for the second or sometimes third time that day. Households in which the PHEV was used by a homemaker, self-employed person who worked at home, or retired person also charged the PHEV multiple times throughout the day. The generalization "EVs are good for commuters" obscures that—stripped of specific location and time of day data—the same pattern of multiple charges per day and charges throughout the day rather than only at night are shown by a wide variety of the demonstration drivers.

Serial Correlation in Driving and Charging?

The National Household Travel Survey (NHTS) [15] data has been used in many analyses of PEV markets, charging, and their impacts in the USA. As there are no PEVs in the NHTS data as yet (the most recent are from 2009), the data cannot describe how PEVs are driven. More typically, the data are used to attempt an answer to how much existing travel done by gasoline-fueled vehicles could be done by PEVs if PEVs engender no changes in travel. Further, the NHTS contains only one day of data per household. The use of such one-day data to represent what are fundamentally behaviors over time is another form of averaging.

One of the essential features of the demonstration household data presented in Figures 3 and 4 is that they are for sequential days from the same households. Interviews of these households reveal how one day's PEV charging can depend on that day's driving, as well as past days' and expectations of driving and charging over future days. Sunday is illustrative: charging on Sunday depends not only on the driving done that day, but on expectations about driving to be done on Monday. Further, and especially for those households who were away-from home Saturday night, Sunday's charging depends on whether they plugged in the prior Friday night, or in anticipation of leaving the PHEV home on Saturday, left it unplugged. The question that arises is the extent to which these dependencies across days are idiosyncratic to individual households or common across households, i.e., observing serial correlation within behavioral units, do we observe the similar correlations across behavioral units. These questions cannot be answered by any data set, no matter how large, that contains only one day of date per behavioral unit.

Weekday Charging Variability Across Households

If the once-per-day assumption masks variation and possibly correlation across days, how well does the assumption and some stricter variants do in describing charging during the course of a day? [1, 6, and 8] assume that all PEVs charge everyday and off-peak. The definitions of "off-peak" vary. Taking one region as a specific example, in the service area of San Diego Gas & Electric (SDG&E) in southern California, the time-of-use electricity price schedule has three levels: highest prices during the "peak" electricity demand period of noon to 8pm, lowest prices during the "super off-peak" period from midnight to 5am, and intermediate prices during the "off-peak" periods between the prior two. For now, we borrow only the definition of the super-off peak period from midnight to 5am. Figure 5 plots the SDG&E definition of the super off-peak period with the data from the 67 PHEV demonstration households (none of whom were in the SDG&E service area). The top of the color area indicates by time of day the maximum percentage of the PHEV demonstration households that had their PHEV plugged into the electrical grid across the five weekdays; the bottom of area is the lowest percentage of vehicles plugged in at that time.

At no point during the amalgamated weekdays in Figure 5 does the percentage of PHEVs plugged into the grid approach one hundred percent of all vehicles at the same time, as required by the assumption that all the PEVs charge everyday during the super off-peak period. The observed variation during this time period is from 60 to 80 percent. A less restrictive assumption regarding time-of-day, i.e., all PEVs charge every day, but don't all charge during the same time period would allow for different lines and areas to be drawn showing more variability, but the observed variability by time of day shown in Figure 5 and the observation in Figure 4 that at least 10 percent of households didn't charge the PHEV at all on any given day clearly show the assumption that all PEVs charge every day, and the special case that they all charge off-peak or super off-peak, is violated.

That far from all households will plug in their PEV every day is borne out by analysis of 1,000s of PEVs in several regions across the USA [16, 17]. For cities across the USA, only a few match as high a percentage of PEVs plugged into an EVSE at the PEV driver residence as shown for the PHEV demonstration households in Figure 5. Averaged across the cities that are part of the EV Project, 50 to 70 percent of the participating PEV drivers have their vehicle plugged into the grid between midnight and 5am on any given weekday night. In some of these cities, the observed low is 35 percent and in none does it exceed 80 percent [17].

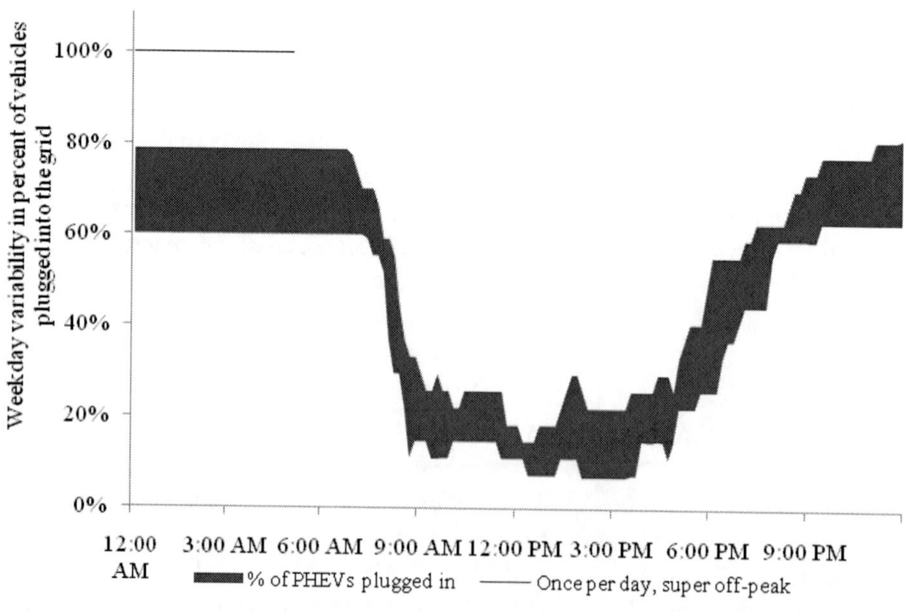

Figure 5. Once-per-day, off-peak recharging vs. Time of day variability in PHEV charging, weekdays (67 one-week data sets from the last week of a one-month household PHEV trial).

ADDITIONAL FORCES ACTING TO INCREASE OR DECREASE THE INCIDENCE OF CHARGING AND TIME-OF-DAY VARIABILITY OF CHARGING

Deploying an away-from-home charging network increases the likelihood of violating the once-per-day charging assumption. On the other hand, imposing time-of use (TOU) electricity pricing and other mechanisms to prompt off-peak charging reinforces the assumption. Initial observations from drivers of PEVs suggests the effects of these countervailing forces may be to increase variability in charging behavior across days and drivers.

Interviews were conducted with buyers and lessees of Nissan's Leaf EV in San Diego, CA during Spring 2012 and focus groups were conducted with different people from the same population in Fall 2012 [18]. During this time (and continuing through 2013) SDG&E is conducting a TOU experiment, randomly assigning a sample of EV drivers to one of three different TOU rate schedules. (SDG&E expects to release their report in early 2014.) Our sample includes some PEV drivers who are participating in SDG&E's experiment and some who are not. These interviews did not measure EV driving and charging behavior, but engaged drivers in a conversation about why they acquired a PEV, how they drive it and charge it, whether they see the EV as an opportunity to enter into a community of like-minded people or to otherwise enact specific values.

EV drivers who have charged their EV away from home generally describe doing so in addition to charging at home. Exceptions included people who could routinely charge their EV at a workplace on weekdays and did not require charging at both home and work to complete their daily travel. In general though, the reports of these EV owners echo some of the charging behaviors reported by the PHEV demonstration drivers: people who charged multiple times per weekday included commuters and non-commuters; overnight, away-from-home trips on weekends may suppress the number of charging events on Saturday; those who stay home on the weekend may charge more often than once per day as they are more likely to be in and out of the house multiple times than they are on weekdays.

Differences from the behaviors reported by the PHEV demonstration drivers include a more concerted effort to charge (at home) everyday and a greater likeliness by more EV drivers in San Diego to explore away-from-home charging opportunities. Their exploration of away-from-home charging is facilitated by the different context from the PHEV demonstration—for which the only away-from-home charging was whatever 110V electrical outlets drivers might find. Coincident with the launch of PEV sales in San Diego, the deployment of Level 2 (up to 7.68kW) charging in public places, workplaces, public parks, and other destinations started, too.

However, the "technological" launch of PEV charging at away-from-home locations has not had as strong effect on increasing the incidence of EV charging as it might have had, and still may have. Though EVSEs continue to be installed at an increasing number of locations, EV drivers report the absence of a concomitant development and practice of "charging etiquette": rules to guide PEV drivers expectations of what to encounter in an away-from-home charging interaction with other PEV drivers [19]. These reports generally conform to stories of charging opportunities missed, not sought, or even avoided, i.e., the lack of etiquette appears to be suppressing away-from-home recharging.

If the deployment of away from home PEV charging infrastructure facilitates increased opportunity for charging throughout the course of a day, the imposition of TOU pricing on electricity for charging PEVs attempts to confine such charging to a specific time period: midnight to 5am, and thus to reduce the number of (home) charge events to one per day. The reports from the EV drivers are varied as to the effectiveness of TOU pricing [18]. For those EV drivers whose home EV charging is on a TOU price schedule, many report they do attempt to strictly adhere to charging during the desired period. In addition to pricing, timers on both the vehicles and the EVSEs allow drivers to plug-in their EV whenever they arrive home, but to initiate charging after midnight. Pricing is also reinforced by long-standing social marketing efforts by electric utilities in California to exhort electricity users to curtail their demand (for all purposes) during the afternoon-to-evening peak demand period. The combined effect of pricing, timers, and exhortation is appears to be broadly, but not universally, successful.

Households who had both an EV and a home solar energy system (SES) are among those EV drivers who reported they did not adhere to the desired pattern of reducing home recharging to once per day, starting at midnight. These households are also on a TOU electricity tariff. However, many of them seem to have a broken mental model of the relationship between their SES and the grid: they believe their SES 1) provides electricity directly from the sun to their home and EV and thus, 2) the grid is insulated from the effects of their EV charging. These people do understand and seem to be sympathetic to the calls for moderating peak period electricity demand. They just think their charging behavior doesn't affect the grid. These EV+SES owners often report charging their EV whenever they return home, including multiple times throughout the day.

IMPLICATIONS

Deviation from "all PEVs charge once-per-day, everyday" has implications for estimating total electricity demand, electric miles driven, gasoline displacement, emissions, and all other outcomes. An example from the USA is the use of the "all PEVs, once-per-day" charging assumption in the Society of Automotive Engineers (SAE) standard J2841. Illustrated in Figure 6, the standard is meant to represent the "utility factor" (UF): the fraction of the total driving done by a PHEV (with a given electric range) that are electric-powered. (The SAE standard also assumes PHEV driving can be represented by the NHTS data.) The California Air Resources Board (ARB) uses the standard to calculate zero emission vehicle (ZEV) credits to award to PHEVs in comparison to the "full credit" given to EVs and Fuel Cell Hybrid Electric Vehicles (FCHEVs).

Analysis of the driving from our PHEV demonstration households' data indicates 1) that even in matched vehicles the achieved UF varies across drivers and within drivers across days [12]. Further, with experience, PEV drivers are likely to learn what electric range capability is suitable for them. The SAE J2841does not allow for consumers to self-select into vehicles that match their travel and charging behaviors. In Figure 6, the black dashed lines show individual household's UF from simulations based on the PHEV demonstration data. J2841 does not provide a good measure of the centroid of the space defined by the PHEV demonstration households—who all drove closely matched PHEVs.

The distribution and slopes of individual household's UF indicate that neither households nor society would benefit equally from additional electric range for the PHEVs; for each household there is a threshold beyond which the marginal effects of adding range decreases markedly. The three small boxes identify specific electric ranges and the households whose driving and charging behaviors match those vehicle ranges (in the sense that additional electric range and thus additional cost bring reduced marginal benefits in terms of electric-miles driven). Fitting a line to the averages of the UFs within the boxes produces what we refer to as a consumer-informed UF.

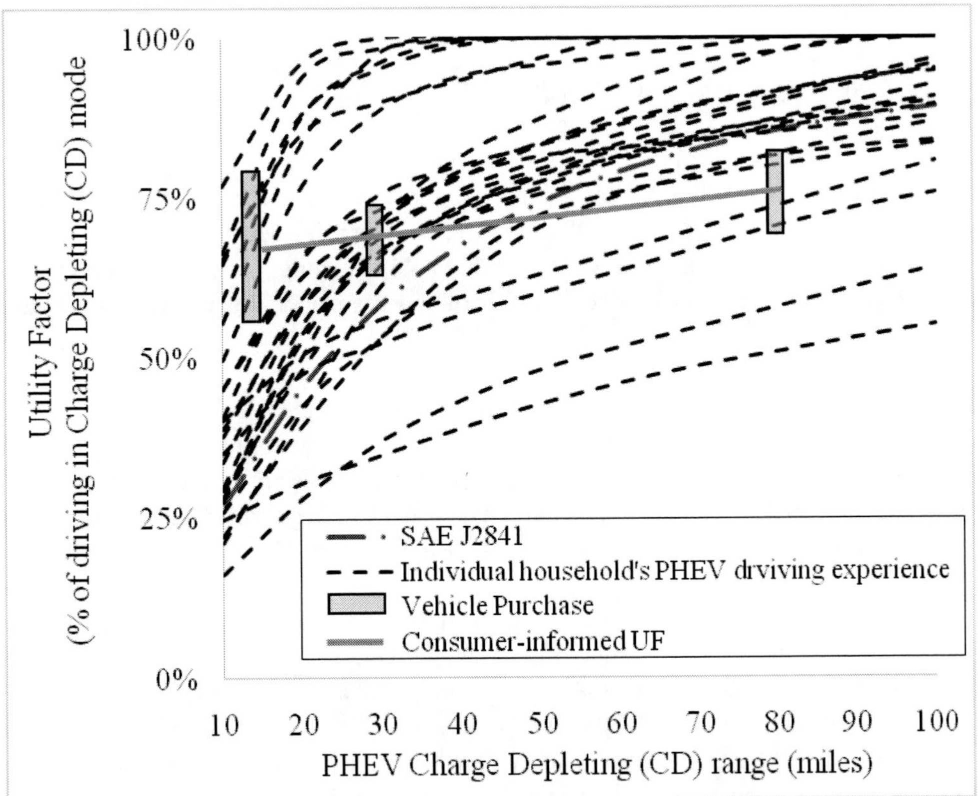

Figure 6. SAE J2841 and potential for self-selection in PHEV purchase range to affect utility factors.

CONCLUSION

Contrary to simplifying assumptions, observations of incipient and initial PEV charging shows wide variation in charging behavior within a single PEV driver's life over time, between PEV drivers, and across types of PEVs. If some drivers routinely charge their PEV more than once a day, some don't charge every day. Frequency and timing of charging throughout a day are affected by PEV electric-driving range capabilities, drivers' (and their households') travel and access to other mobility tools, fuel prices (gasoline and the cost to charge), the localized recharging infrastructure in the region in which they live, and the efforts (or lack thereof) of their local electricity provider to confine charging to a specific time of day (and through the correlation between time-of-day and most people's diurnal patterns, to a specific location, i.e., home). The continued introduction of more PEVs and more kinds of PEVs are likely to increase variability in charging behavior beyond what is already observable. The continued deployment of away-from-home infrastructure is likely to increase both variability and the number of charge events per PEV per day. These effects will be amplified if 440v "quick charging" is widely deployed. Further, if quick charging differentially facilitates long distance travel, then we can expect the effects to be more pronounced on weekends when this travel is more likely to occur. Reservation systems could alleviate some of the reluctance many PEV drivers show to rely on away-from-home charging—further increasing the effect of such infrastructure on charging frequency and variability.

If PEV charging is more frequent and variable than has commonly been assumed, what are the implications for estimates of social, economic, and environmental impacts—good or bad—of PEVs? What else might analysts do than make simplifying assumptions? The implications for impacts estimates are that spatial and temporal electricity demand to charge PEVs has been mischaracterized. Furthermore, total electricity demand for charging PEVs has been underestimated because either or both 1) charging events are simply being undercounted (as the mean of observed number of charges per PEV per day has been higher than 1.0 in both demonstration households and households of buyers and lessees of PEVs) and 2) calculations of correlations between charges per day and miles driven are likely to be positive. The ongoing tension between increasing PEV charging by building more away-from-home infrastructure and limiting charging to once per day at home during a few hours late at night is correlated both with emissions effects—as emissions in many electrical grids are dependent on time of day—and the additional costs electricity providers face from having to upgrade local distribution and the additional benefits electricity producers stand to gain from increasing the efficiency of their production base.

A prior condition for improvement in the analysis of the effects of widespread sales of PEVs and deployment of PEV charging infrastructure is improved quality of data during this dynamic period. At present, in the USA there is no widely available, comprehensive view of PEV driving and charging. The EV Project data cited in this chapter is helpful, but it focuses on the PEVs and the charging infrastructure within the project. The accrual and availability to analysts of comprehensive (across household, vehicle and infrastructure contexts) data are central to testing assumptions and building understanding of the societal, environmental, and economic implications of PEVs.

As described in [20] and again here, unless and until we have such data, and if the limits of our analytical scenarios are the analytical community's imagination and creativity, then in times such as this early phase of a PEV market launch, the authors encourage more complicated assumptions about PEV driving, charging and purchase behaviors: rather than independent point values, sampling from correlated distributions of behaviors and in the case of charging behavior, distributions of temporally correlated behavior. In such a way, estimates of social, economic, and environmental impacts can be tested for their robustness in the face of our understandable but undeniable ignorance of the nascent nexus of PEV driving and charging behaviors of future PEV drivers. Premature simplification is already having long-term consequences for standard setting and designs of performance based incentive programs.

REFERENCES

[1] Hadley, S., and A. Tsvetkova (2008) "Potential impacts of plug-in hybrid electric vehicles on regional power generation." Oak Ridge National Laboratory. Report ORNL/TM-2007/150.

[2] EPRI (2007) "Environmental Assessment of Plug-in Hybrid Electric Vehicles Volume 1: National Greenhouse Gas Emissions." Electric Power Research Institute. Report 1015325.

[3] Kintner-Meyer, M., and Schneider, K. Pratt, RG (2007) "Impacts Assessment of Plug-In Hybrid Vehicles on Electric Utilities and Regional U.S. Power Grids. Part 1: Technical Analysis." Pacific Northwest National Laboratory. Online journal of EUEC 1: paper #4.

[4] Society of Automotive Engineers J2841 (2009) "Utility Factor Definitions for Plug-In Hybrid Electric Vehicles Using Travel Survey Data."

[5] Khan, M., and .K Kockelman (2012) "Predicting the market potential of plug-in electric vehicles using multiday GPS data." Journal of Energy Policy. Volume 46, pages 225 – 233.

[6] Parks, K., P. Denholm, and T. Markel (2007) "Costs and Emissions Associated with Plug-In Hybrid Electric Vehicle Charging in the Xcel Energy Colorado Service Territory." Report NREL/TP-640-41410 National Renewable Energy Laboratory.

[7] Weiller, C (2011) "Plug-in hybrid electric vehicle impacts on hourly electricity demand in the United States." Journal of Energy Policy. Volume 39, pages 3766 – 3778.

[8] Tate, E., and P. Savagian (2009) "The CO2 benefits of electrification E-REVs, PHEVs and charging scenarios." SAE world congress and exhibition. Report 2009-01-1311.

[9] Ecotality (2012) "EV Project Nissan Leaf Vehicle Project Summary Report." US Department of Energy, Energy Efficiency & Renewable Energy, Vehicle Technologies Program. Report INL/MIS-11-21904. http://avt.inel.gov/pdf/EVProj/EVProj NissanLeafQ42012.pdf

[10] Ecotality (2012) "EV Project Chevrolet Volt Vehicle Project Summary Report." US Department of Energy, Energy Efficiency & Renewable Energy, Vehicle Technologies Program. Report INL/MIS-11-24041. http://avt.inel.gov/pdf/EVProj/ EVProjChevrolet VoltQ42012.pdf

[11] Schey, S. (2012) Crunching the Big Data of EV Charging: A Q&A with ECOtality on what 1 million "charge events" in the DOE-backed EV Project can tell us." October 8. http://www.greentechmedia.com/articles/read/crunching-the-big-data-of-ev-charging

[12] Kurani, K.S., J. Axsen, N. Caperello, J. Davies, and T. Stillwater (2010). Learning from Consumers: Plug-in Hybrid Electric Vehicle (PHEV) Demonstration and Consumer Education, Outreach, Market Research Program. Volumes I and II. Institute of Transportation Studies, University of California, Davis, Research Report UCD-ITS-RR-10-21.

[13] Project Better Place www.betterplace.com

[14] Tesla Motors http://www.teslamotors.com/charging#/onthego

[15] United States Department of Transportation (undated) National Household Travel Survey. http://nhts.ornl.gov/index.shtml

[16] Schey, S., Scoffield, D., and J. Smart (2012) "A First Look at the Impact of Electric Vehicle Charging on the Electric Grid in The EV Project." EVS26. Los Angeles, CA.
[17] Ecotality (2012) "EV Project Electric Vehicle Charging Infrastructure Summary Report." US Department of Energy, Energy Efficiency & Renewable Energy, Vehicle Technologies Program. Report INL/MIS-10-19479. http://avt.inel.gov/pdf/EVProj/EVProjInfrastructureQ42012.pdf
[18] Kurani, K.S., J. TyreeHageman, and N. Caperello (2013) Potential Consumer Response to Electricity Demand Response Mechanisms: Early Plug-in Electric Vehicle Drivers in San Diego, California (Draft). Plug-in Hybrid & Electric Vehicle Center, University of California: Davis. CA.
[19] Caperello, N., K.S. Kurani, J. TyreeHageman (2013) "Do you Mind if I Plug-in my Car? How etiquette shapes PEV drivers' vehicle charging behavior." Presented at Transportation ResearchBoard 2013 conference proceedings. Paper number 13-2816.
[20] Davies, J., and K.S. Kurani (2010) "Recharging Behavior of Households' Plug-In Hybrid Electric Vehicles: Observed Variation in Use of Conversions of 5-kW-h Blended Plug-In Hybrid Electric Vehicle." Transportation Research Record.

In: Grid Electrified Vehicles
Editor: Carla Alexandra Monteiro da Silva

ISBN: 978-1-62808-839-7

Chapter 6

Analysis of Large Scale Adoption of Electrical Vehicles with Wind Integration in Nord-Trøndelag

Åshild Vatne*[*]*, Marta Molinas and Jan A. Foosnæs
Department of Electrical Engineering,
Norwegian University of Science and Technology, Trondheim, Norway

Abstract

With the Agreement on Climate Policy (Klimaforliket) signed by the Norwegian government on January 17th 2008, Norway has set a goal to reduce emission caused by transportation with at least 2.5-4 million tons CO2 equivalents compared with the reference for 2020 (EBL, 2009). To reach this goal, high penetration of electrical vehicles is essential, and new technologies and solutions for the infrastructure must be cleared early in the process. With the aim of triggering a discussion on the topic, this chapter presents a methodology for analysing the impact of large scale adoption of EVs on the electrical grid. A specific portion of a real network was selected and two charging modalities for the electrical vehicles were investigated. The analysis will focus mainly on chargers located at residences, to then explore how the utility can put forward a system for smart charging strategies ("dumb" vs. "smart" charging). Data from a low voltage network was provided by Nord-Trøndelag Elektrisistetsverk (NTE), located in Steinkjer in Nord-Trøndelag. Three different scenarios were analysed. Scenario 1 was given as the base scenario, were the share of EVs where 0%. This was simulated to get a proper comparison. In scenario 2, a share of 10% EVs was implemented in the grid. The share of EVs in scenario 3 was decided to be 60%. The result obtained in the analysis, verified that the smart charging approach causes less strain on the grid. The low voltage network was not capable to handle a large share of EVs (>60%) without any charging scheduling. The smart charge strategy did not cause any extra strain at the grid during peak hours. In addition, the smart charging can introduce the Vehicle-to-Home solution (V2H). The EVs can provide ancillary service and support the network with matching supply/demand and reactive power support. A simplified analysis of V2H and reactive compensation was carried out to demonstrate how the grid could benefit from an implementation of EVs.

*Email:ashvatne@gmail.com

The second part of the analysis, a series of wind measurement was included into the simulation in order to see if wind power can supply the load of the entire residential area. A design for suitable energy storage was also proposed in order for the system to operate as a stand-alone system. Grid stability and power quality was not included in the analysis. The result from the wind integration shows that in order for the network to operate as a stand-alone system in the worst-case scenario, there is a need of an enormous storage. It is assumed based on the results, that the system is self-supplied most part of the year. This chapter proposes a storage consisting of 7 battery-packs from old vehicles, with the capacity of 50 kWh each. This will result in a ≈30% reduction of the peak demand from the grid, when wind power is integrated.

The case study addressed in the chapter, present a methodology for analysing the impact of a large adoption of EVs on the distribution network. The results obtained from this analysis, are considered transferable to similar networks. In order to achieve smart charging, there is need for further research on scheduling algorithms.

Keywords: Electrical Vehicles, Smart Grid, Smart Charging, Wind Integration

INTRODUCTION

The transport sector is accountable for more than half of the world's consumption of oil, and a large amount of this are consumed by passenger cars (P. Suwanapingkarl, 2009). A large scale adoption of electric vehicles would reduce the greenhouse gas emissions, as well as the dependency of oil. Nevertheless, the environment benefit is dependent of the generation mix. The higher percentage of renewable in the generation, the more beneficial is the integration of EVs. Renewable resources, such as wind energy, tend to have a stochastic production and will cause surplus energy in certain periods. To have the most efficient usage of the power, a good solution is to implement EVs in the grid. The EVs can also provide ancillary service and support the network with supply/demand matching and reactive power (Statistisk Sentralbyrå, 2011)support (P. Suwanapingkarl, 2009). The solution where the vehicle delivers power back to the grid is called Vehicle-to-grid (V2G), and Vehicle-to-home (V2H) where the vehicle delivers power to the owner's home. Although the integration of EVs is an intelligent solution to use the energy surplus, the implementation of a large EV fleet causes a lot of challenges. This chapter investigates the impact of EVs on the distribution grid where the first bottlenecks are likely to occur.

The 17th of January 2008 the Norwegian government signed Klimaforliket (Agreement on climate policy) declaring that the state of Norway will be carbon-neutral no later than 2030. This agreement stated that Norway have to reduce the emission with 15-17 million tons CO^2 equivalents, a reduction of 25% compared with the 2007 reference (EBL, 2009). The report also states that the transport sector needs to reduce the emission with 2.5 - 4 million tons CO^2 by the year 2020. The transport sector in Norway is accountable for 19% of the emissions. As a part of the obligation given in Klimaforliket a resource group were formed in December 2008 to elaborate a plan for electrification of the transport sector in Norway. The result from this project was presented in "*Handlingsplan for elektrifisering av veitransport*" (Plan of action to electrification of the road transportation) (EBL, 2009). To reach the goal of a reduction of 25% of emission, the resource group suggests that the traditional vehicle become more efficient as well as an integration of EV and vehicles that runs on bio fuel. A share of at least 10% electric vehicles is adequate by 2020 to meet the above goal. It was also stated in the same report that a share of 50% of EVs would cause a 36% reduction in emission compared to a vehicle-fleet with only efficient traditional gasoline vehicle. An efficient vehicle is calculated with 95 g CO^2 per km. If Norway wants to achieve the ambition of being carbon-neutral, the report proposes a large (>50%) implementation of EVs by 2030 (EBL, 2009).

The potential for regional use of wind power for EV charging is good, and already in place. Green certificates were a centralized theme in the Agreement on Climate Policy (2008). The certificate is now called electricity certificate. As of January 1st, 2012 Norway is a part of a Norwegian-Swedish electricity certificate market. The purpose of implementing the certificate is to increase the development of renewable energy production. Together with Sweden, Norway plans to expand the production with 26.4 TWh by 2020 (Norsk Vindkraftforening). One of the prior resources for new development is wind power.

There are, as of today, installed 525.8 MW wind power in Norway. These turbines will deliver approximate (Society of Automotive Engineers, SAE, 2010) (International Standard, IEC, 2010) (M C. Kisacikoglu)ate 1.2 – 1.3 TWh, which equal 1% of the energy supply in Norway. In addition to this, there are five wind farms that are under construction with a total capacity of 228 MW (Norsk Vindkraftforening). There are also several wind power projects under evaluation at Norwegian Water Resources and Energy Directorate (NVE). The total capacity of the projects that have applied for license is 5500 MW.

ELECTRICAL VEHICLE

There are several standards for charging electric vehicles. In this chapter it is chosen to look at the SAE J1772 and IEC 61851-1. These are the standard used by the manufacturers. IEC used SAE J1772 as the background for IEC 61851-1. The American standard, SAE J1772, classifies three levels for charging for EVs. These three levels for charging are and will be used as standards in the USA(Society of Automotive Engineers, SAE, 2010). Europe has in comparison, classified the different charging levels as modes. The charging modes are defined in IEC 61851-1.

Charging Standards: SAE J1772 vs. IEC 61851-1

Mode 1 and level 1 is more or less the same solution, where one charges the EV from a standard household-type socket-outlet. It is the solution that is mostly used today. As for the future, it is expected to use the other modes and levels. Mode 2 and level 2 refers to charging for a fixed station, with a higher current than for mode 1/level 2. This results in a lower charging time. Level 2 and mode 2 and 3 chargers are the solution that will be the most common.

The chargers will be stationed at residences, workplace and public sites. The difference between mode 2 and 3 is that mode 3 requires more communication and control in on the vehicle side. One can expect that mode 3 will become the standard solution for residence chargers if the V2H is introduced. Mode 4 and level 3 charging is also called "DC Fast

Charge". To be able to replace a large share of the traditional ICE vehicles with EVs, it is necessary to build the infrastructure for it. This includes fast chargers at shopping malls, at gas stations and in the streets and at rest stops along the highway.

Charging Operations

If one consider $v_c(t)$ as the voltage at charger and $v_s(t)$ as the grid voltage, the relationship between the two terms is expressed by $v_s(t)$=$v_c(t) + jX_c \cdot I_c$ (M C. Kisacikoglu), where $jX_c \cdot I_c$ is the voltage drop in the charger. This equation is used to form vectors to express the different charging modes. Vector diagram are presented in Figure 1a presents the "normal" operation. This mode is used in the case study. P is positive and Q is zero. When P is negative, the battery delivers power to the grid. This is further called Vehicle-to-Grid or Vehicle-to-Home, and is explained in the section below. As seen in the figure, there are four modes where reactive power, Q, is "active" - not equal to zero. This means that the charger provide reactive compensating. By controlling the charger voltage $v_c(t)$ and its phase angle, δ or the charger current $i_c(t)$ and its phase angle, θ, one can control the magnitude and direction of P and Q.

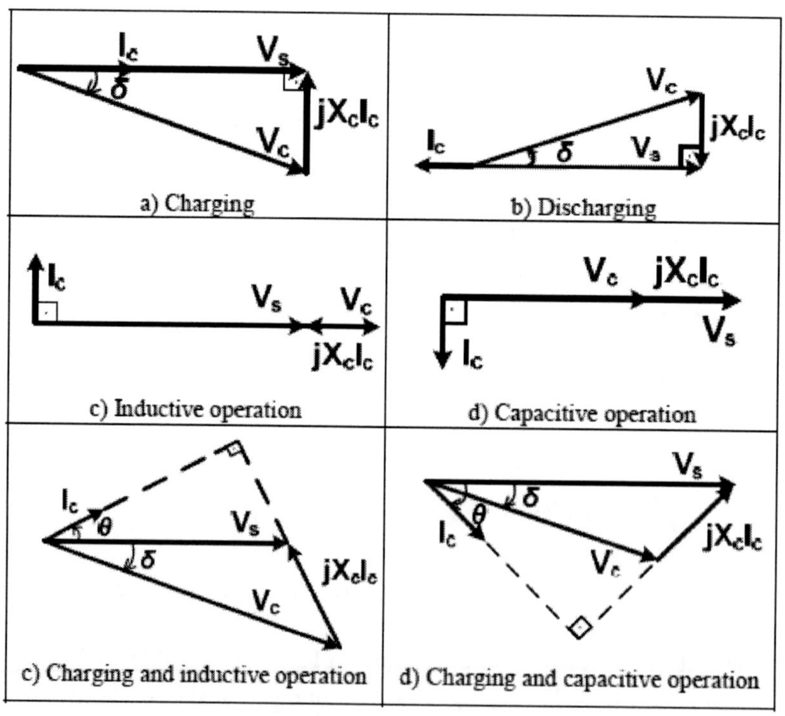

Figure 1. Vector diagram for operation modes (M C. Kisacikoglu).

A small analysis of reactive compensating will be carried out in the case study addressed in this chapter. By changing the magnitude of Q, one can control the voltage drop ΔV. The purpose is to make a more uniform voltage curve.

Vehicle to Grid (V2G) and Vehicle to Home (V2H)

The terms V2G and V2H cover the solution were the battery in the electric vehicle delivers power back to the source. The intention is to save renewable unregulated energy in the battery and save it to feed it back to the grid in peak hours. The introduction of electrical energy storage, such as the EV, could improve the efficiency and reliability. The demand on the grid will be smoothed. However, V2G is associated with challenges such as the availability of the vehicles and state of charge (SOC) of the different vehicles. How many vehicles are required to stand by as storage? How can we obtain the reliability, and at the same time the EV owners satisfied? Due to the fact that the V2H solution avoid the infrastructure and tariff problems raised by the V2G solution, the V2G solution has been neglected from this project(G Haines, 2009). In the V2H solution, the load is geographically close to the source that makes transmission minimal, reducing losses. The EV supplies the home with electricity during peak hours, causing less strain on the grid. This may reduce the cost of the transmission grid, and may prevent building new lines. A study made by Haines/ McGordon/ Jennings at this topic present that the V2H solution reduces the peak demand(G Haines, 2009). This will be analysed in the second part of the analysis, where wind turbine and energy storage delivers the energy needed to cover the demand and the charging.

Case Study

For the purpose of this case, a low voltage distribution network was chosen to be simulated. The real data is provided by NTE (NTE, 2011). This low voltage network is located in Steinkjer, Nord-Trøndelag.

The schematic sketch Figure 2 shows the layout of the low voltage network in Steinkjer. The provided system includes a distribution network with primary voltage level of 22 kV transfer down to the low voltage level of 230 V. From the substation there are six outgoing feeders, supplying together the load from 35 residences. There are also 3 residential loads that are supplied directly from the substation. This gives a total load of 38 residences. NTE provided hourly consumption from *one* of the residences (residence number 020 Figure 2), and yearly demand from the other residences in the network in Steinkjer.

To be able to make a 24-hour load curve for the entire system, a power coefficient was introduced. The coefficient κ is the relationship between the yearly consumption for any other residence in the system and the yearly consumption for residence 020, as seen in equation; $\kappa = T_\eta / T_{020}$.

Since there was no information about the hourly reading from the other residences, a 24-hour consumption was calculated for each residence. The momentary demand for any residence, η, is scaled from the momentary demand for residence 020 ($P_\eta = P_{020} \cdot \kappa$). It is therefore assumed in this case, that every residence has the same load curve throughout the day. Due to the fact that demand is highest in the winter, it is beneficial to analyse this case. It is therefore chosen to use the load from December 2010 from the given residence as the simulation data. The hourly real time data from December 2010 for the residence in Steinkjer was used to calculate the daily hourly average demand for the residence. This value is shown Figure 3. The grey line is the average demand throughout December. The dotted line in the figure is the demand for December 22nd. It was decided to use this date as the "simulation date", because it would represent the worst case scenario as the demand was highest at this date. The figure shows the difference between the average demand for the month, and the demand on the simulation date. The demand peaks are, as expected, in the afternoon. However, it was expected that the highest peak would occur later. For the chosen simulation date, the highest peak occurs at 15.00. It is expected that for a different day this peak would occur later, like shown on the average where the peak occur around 18.00

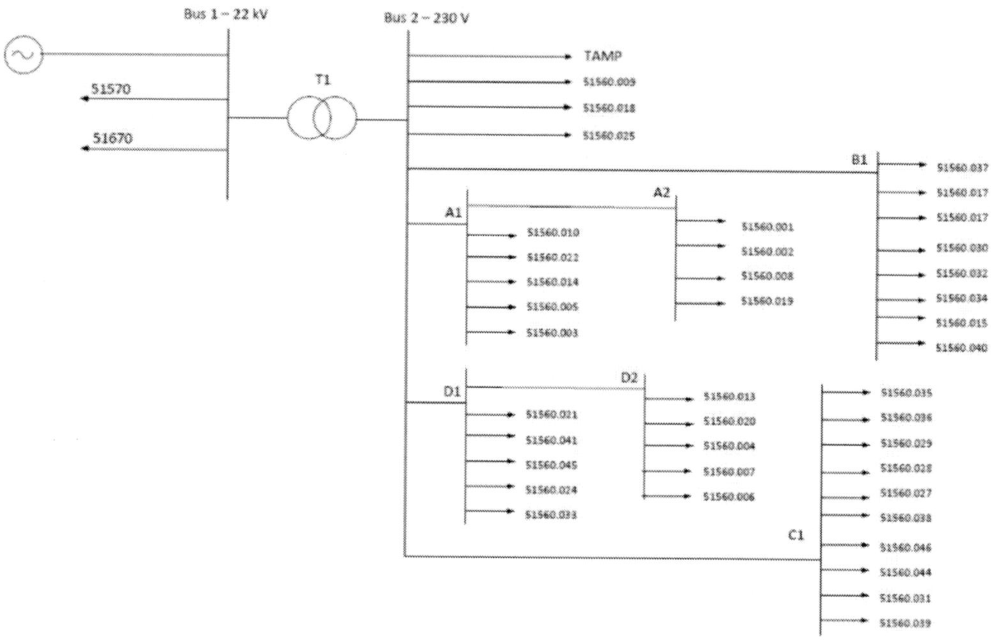

Figure 2. Overview of case system.

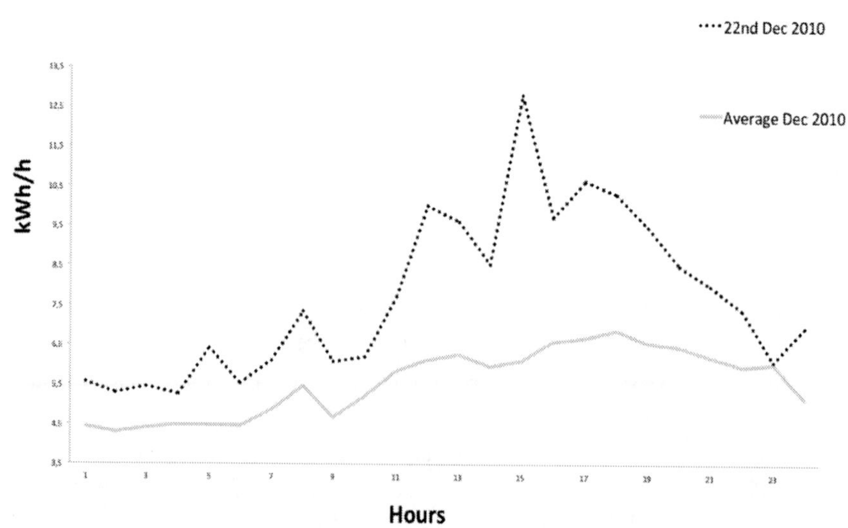

Figure 3. Average daily load for residence 020.

EV Charging Model

In the study addressed in this case, a simplified charging model will replace the actual charging process. The EV is considered as a constant power demand.

Further, the following is assumed:

- All EVs are charged at rated power
- Initial SOC is assumed to be 0%
- EVs will be charged to full capacity

To determine the load and recharge time for the EVs in the simulation, Nissans electric vehicle, *Nissan Leaf* is used as a basic for the electric vehicles in the analysis. The Nissan Leaf has a Li-ion battery with a capacity of 24 kWh. The recharge time depends on which charging level or mode is used. As of the solutions used today, a charging dock 220/240V and 40A has a recharge time of 3.5 hours (level 2 –USA) (Nissan USA, 2011), while a 220/240V and 16A the recharge time will be 6.5 hours (mode 1 –EU) (Nissan Norway, 2011). In this analysis it is preferable to assume a 4 hours recharge time. In the American solution mentioned above (level 2) the dock has a 40A charging (J. A. Peças Lopes, 2009)current; mode 2/3 allows a maximum current of 32A. It is therefore assumed that the recharge time in mode 2/3 docks will be equal to 4 hours. As stated previously, mode 2/3 will be the preferable level in the future in Europe. The assumption of a 4-hour recharge time is also used in the previous publications(J. A. Peças Lopes, 2009).

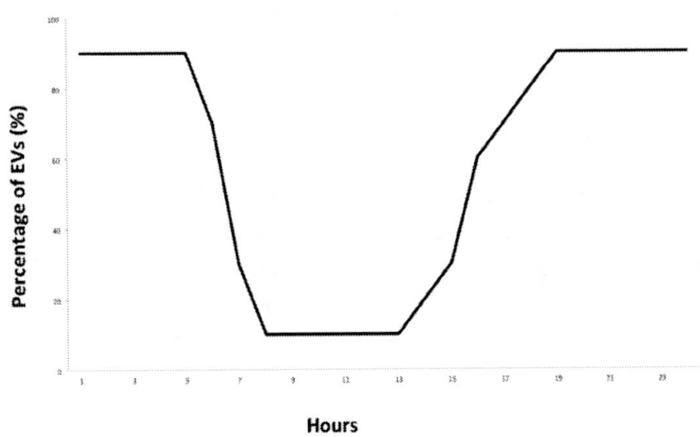

Figure 4. Percentage of EVs connected to the grid.

To simplify the analysis, it was assumed that the modelled EVs would consume a linear load. As mentioned, the Nissan Leaf has a 24kWh battery and a recharge time of 4 hours. It is determined that the EV, during charging, will consume 6 kW per hour based on; $P_{EV} = W_{batt}/T_{charge} = 24\ kWh/4\ h = 6\ kW$. It was determined that average value of vehicle per household is equal to 1.5 vehicle, which gives a total number of vehicles enclosed in the area of the network, are calculated to be 57 vehicles. This was based on that the vehicle density in Nord-Trøndelag is, according to Statistics Norway (J. F Manwell, 2009)566 per 1000 citizen (Statistisk Sentralbyrå, 2011). It is further assumed, in this case study that all the EVs will charge during the 24 hours analysis. Figure 4 shows the assumed percentage of EVs connected to the grid in the following analysis.

Scenarios

According to the Norwegian government's plan for electrifying the transport sector, it is determined that at least a share of 10% of the passenger vehicle must be chargeable by 2020. Further it was stated that if the share of electrical vehicles reached 50% in 2050, it would cause a reduction in CO^2 emission with 36% (EBL, 2009). In consideration to this, three different scenarios were analysed. The scenarios are presented in Table 1.

Table 1. Overview of Scenarios

Scenario	1	2	3
No of vehicles	57	57	57
No of EVs	0	6	36
Percentage of EVs (%)	0	10.53	63.16
Total Demand (MW)	4.77	4.91	5.63

The numbers in the table is based on assumptions made earlier in this chapter. The first scenario is an analysis of the system without EVs. This was done to have it as a base to compare the other two scenarios with. In scenario 2, a small scale integration of EVs was simulated. The share of vehicle is approximate 10% of the total number of vehicle, and reflects the government's goal for 2020. Six EVs were added into the grid. These EVs were placed in six different residences. The extra demand caused by the charging of EVs was added to the hourly measurement. The demand was added during 14.00 - 21-00 as a direct function of Figure 4. It is expected that the grid will be able to handle a small share of electrical vehicle. For the third scenario it was decided to exceed the government's ambition of a share of 50% EVs by 2050. It was chosen to see what consequences a >60% share of EVs would have in the LV-grid. Thirty-six EVs were added in the system, leaving 95% of the residences with one electrical vehicle. The extra demand was, like in scenario two, added to the hourly measurement.

Wind Power Model

Further it was decided to look at a case where the small residential area was supplied with wind power. A series of wind-measurement data was also provided by NTE. The measurements are collected from NTE's wind farm at Hundhammarfjellet, Nord-Trøndelag. The total rating of the wind farm is 53.7 MW and it has a yearly production of 170 GWh. NTE provided wind measurement for December 2010 and March 2012. The mean wind speed throughout the months is respectively 7.8 m/s for December 2010 and 11.34 m/s for March 2012.

The typical cut-in wind speed for small-scale turbine is 3-4 m/s (J. F Manwell, 2009). The wind speed is lower than the cut-in wind speed 22.55 % of the time in December 2010. The wind speed does not exceed the cut-out speed of typically 25 m/s. This means that the turbine cannot produce 22.55 % of the time due to low wind speed. As for March, the percentage of time where the turbine can produce is 94.08 %. The wind speed exceeds the cut-out speed 0.404 % of the time and below the cut-in speed 5.518 % of the time. The analysis will compare a 24-hour power analysis for a day in December 2010 and a day in March 2012. It is expected that there will be produced more during the day in March, due to the difference in the wind measurement. In addition to more production, the demand is not nearly as high in March than December. It is therefore expected an energy surplus in March, and a need for energy storage.

In the case addressed in this study, one turbine was assumed connected to the high voltage side. The size of the turbines was determined on the bases of calculations made in a previous study. It was determined to choose a turbine with a power rating that stayed under the respective capacity of the system. The limit is the thermal capacity of the cables, 380 kW. It was decided to use Enercon turbine models for the analysis. The 300 kW turbine was modelled after Enercon's E_{33} (330 kW). The hub height for the turbine varies, but it was chosen to use a hub height of 50 m in this study. The wind data provided by NTE was measured in the middle of the wind farm at a height of 82.6 m. In order to simulate the wind speed at 50 m, the wind profile power law were applied. The law represent a simple model for vertical wind speed (J. F Manwell, 2009).

$$\frac{U(z)}{U(z_r)} = \frac{z}{z_r}\alpha$$

where U(z) is wind speed at the height z, $U(z_r)$ is wind speed at reference height z_r, and α is the power law exponent. In simplified studies, like this analysis, the power law exponent $\alpha \approx 1/7, which$ indicates a correspondence between wind profile and flow over flat plates. The following equation shows how the wind speed is adjusted for the turbine height.

$$U(z) = U(z_r)(\frac{z}{z_r})^{\frac{1}{7}} = U(z_r)(\frac{50}{82.6})^{\frac{1}{7}}$$

In practice α is affected by terrain, temperature, time of day and season. In a more detailed analysis it had to be accounted for that α is a variable parameter. For the work addressed in this chapter, the assumption $\alpha = 1/7$ is adequate.

To calculate the produced power for the given time period it is assumed a linear power curve between cut-inn wind at 3 m/s and 5 m/s, and between 5 m/s rated wind speed at 13 m/s. It was determined that for wind speed between rated wind speed and cut-out speed at 25 m/s is it assumed that the E_{33} produces 335 kW.

Figure 5 shows the simplified calculated power curve (blue line) and the power curve given from the producer Enercon (red line). The figure shows that the calculated production will be a little bigger than the estimated production for wind speed between 5 - 9 m/s and a little smaller for wind speed between 9 - 13 m/s.

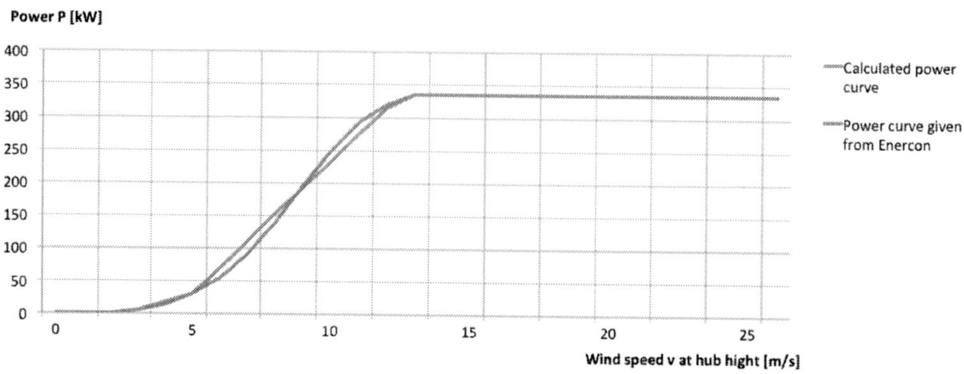

Figure 5. Simplified and actual power curve E33.

Charging Strategies

As it is previously mentioned, this chapter will focus on the impact electric vehicle will have in a distribution network. Figure 3 and Figure 4 in `Case Study` illustrates that implementation of EVs in the grid *can* cause a capacity problem in the afternoon. There is an existing peak in the afternoon and when a large share of EVs is implemented, the peak will increase. That may result in a capacity problem, a situation where the grid is not able to handle the heavy load. For the case study addressed in this work, two different charging strategies for EV charging were analysed:

Strategy 1 – "Dumb Charging"

In the first approach it is assumed that the electric vehicle owners can connect the vehicle to the grid whenever they want to, and that the charging will start immediately. In worst case the owners will plug their vehicles at the same time. This situation may occur when the users come home from work, which are typically existing peak hours. The load from the vehicles will be added to the excising peak and this may result in a capacity problem. This approach is described as a non-controlled strategy, leaves the power company to almost guess the production planning. In the dumb charging strategy the following assumption have

Strategy 2 - "Smart Charging"

In the second approach, the smart charging strategy is introduced. It is assumed that there is an active control and manage system that continuously monitor all the elements connected to the grid. Then the charging schedules will be phased in. It is expected that the demand curve will be made more uniform.

The vehicle will communicate to the grid when it is connected, and will start to charge when the demand is low. The smart charging strategy requires technology that can measure and communicate. It will also require a commitment from the vehicle owners. This approach will provide a beneficial usage of available resources, and also prevent damaging the grid. The V2H can also be introduced as a part of the smart charge. Due to the continuously monitoring of the elements connected to the grid, the EVs can operate as ancillary services and support the grid. The EVs can both deliver and store electric energy from the battery to the supply in situation with surplus/deficit of energy.

RESULTS

Power Analysis

The first part of the analysis consisted of looking at the capacity of the system. This was primarily done in Excel. As mentioned, the dumb charging was added to the base demand as a direct function of Figure 4. The dumb charging lasted for four hours and one EV consumed 6 kW each hour. In the second approach - the smart charging, the load from charging the vehicles, were added to the demand as a function of when the demand was low. It is expected that in a smart manage system, the vehicle can be charged at a lower power and over a longer time period, but to simplify this analysis, the smart charging will as the dumb charging last for four hours and each vehicle will consume 6 kW.

Scenario 1

The demand from the rest of the residences was calculated and a 24-hour demand profile for the system was made. This profile is shown in Figure 6. The capacity of the substation is, as previous mentioned, exceeded in scenario 1. It is therefore chosen to set the capacity limit in this analysis as the cables thermal limit. The power capacity for each cable was calculated based on the nominal voltage and I_{th}, the maximum continuous operating current. I_{th} was given from Quire for power grid, Technical data provided by SINTEF (SINTEF , 2010). Further it was calculated for the entire system. The thermal limit was calculated to approximate 380 kW.

Scenario 2

The demand profile in scenario 2 is shown in Figure 7. The extra demand of the six electrical vehicles does not exceed the capacity at any point. Figure 7 represent the worst case scenario, when all the vehicles are connected during peak hours. As seen in the figure, the highest peak of 366 kW occurs at 15.00.

Except for the peak hours, the dumb charging, represented as the red line, is consistent with the base demand. The smart charging is, as shown in the figure with a blue line, slightly higher during the night. This represents the charging of the vehicles. It is assumed that all the vehicles are disconnected between 06.00 and 14.00.

As expected, the extra demand from charging in the dumb charging approach does not exceed the limits of the system. It does however cause an extra strain during peak hours of 36 kW. Since the charging is moved to nighttimes in the smart approach, there are no additional strains in peak hours.

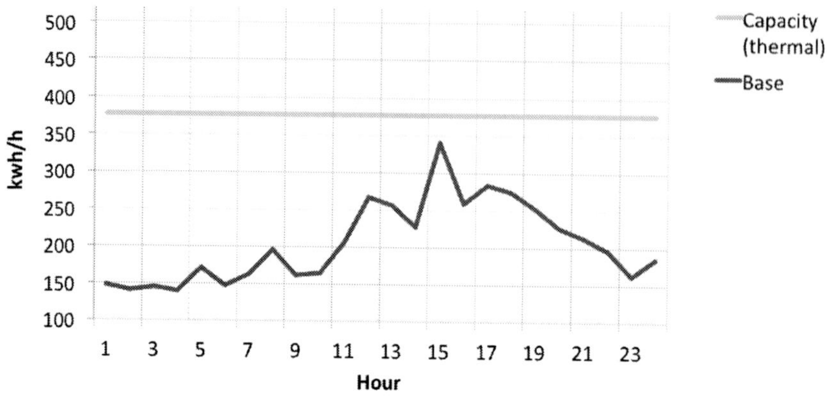

Figure 6. Scenario 1.

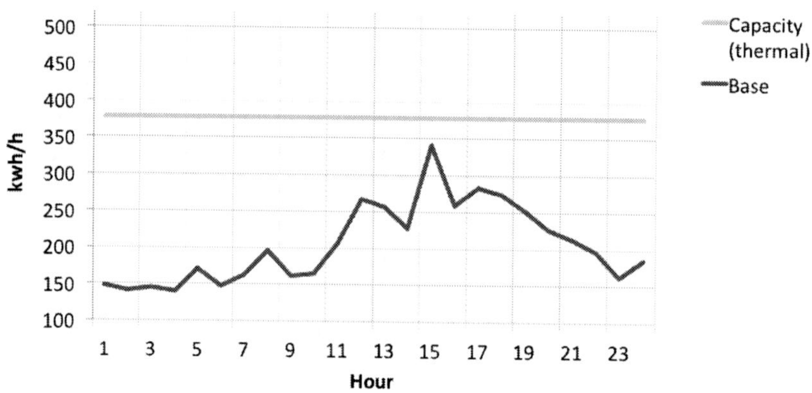

Figure 7. Scenario 2 - 10% EVs.

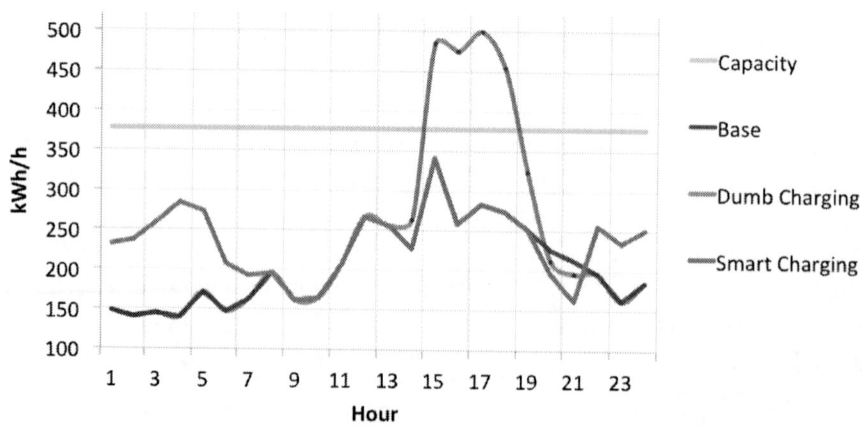

Figure 8. Scenario 3 - 63% EVs.

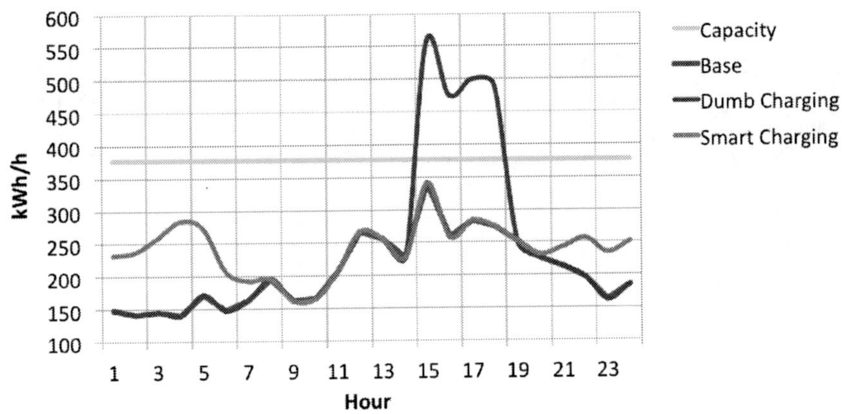

Figure 9. Scenario 3 - Worst case.

Scenario 3

In the third scenario, the load for charging 36 vehicles is added to the power profile. This is shown in Figure 8. The demand for the dumb charging will, as Figure 8 presents, follow the base demand until 14.00 - when the vehicles are connected to the grid. The consumption peaks at 17.00 with a load of 454.5 kW, which exceed the limits of the system with 76,1 kW. The demand exceeds the capacity limits between 15.00 and 19.00. During the second approach, the smart charging, the charging is moved to late evening and night. The demand profile is consistent of the base demand during peak hours, leaving the highest peak to 330 kW at 15.00. The low demand between 05.00 and 07.00 is due to that the vehicles are done charging. In a case with a higher EV percentage this time could also been used for charging.

In scenario 2, the dumb charging represented the worst-case scenario. In the figure shown here, this is not the case. The worst-case scenario would occur if all 36 vehicles started to charge at 15.00. In that case the load would peak at 15.00 with a total load of 546 kW, an increase of 91,5 kW compared to scenario 3 (ref Figure 8. This case is shown in Figure 9. This is not simulated further since it was assumed that the percentage of EVs connected to the grid at that point is approximate 20% (ref: Figure 4).

Voltage Profile Analysis

In the second part of the analysis, the voltages in the system were measured. This was done in Simpow. According to the government's Regulations on Delivery Quality in the Electric System the voltage shall not exceed ±10% (FOR-2004-11-30-1557, 2004). In addition to the general U_n±10% requirement, the quality of voltage at the point of end user is also included in the regulations. The voltage at the point of end user shall not exceed the limits given in Table 2 when the nominal voltage is $\leq$1kV .

Table 2. Voltage limits

	Variation	Per Unit	Volt
Max voltage	$U_n + 8\%$	1.08	248.4 V
Min voltage	$U_n - 6.5\%$	0.935	215.05V

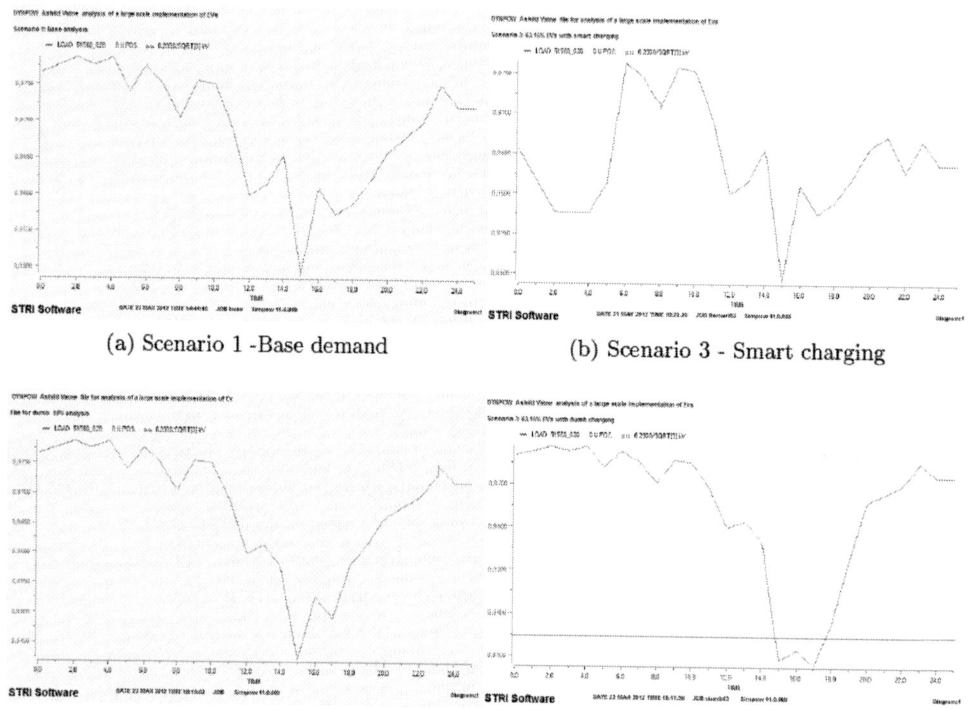

(a) Scenario 1 -Base demand

(b) Scenario 3 - Smart charging

(c) Scenario 2 -Dumb charging

(d) Scenario 3 - Dumb charging

Figure 10. Voltage measurement at 020 - from SIMPOW.

Figure 10 shows the voltage during the different scenarios and charging approaches at residence 020 in the system. This was the point of measurement provided by NTE. Figure 10a present the voltage during scenario 1 - the base demand. The voltage is decreasing as the load is increasing. The lowest voltage dip measured in scenario 1 is 0.9491 p.u, which occur at 15.00. Figure 10b present the voltage at the residence in scenario 3 with smart charging. The lowest measured voltage is as for scenario 1, 0.9491 p.u. The only difference from Figure 10a is a reduction in the voltage between midnight and 06.00 due to charging of the vehicles. The voltage measurement for smart charging in scenario 2 was not included here due to the result is the same as in scenario 3, just with smaller values. In Figure 10c, the voltage at the residence is scenario 2 during dumb charging is presented. As explained previous in this chapter, the value will follow the base demand until the vehicle is connected around 14.00. Due to the extra load caused by the charging the voltage dips, and the lowest measured voltage is 0,9424 p.u. This does not exceed the limits in table.

Figure 10d shows voltage at residence in scenario 3 during dumb charge. As seen in the figure, there are no change in the voltage compared to scenario 1 between 22.00 and 13.00.

During the charging of the vehicle the voltage dip below the limits of $U_n - 6.5\%$. The limit is marked in the figure. The voltage is below the limit between 15.00 and 18.00, with its lowest value at 17.00 when the voltage is 0.9283.

Residence 001 is the point furthest away from the main supply. It is therefore expected that this point will face voltage problem first. Figure 11 present the voltage measurement for residence 001. The voltage in scenario 1, the base demand, is shown in figure Figure 11a. Due to the up scaling of the system from the consumption from residence 020, the consumption-curve for residence 001 will have the same form as the demand for residence 020, shown in Figure 10d. (Plassholder2) (Vågane, 2006)The lowest measured value is 0.09419 p.u. in scenario 1. The same applies for scenario 3 - smart charging, this can be seen in Figure 11b. The dip in voltage from midnight to approximate 04.00 is due to the charging of the vehicles.

Figure 11c and Figure 11d shows the voltage for respectively scenario 2 and 3 during the dumb charging. In this approach the demand is consisted with the base demand from scenario 1, with the exception of peak hours. In both figures, the voltage drops below the margin of $U_n - 6.5\%$. This limit (0,935 p.u) is marked in both figures. In scenario 2 the voltage drops below the limit at 15.00, with the lowest measured voltage as 0.9332 pu.

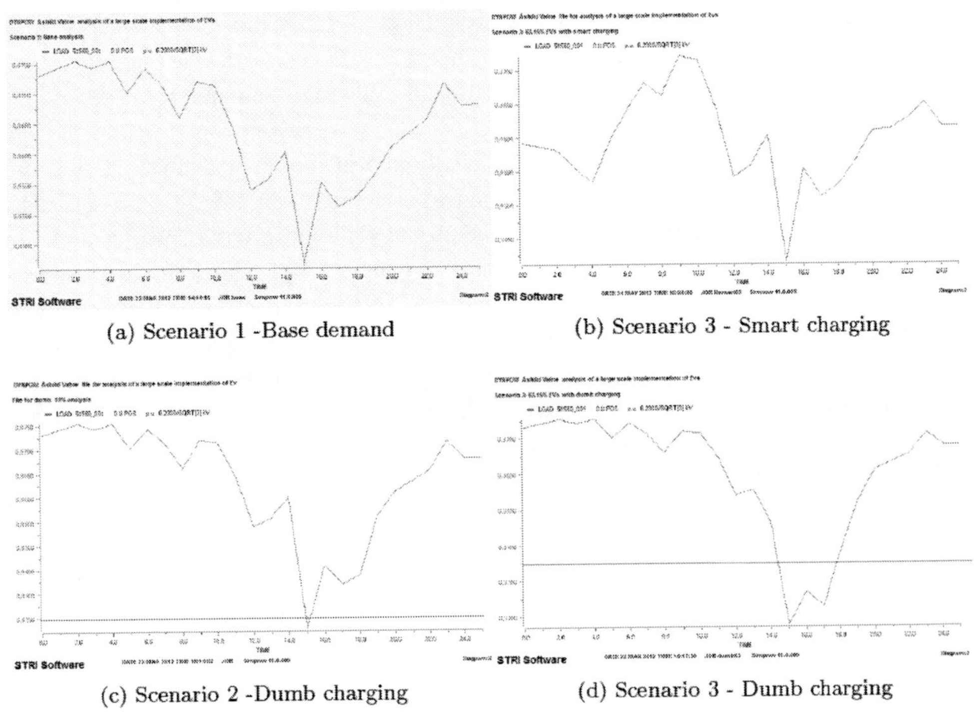

(a) Scenario 1 -Base demand

(b) Scenario 3 - Smart charging

(c) Scenario 2 -Dumb charging

(d) Scenario 3 - Dumb charging

Figure 11. Voltage measured at the node farthest away - 001 (SIMPOW).

In the case of scenario 3, Figure 11c shows that the limit is exceeded between 15.00 and 18.00. The lowest voltage is measured to be 0.9179 p.u., which occur at 15.00. This differs from the measurement from residence 020, which had the lowest voltage dip at 17.00 in scenario 3. The voltage in this case is 1,71 % below the limit.

Wind Power Profile Analysis

In the third part of the analysis, it was determined to look at the correlation between demand and wind power production in the distribution network in Nord-Trøndelag. Can wind power solely supply the low voltage network? What capacity of energy storage is needed?

It is assumed that smart charging will implement production as a parameter for the scheduling. In the scheduling shown previous in the report, the scheduling is based on low demand only. The charging of the vehicle has therefore been neglected in the following analysis.

Figure 12 shows the power flow for December 22nd 2010. The demand for the given low voltage system is presented as the green line. The produced power shown as the purple line in the figure is the estimated wind power production, calculated as described in the sections above.

The demand is higher than then production throughout the day. The deficit of energy is calculated to approximate 2.46 MWh. The only time the production curve reach the demand curve, is a 04:00, where the production peak with 5 kW over the demand curve. In order to be able to determine the available power in storage before the 22nd, an analysis of data from the 15th and until the 25th were carried out. The production and demand for these days can be seen in Figure 13. The demand for the given days has a higher average value than the average value for the month. The value is 5,91 kW, while the average demand for the month of December is 5,52 kW.

The demand in the winter of 2010 was, as mentioned in the other analysis, higher than usually due to an extreme cold winter. The wind measurement for the 10 days has also a higher average wind speed than for the rest of the month. The mean wind speed was 8,5 m/s in comparison to 7,5 m/s as it was for the month of December. The figure shows that there is production on the 15th, from noon and until midnight, and then there are not produced any energy until the morning of the 17th. The wind is steady until the middle of the day at the 20th, except for a period with lower wind speed the 19th. From noon the 20th, the demand increases some while the wind speed decreases. There is a small surplus of energy at the night of the 24th. During the day of the 24th, the wind goes below the cut-inn wind speed at there is no available wind power.

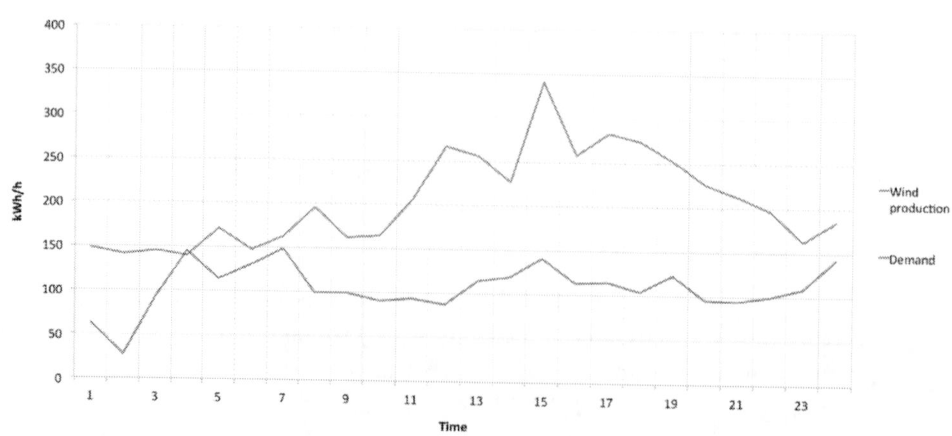

Figure 12. Production/ demand - Dec 22.

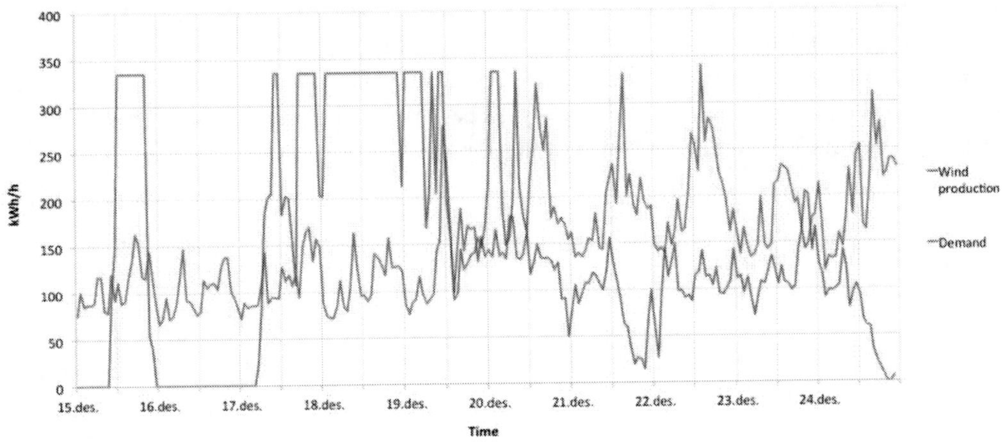

Figure 13. Production/ demand - week Dec.

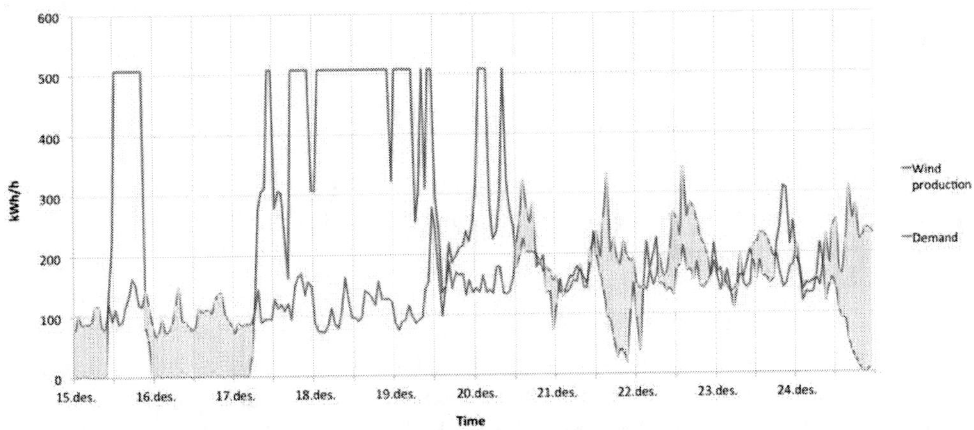

Figure 14. Production/ demand (with 500kW turbine) for 10 days in Dec 2010.

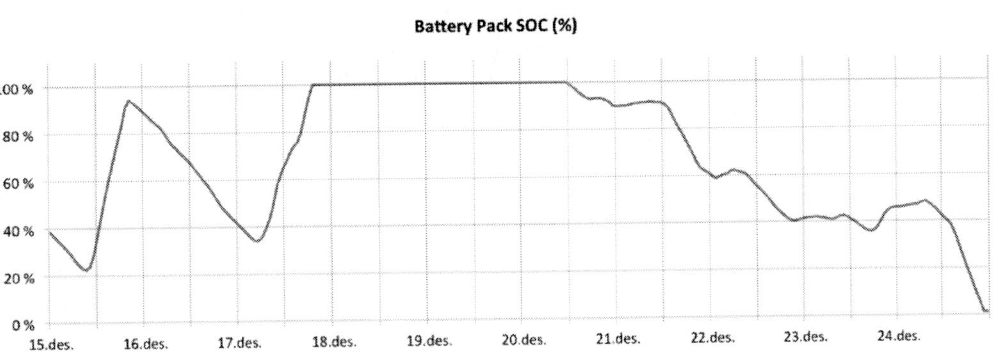

Figure 15. State of Charge with battery-pack of total 5 MW.

In these periods, an external power supply or local energy storage need to deliver energy to the system. The sum of deficit power is calculated to 14,83 MWh. In the periods where then production is more than the demand, the surplus must be stored or delivered to the rest of the grid. Surplus energy is calculated to only 12,68 MWh. The production from the 330 kW turbine is not enough to cover the demand for the period.

Energy Storage Needs

Since the case study presented in this chapter is discussing a large scale adoption of electrical vehicle, it is therefore chosen to look at vehicle batteries as preferable storage. As previous stated, Nissan Leaf is delivered with a 24 kWh Li-ion battery pack. Tesla delivers their vehicles with larger battery-packs. Tesla S will be delivered with a Li-ion battery-pack of 42 kWh. Tesla Roadster can be delivered with 53 kWh. For Tesla's newest model, Tesla Model X, Tesla has developed state of the art battery-pack with the total capacity of 60 kWh and 85 kWh (Zero). The energy storage is build up by a vehicle fleet. A fleet with 50 Tesla X - or other models with a Li-ion battery of 85 kWh, can have the total capacity of 4.25 MWh.

It was desirable to try to operate the low voltage network as an "island grid", when neglecting stability and power quality issues. Since the demand peak the production from the 330 kW turbine, it was decided to use the production from a 500 kW turbine in the storage design.

The deficit of energy is marked in green in Figure 14. As seen in the figure, a large share of this deficit is due to the demand on the 15th and 16th. Therefore the required minimum SOC t the beginning of the time period is 40%. In order to cover the total consumption for the time period, there is a need for a minimum capacity of 5,0 MWh. The lowest SOC will in that case be 1,2 %. This occurs in the end of the time period, as shown in Figure 15. It is expected that a longer analysis would require a larger storage. Figure 16 shows the surplus energy when the battery is charged to 100 %. This power can be delivered out to the rest of the grid.

A total capacity of 5 MWh, requires a huge number of vehicles. The Nissan Leaf fleet would consist of 209 vehicles, while a Tesla X fleet would require "only" 84 (60 kWh) or 59 (85 kWh) vehicles. For the given low voltage network, the vehicle fleet is 57 vehicles. If one considers an implementation of EVs as described in scenario 3, there are 36 battery-packs available. It is expected that the battery-pack in future EVs will be Li-ion batteries, with larger capacity for lower cost (Ferdowsi, 2008). With that in mind, an energy storage consisting of 36x85 kWh Li-ion batteries, gives a total capacity of 3060 kWh. The designed energy storage described in the section above, consist of a 36 Li-ion 85 kWh batteries. The number of vehicles suits a medium sized parking lot in Norway. However, the battery is the state of art Li-ion battery and only delivered with Tesla X. The first Tesla X will be delivered late in 2013, and the price will vary between $60.000 and $100.000 (Zero). One can therefore assume that there is not possible to design this storage before 2020. The Nissan Leaf was chosen as the model for simulation - a energy storage consisting of a fleet of 36 Nissan Leafs would have a total capacity of 864 kWh. Due to a smaller storage, an external source must supply in the periods when demand is higher than the production and storage can deliver.

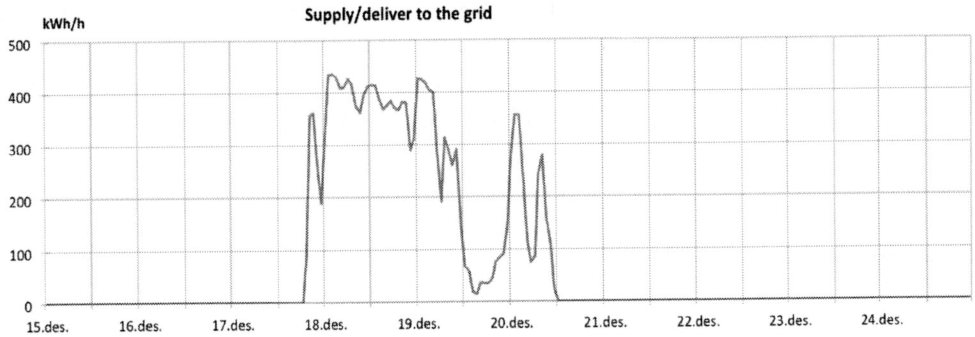

Figure 16. Surplus power delivered to the grid when storage is sized 5 MW.

This is marked in green in Figure 17. It is in this case assumed that the 22 kV grid will match the power with supply and receive power as shown in Figure 17. The purple field in the figure is surplus energy.

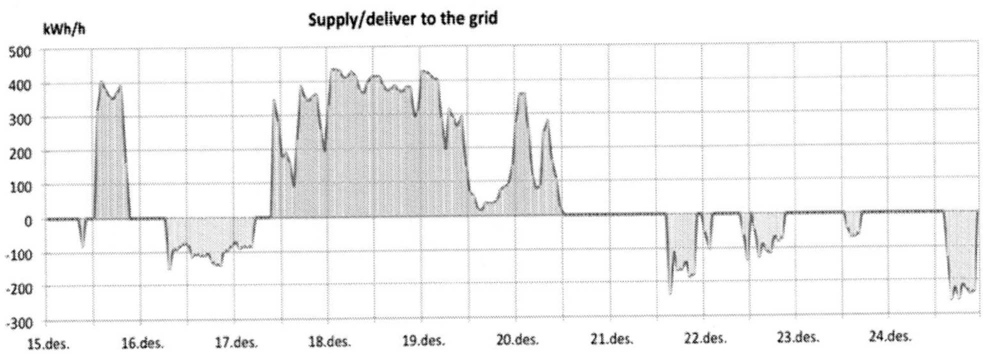

Figure 17. Grid support for storage sized 864kWh.

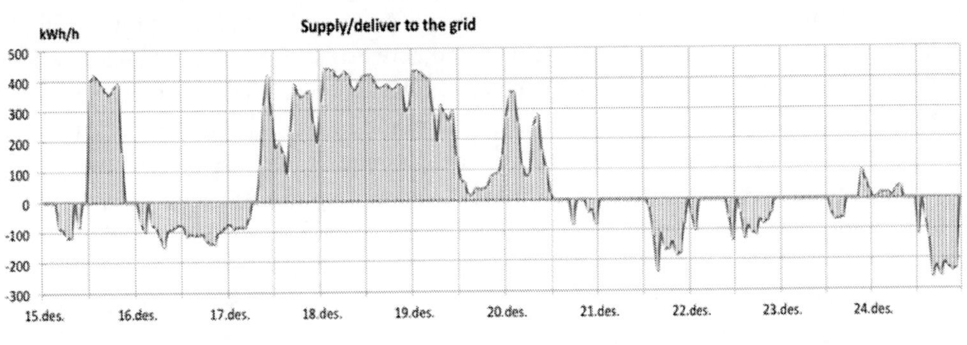

Figure 18. Grid support for storage sized 350kWH.

Energy storage of 36 Li-ion batteries with the capacity of 24 kWh, is not large enough to store the surplus power or to supply the system in low production periods. In order for the vehicles to operate as storage, they must be part of an organized fleet, such as a taxi fleet or a parking lot. This solution would be possible to conduct in larger cities, but is not suitable for

Steinkjer and similar smaller cities. Figure 18 represents the power supplied from and delivered to the grid when the storage is set to 350 kWh. As seen in the figure, it does not differ significant from Figure 17. The designed storage in Figure 18 is a bank of 7 battery-packs with a capacity of 50 kWh each. The battery-packs consist of Li-ion batteries from old EVs.

Vehicle-to-Home

A different approach than one large energy storage is to use the solution of V2H. The vehicle will be charged from a mode 3 chargers at the residence when the demand is low or available renewable production. The power flow shown in Figure 20 shows the correlation between one Nissan Leaf and the consumption from residence 020.

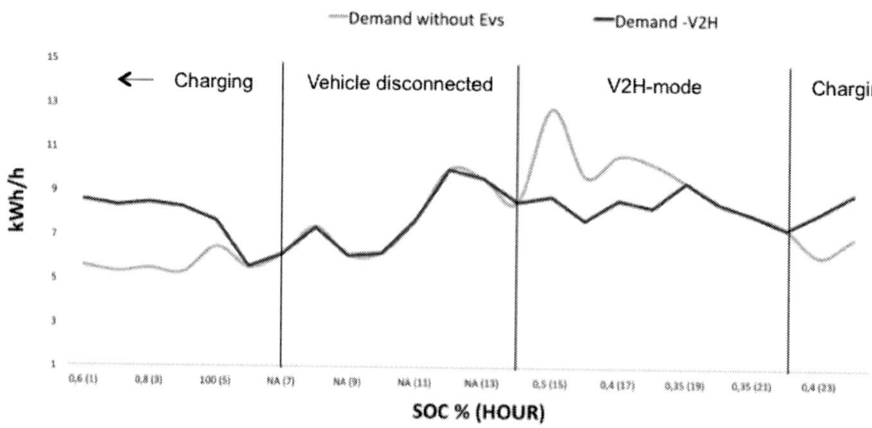

Figure 19. V2H - One EV connected to residence 020.

The x-axis is the state of charge - and the hours of the day is in parenthesis. It is assumed that the vehicle will be charged from 22:00 and until 06:00, and it will be charged with 3 kW each hour. As previous, the vehicle is disconnected between 07:00 and 14:00, and the SOC is therefore applied as not applicable (NA). When the EV is connected to the charger again at 15:00, it is assumed a SOC of 50%. This is based on the average distance a person drives in Norway per day and the range of the Leaf, it is also assumed some loss in the battery. During the peak hours the vehicles battery supply the home with power as seen from 14:00 until 19:00 in Figure 20. The purple curve is the demand supplied from the grid when V2H solution is engaged. The load will be flatten and more even as seen in the figure.

It is assumed that the utility or a third party will continuously monitor all the elements in the grid, and phase in a charging schedule. One of the problems for this issue is that some of the vehicle manufacturers have indicated that they will not allow their vehicle to operate in V2H or V2G due to warranty, and other problem is the consumer's willingness for peak shaving (Zero) (Seljeseth). Are they willing to postpone charging or "loose" range at their vehicle?

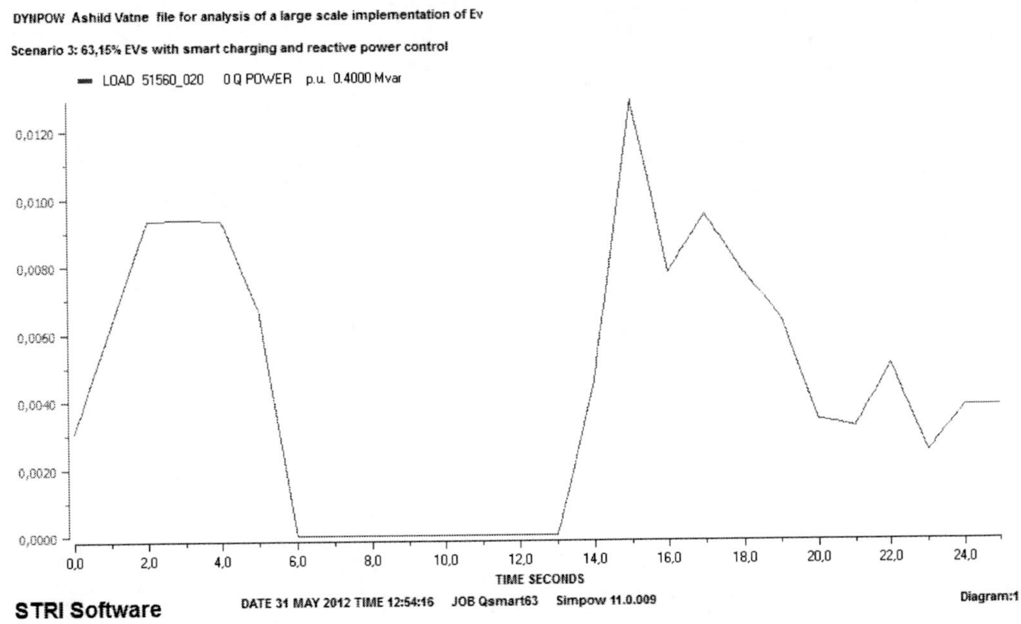

Figure 20. Reactive power at residence 020.

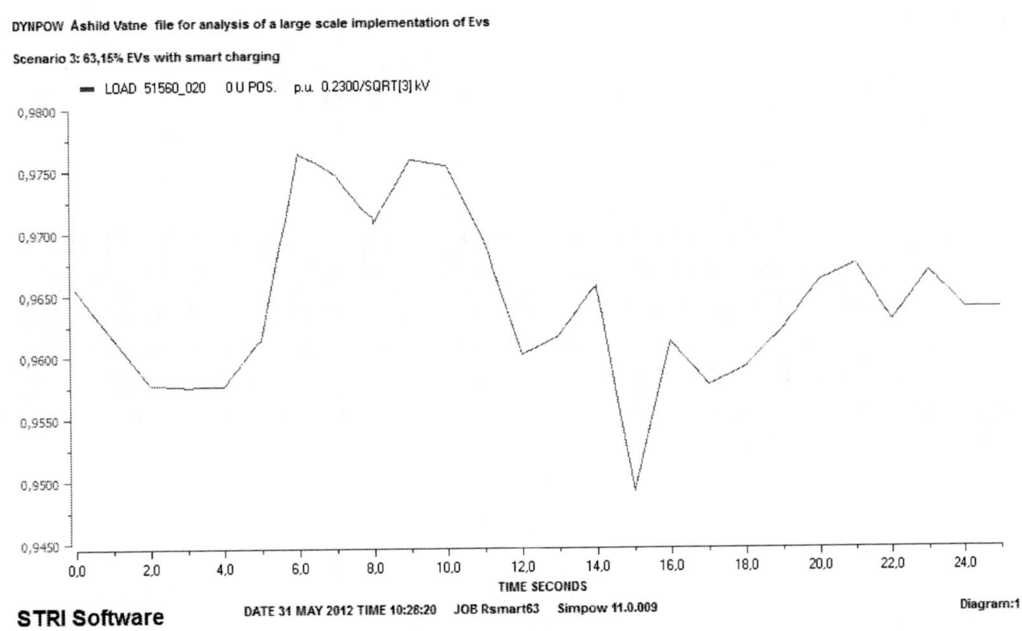

Figure 21. Voltage at residence 020 without Q-compensation.

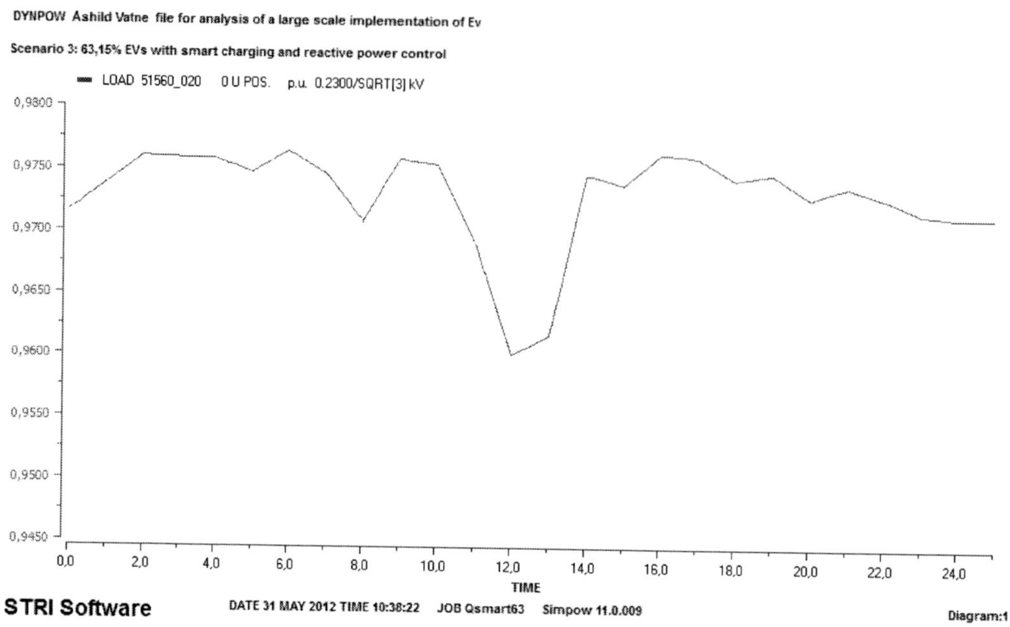

Figure 22. Voltage at residence 020 with Q-compensation.

Reactive Compensation

EV chargers can provide several services that are included in the V2H, to the grid. Voltage support, harmonic filtering and reactive power compensation, are some of the services. The reactive power on the nodes were controlled as the same way as the active power, where $Q = Q_{max} \cdot Q_{TAB}$

Q_{max} was set to 0.002 p.u as a capacitive load in optpow, S_{ref}=0.4 MVA, the value of the reactive maximum power is 0.8 kVAr. Figure 20 shows the changing value of Q throughout the day of the 22nd of December 2010. The case simulated is scenario 3 with smart charging. Figure 21 and Figure 22 shows the voltage at the charger at residence 020. The voltage without Q-compensation is shown in Figure 21. In Figure 22 one can see how the reactive compensating smoothens the voltage. The compensation shown here is a very simplified analysis, but the result show the trend of the possibility.

Discussion of Results

The purpose of the work addressed in this chapter was to show consequences of a large-scale adoption of electrical vehicles. A low voltage network was simulated. The specific network was provided by NTE, and was located in Steinkjer, Nord-Trøndelag. The network data is not compared with other network, but it is expected that the result of the report can be used for similar networks. The average annual consumption for a residence in Norway is 20 MWh. The average for the 38 residences in the system is 21.45 MWh. If one consider that the consumption was extremely high in 2010 with 130,4 TWH, the average value for the residence is a typically residence area in the Norwegian power-market. The annual consumption for 2009 is 123,7 TWh while the value was 121,8 TWh in 2011 (Statnett). According to the Norwegian Meteorological Institute, December 2010 was the 4th coldest since 1900 (Meteorologisk institutt). This caused a very high demand. One can expect that the analysis of December 2011, would give different results than stated here. Nevertheless, it is important and interesting to look at the worst case scenario as December 2010 represent.

The hourly consumption for each residence was scaled up from one residence. A power profile with hourly measurement from each residence may differ from the scaled version. The power profile used in the analysis does however resemble a typical 24-hour load profile. The highest peak will normally occur later in the afternoon in a typically day, but as the simulation date is set to the 22nd of December one may assume that the peak occur earlier due to residents arrive earlier for Christmas.

The scaling of the systems demand profile caused the demand to peak the substation. The substation was therefore set to 400 kVA. This is not a typical size in the Norwegian power-grid, but since the cables thermal limit was calculated to 378 kW, it was decided that a 400 kVA substation would be suitable.

The charging model for the electrical vehicles is very simplified. A simulation with a more detailed charging model may cause different results. Shouxiang Wang presented the difference between the actual charging and a simplified charging(Wang, Han, Wang, Li, & Shahidehpour, 2012). The model is used for lithium batteries. The study supports that the simplified charging is conformable for the work addressed in this chapter.

It was assumed that all the EVs were charged during the 24 hours time period. According to The Institute of Transport Economics (Transportøkonomisk institutt) the average distance a person drives each day in Norway is 43 km(Vågane, 2006) (Wang, Han, Wang, Li, & Shahidehpour, 2012). Considering that Nissan Leaf has a driving range of 150 km (Nissan Norway, 2011) (Statnett) the average charging frequency *could* be that the owners charged their vehicle every third night. However, the worst case scenarios where all the owners plug their vehicle in the grid at the same day will be analysed in this study due to the insecurity in the owners charging behaviour.

The power flow for the three different scenarios is the result of manual calculation. The charging scheduling was based only on low demand. A preferred method for the scheduling would be a smart algorithm based on price, low demand, available renewable production and available energy storage. Nevertheless, the manual scheduling shows of the power flow and reflect a smarter charging. The power profile for scenario 3 dumb charging exceeds the systems limit. It was expected that the high power would cause a large voltage drop in the nodes far away from the supply. It was expected the voltage would drop below 0.9 p.u. Voltage simulations shows that dumb charging cause a voltage drop of 2.55% more than the smart charging. The LV-network is however connected to a relative stiff network. The 22 kV side had a $\cos\Phi \approx 0.97$. It is therefore expected that in a low voltage network further away from production, would have larger voltage drops.

Even if the result did not give a voltage drop below 0,9 p.u, the dumb charging from scenario 2 showed some interesting found. Table 3 lists the lowest measured voltage for each case. As seen in the table, the node furthers away from the supply, residence 001, will face voltage problems also in scenario 2 with only a 10% share of EVs. This means that the system is not capable to handle a share of 10% of electrical vehicles without any restrictions in charging.

Table 3. Lowest voltage for the different scenarios

		Scenario 1	Scenario 2	Scenario 3
Residence 020	Dumb charge	0.9491	0.9424	0.9283
	Smart charge	0.9491	0.9491	0.9491
Residence 001	Dumb charge	0.9491	0.9332	0.9179
	Smart charge	0.9491	0.9491	0.9491

The wind measurement was collected from a met-mast at 80 m high. The wind speeds were adjusted for 50 m with the power profile law. The power law exponent α was set to 1/7. This assumption is not adequate if designing a wind farm, but due to the simplicity of the wind power analysis it is sufficient. Loss in the production has not been accounted for either. The Norwegian Water Resources and Energy Directorate (NVE) use a loss of 20% of the total production. Wake loss was given as the largest factor causing loss. Other factor is named as transmission loss, icing, etc. Considering that the analysis only covers one turbine, the wake loss is neglected. The electrical loss is not included either. The resistance in the lines are very small - which gives small loss in active power. Reactive power has been neglected throughout the analysis.

The storage was designed as a number of batteries in parallel. The total capacity of the storage was the sum of the capacity. The availability of the vehicles as a storage has not yet been discussed, but is an important factor for the solution. Can one secure the capacity of the storage when considering of the indeterminacy of the owners behaviour? It is not assumed that storage will consist of several private vehicles due to these issues, but a taxi-fleet or an organized parking lot connected to renewable power resources is reasonable choices.

A different approach is also to use batteries from old vehicles. There is an on-going project on turning the batteries from old Nissan Leafs into storages. Even if the storage can both supply and receive the entire power flow, the system cannot operate as a stand-alone system without power application storages or grid support. A common problem in wind power production is flicker and harmonic disturbances.

Harmonics are AC voltages or currents with a frequency that is in an integer multiple of the fundamental frequency (grid frequency). The harmonic disturbance or distortion is the effect on the fundamental waveform cased by electrical switches. In a circuit with several power electronic components, it is more likely to experience more harmonic disturbance. The harmonics are mainly caused by inverters, industrial motor drives, electronic appliances, light dimmers, fluorescent light ballasts and personal computers (J. F Manwell, 2009). The harmonics disturbance is measured by THD (total harmonic distortion), the worse waveform, the higher THD. One can reduce these disturbances by placing a sinusoidal-filter after the inverter. In *Wind Energy Explained* by Manwell, flicker is defined as "disturbance to the network voltage that occur faster than the steady state voltage changes and which are fast enough and of a large enough magnitude that light noticeable change brightness" (J. F Manwell, 2009). The disturbance occurs due to the torque fluctuations in the turbine or when the turbine is connected/ disconnected. Flicker is noticeable in weak grids. There was not carried out a stability analysis in the case study. The wind power production needs to be balanced, both immediate balancing for fast response with propped inertia and load balancing as renewable backup.

CONCLUSION

In the study addressed in this chapter, an analysis of a large share adoption of EVs in a low voltage network has been carried out. The analysis was based on the worst case scenario, using demand from December 2010 - a month of extreme high demand. The result from the analysis shows that the utility companies need to forward a smart charging strategy. The system was not capable to handle a large share of EVs (>60%) without scheduling or restriction for charging. The smart charging did not cause any extra strain on the grid during peak hours. In order to reach the goal of 25% reduction in emission, it was stated at least a share of 10% of the vehicles were electrical by 2020 (EBL, 2009). One can see in the result from the power flow, that the system provided here is capable to handle a 10% EV-share. However, the voltage simulations resulted in that the node furthest away from the supply, residence 001, will face voltage problems in this case, where only 10% of the vehicles are electrical. One can therefore conclude that system is not capable to handle a smaller implementation without charging scheduling or restrictions.

In the second part of the analysis, a series of wind measurement was added to the analysis. The results show that the system *can* operate as a stand-alone system when the designed storage consists of 59 Tesla X vehicles with 85 kWh batteries. Storage like that would be extremely expensive, and accounted for as not realistic. A more realistic storage would consist of 24 kWh Li-ion batteries from ≈40 EVs. Due to the uncertainty of the private EV owner's behaviour, the vehicles should be a part of a taxi fleet or an organized parking lot. Another storage possibility is the V2H, but the vehicle manufactures is not willing to engage this solution into their vehicles. It is therefore concluded to use battery-packs from old vehicles as storage in this low voltage network. It is also concluded that the storage will not be designed as the backup for a stand-alone system. A storage bank of ≈7 battery-packs of 50 kWh each was designed as the most suitable storage based on vehicle availability, storage cost and location and size of network. The analysis from December 2010 was given as the worst case scenario, and the result from March 2012 shows that the area will be able to be self-supplied most of the time. Nevertheless, the connection to the grid is still needed to receive surplus energy and also supply stability services from the grid.

Large share adoptions of EVs require a beneficial infrastructure. In addition to chargers, both from suitable docking station (mode 2/ mode 3) and fast chargers (mode 4), communication are essential in the implementation for smart charging. The smart charging and the smart grid in general, forward large opportunities for new markets and new technology. The smart grid infrastructure gives opportunity for aggregation services, and it is therefore important to establish suitable standard for the equipment and solutions. There is also a need for further research on optimization algorithms and scheduling algorithms is essential in order to achieve smart charging.

REFERENCES

EBL. (2009). Handlingsplan for elektrifisering av veitransport. Energi Norge.

Executive Champers Honolulu. (2008). State and Hawaiian Electric Strike Sweeping Agreement For Hawaii`s Energy Future.

Ferdowsi, J. P. (2008). Comparison of NiMH and Li-ion Batteries in Automotive Applications. Missouri University of Science and Technology. IEEE.

FOR-2004-11-30-1557. (2004). *Forskrift om leveringskvalitet i kraftsystemet*. Olje- og energidepartementet. Lovdata.

G Haines, A. M. (2009, 03). The Simulation of Vehicle-to-Home Systems - Using Electric Vehicle Battery Storage to Smooth Domestic Electricity Demand. EVRE Monaco 2009.

International Standard, IEC. (2010). IEC 61851-1. Norsk Elektronisk Komite (NEK).

Kisacikoglu, M. C., B. O. (n.d.). Examination of a PHEV Bidirectional Charger System for V2G Reactive Power Compensating. Retrieved 06 8, 2012, from http: //ieeexplore.ieee.org/ielx5/5426413/5433335/05433629.pdf?tp=&arnumber=5433629&is number=5433335

Manwell, J. F,; J. G. (2009). Wind Characterristics and Resources. In Wind Energy Explained. John Wiley & Sons.

Meteorologisk institutt. (n.d.). Desember 2010 den 4. kaldeste. Retrieved 11 2011, from http://met.no/Desember+2010+den+4.+kaldeste.b7C_w7vSZm.ips

Nissan Norway. (2011). Retrieved 11 11, 2011, from http://www.nissan.no/NO/no/ vehicles/electric-vehicles/electric-leaf/leaf.html

Nissan USA. (2011). Nissan Leaf Technology. Retrieved 11 11, 2011, from http://www. nissanusa.com/leaf-electric-car/index

Norsk Vindkraftforening. (n.d.). Infoside om vindkraft. (E. Norge, Producer) Retrieved 05 28, 2012, from http://www.vindkraft.no

NTE. (2011, 10). Information and data from the LV-grid from NTE.

Peças Lopes, J. A., F. J. (2009). Smart Charging Strategies for Electrical Vehicles: Enhancing Grid Performance and Maximizing the Use of Variable Renewable Energy Resorces. IINESC Porto and FEUP, Porto, Portugal. EVS24, Stavanger.

Petterteig A., O. M. (2006). Tekniske retningslinjer for tilknytning av produksjonsenheter. SINTEF Energiforskning AS.

Seljeseth, H. (n.d.). Nettutfordringer (spenningskvalitet) ved storskala lading av elbiler. Retrieved 06 2011, from http://www.energinorge.no/getfile.php/ FILER/KALENDER/ Foredrag%202011/Brukerm%F8te%20spenningskvalitet/1409_1435_Nettutfordringer%2 0ved%20storskala%20lading%20av%20elbiler.pdf

SINTEF . (2010). Planleggingsbok for kraftnett, Tekniske data. SINTEF ENERGI AS.

Society of Automotive Engineers, SAE. (2010). SAE J1772-2009. Retrieved 10 27, 2011, from http://standards.sae.org/j1772_201001/

Statistisk Sentralbyrå. (2011, 04). Bilbestand of folkemengde (F). Retrieved 2011, from http://www.ssb.no/bilreg

Statnett. (n.d.). Nøkkeltall 1974-2011. Retrieved 06 2012, from http://www.statnett.no/ no/Kraftsystemet/Produksjon-og-forbruk/Historikk/

Suwanapingkarl, P., D. J. (2009). Impact of Electric Vehicles on Power Distribution Networks. Energy System and Advanced Materials Research Group.

Vågane, L. (2006). Bilhold og bilbruk i Norge. TIØ.

Wang, S., Han, L., Wang, D., Li, Z., & Shahidehpour. (2012). ierchical Charging Management startegy og Plug-in Hybrid Electric Vehicles to Provide Regulation Service.

Zero. (n.d.). Klimabiler. Retrieved 06 2012, from http://www.klimabiler.no

In: Grid Electrified Vehicles
Editor: Carla Alexandra Monteiro da Silva

ISBN: 978-1-62808-839-7

Chapter 7

SIMULATION OF ELECTRIC VEHICLES MOTION USING A STOCHASTIC MODEL: ASSESSMENT OF ENERGY REQUIREMENTS AND ENVIRONMENTAL IMPACTS

F. J. Soares, **P. M. Rocha Almeida and J. A. Peças Lopes***
INESC TEC - INESC Technology and Science
(formerly INESC Porto), Porto, Portugal

ABSTRACT

In this chapter an innovative methodology for determining the energy requirements of Plug-in Electric Vehicles (PPEV) in a given area or region is presented, in order to be used in PEV integration planning. PEV movement patterns depend on the needs of PEV owners for commuting and so to accurately estimate PEV energy needs, it is necessary to describe in the best possible way PEV journeys. The proposed methodology allows not only to estimate energy needs of PEV individually and globally but also to determine the impacts they may have in the global reduction and/or displacement of greenhouse gases emmissions and other environmental indicators. Additionally, this methodology can be used for evaluating other types of impacts relative to the presence of PEV in the vehicle fleet, more precisely on the electricity distribution grids. In the final part of the chapter, some application examples of this methodology will be presented. The case study refers to the Portuguese reality and the data used was collected from a statistical study, whose main goal was the characterization of the common traffic patterns in a region in the north of Portugal, covering the city of Porto and other smaller surrounding cities.

Keywords: CO_2 emissions, distribution grids, electric vehicles, energy consumption, Markov chain, monte carlo simulation, traffic patterns

---

* Corresponding Author address, Email: fsoares@inescporto.pt.

INTRODUCTION

With the expected near future integration of plug-in hybrid and pure electric vehicles in the light-duty vehicle fleet, herein referred to as Plug-in Electric Vehicles (PEV), the transportation sector will undergo deep changes. The main consequence of this change is a paradigm shifter. PEV will promote the diversification of final energy consumption sources in a sector where most of the fleet would run on petrol derivatives, diesel and gasoline. This diversification of fuel sources is very welcomed by the society but carries great challenges to the energy sector. PEV will need to be refueled and, currently, there is no actual infrastructure deployed to meet their needs. Ergo, in order to successfully deploy this new technology an adequate recharging infrastructure must exist. Yet, it should not pressure capital and operational expenditure costs for users and investors in infrastructure assets that would oppose to the uptake of PEV.

To promote the successful deployment of PEV careful planning actions should be performed and to do so it is necessary to understand the energy needs in space and time. Due to current battery costs it is likely that the first PEV will be acquired as secondary vehicle for commuting mainly in urban routes. It is then straightforward to infer that the initial infrastructures will cover cities. As technologies mature and PEV get more affordable and with a wider range of action the charging infrastructure should be expanded onto other areas until it gets country-wide.

In this chapter an innovative methodology for determining the energy requirements of PEV in a given area or region is presented, in order to be used in PEV integration planning. PEV movement patterns depend on the needs of PEV owners for commuting and so to accurately estimate PEV energy needs, it is necessary to describe in the best possible way PEV journeys. To do so, it is necessary to use historical data available, for instance, in statistics institutes of regions or countries, such as Eurostat. This data must be processed and fed into a mathematical model that will enable mapping individual PEV movement patterns in time and place. Due to the inherent uncertainties that the model has to address, a stochastic approach based on a Markovian process was adopted.

This tool allows not only to estimate energy needs of PEV individually and globally but also to determine the impacts they may have in the global reduction and/or displacement of GreenHouse Gases (GHG) emmissions and other environmental indicators. Additionally, this tool can be used for evaluating other types of impacts relative to the presence of PEV in the vehicle fleet, more precisely on the electricity distribution grids. The electricity sector will also experience great modifications due to the presence of PEV. PEV present a new load for the system that was not prepared for in the planning stage, thus it will be necessary to understand their impacts and also develop new tools for the planning of future electricity distribution networks.

In the final part of the chapter, some application examples of this methodology will be presented. The case study refers to the Portuguese reality and the data used was collected from a statistical study, whose main goal was the characterization of the common traffic patterns in a region in the north of Portugal, covering the city of Porto and other smaller surrounding cities. (S. Bending, 2010)

CHARGING SCHEMES CONSIDERED

There are several types of PEV charging solutions being currently adopted, (S. Bending, 2010), which involve distinct power levels:

- Level 1 – Around 3 kW charging power that can be obtained through common domestic outlets;
- Level 2 – 10-20 kW charging power that can only be obtained through dedicated charging outlet and wiring;
- Level 3 – More than 40 kW charging power that can only be obtained through dedicated charging outlet and wiring and using a dedicated off-board charger for DC fast charging.

The charging type commonly classified as slow refers to level 1, while the fast charging refers to level 3. Level 2 is an intermediary charging level. All the three levels were considered in this work, being assumed that slow charging corresponds to level 1 – Low Voltage (LV) connections, while fast charging includes level 2 and 3 – Medium Voltage (MV) connections.

Depending on the type of application, PEV controllability may vary and, therefore, several control schemes may be adopted. In the solutions involving fast charging (level 2 or 3), a full charge might take less than 1 h (S. Bending, 2010). Due to the urgent needs from the user of these types of services, especially level 3 clients, no controllability is envisaged.

On the other hand, depending on the PEV battery State-of-Charge (SOC) and capacity, full charge solutions involving level 1 might take up to 12 h (S. Bending, 2010). Within this charging alternative, it is assumed that PEV owners can choose between a set of three charging options: two passive or non-controlled (dumb charging and multiple tariff) and one active or controlled (smart charging).

In the dumb charging, the PEV owners are completely free to connect and charge their PEV when they want and the charging starts automatically after the PEV being plugged-in (J. A. P. Lopes, 2011).

As in the previous approach, the dual tariff policy assumes that PEV owners are completely free to charge their vehicles. Yet, electricity cost is assumed not to be constant during the day, existing some periods when its cost is lower (J. A. P. Lopes, 2011), which may lead PEV drivers to prefer charging during these periods. The lower electricity price period assumed in this work was that of the dual tariff policy currently implemented in Portugal: 22h to 8h (EDP Serviço Universal);

The smart charging envisions an active management system where there exists a hierarchical control that is used to control the PEV charging rates. Yet, as referred previously, PEV will in this case be exclusively managed and controlled by the PEV aggregators when the grid is operating in normal conditions (J. A. P. Lopes, 2011).

The Vehicle-to-Grid (V2G) mode of operation, (Tomic, 2005), was not considered in this work, as it is not likely to be a reality neither in the short run nor in the medium term. As it is a very aggressive PEV charging mode, due to possible implications with PEV batteries lifecycle, only in the long term this option is likely to be adopted, when battery technology has reached a high maturation stage.

METHODOLOGY FOR SIMULATING ELECTRIC VEHICLES MOTION

This section will describe a methodology for evaluating electric vehicles motion within a given area or region. It allows building a rather versatile tool for planning PEV deployment in the upcoming years. Its major strengths include the generation of individual PEV movement patterns, track them throughout the simulation period and output indicators, such as individual and global energy needs, GHG impacts and the impacts on electricity grids.

This method is built for assessing simultaneously the commuting patterns and electricity grid issues, as these are highly correlated. Generally speaking, the electrical load is higher where more people gather and in a simulation tool that tracks the movement of PEV owners it is necessary to track people to understand their actual destinations. Therefore, it is henceforth assumed that people with commute more frequently to destinations with greater electrical load. Further on this chapter it will be detailed how in accordance to the load of a particular location a probability of commuters to stop in that place is defined.

Markov Chain Description

The PEV movement over a given period of time, for instance in a one year period, may be simulated using a discrete-state, discrete-time Markov chain to define the states of all the PEV at each time step of 30 minutes. It may be assumed that, at every unit of time, one and only one event from a set of a finite number of events can occur to a given PEV. Four event were considered: E_M, E_R, E_C and E_I. To these events one may callevent states. When the event E_k $(k = M, R, C, I)$ occurs, the tracked PEV changes to state E_k. In practice, these four events correspond to four PEV possible states in real operating environment "*in movement*", "*parked in a residential area*", "*parked in a commercial area*" and "*parked in an industrial area*". These states may characterize to a reasonable extent the quotidian commuting patterns of an average driver. More granularity could be conferred to the method if further states would be considered. For instance, if professional drivers, such as taxi drivers or delivery truck drivers were to be tracked individually, then specific states that would allow simulating their behavior would have to be considered in a similar way to what will be further explained in this section. Also other popular states could be added, such as *"parked in a school"*, since it is a common intermediate destination in daily commuting patterns.

From this point onwards, time terminology will be used, i.e. it is considered that one trial is performed at every discrete and periodic unit of time. When the event E_k occurs at the moment t, it is represented by E_k^t. It is assumed that the initial moment occurs at $t = 0$. Therefore, E_k^0 denotes that the initial state of a PEV is E_k.

This method is classified as a discrete-time process, given that t is finite and can be enumerated. As the purpose of this work is to simulate PEV movement during one week plus their initial state, 337 time steps of ½ hour are considered. Thus, $t \in [0,336]$.

One trial is performed initially to define every PEV state when $t = 0$. In this trial, a PEV may be in the state E_k with probability $P(E_k)$.

The conditional probability that at the moment t, for $t \in [1,336]$, a given PEV passes into the state E_k is denoted by $p_{j\to k}^t$ provided that after the trial $t-1$ it was in the state E_j $(j = M, R, C, I)$:

$$p_{j\to k}^{t} = P\left(E_k^t \middle| E_j^{t-1}\right) \tag{1}$$

As mentioned above, this sequence of trials forms a Markov chain, given that for any j and k and for any $t \in [1,336]$, the equalities

$$p_{j\to k}^{t} = P\left(E_k^t \middle| E_j^{t-1}\right) = P\left(E_k^t \middle| E_j^{t-1} \cdot E_j^{t-2} \cdot \ldots \cdot E_j^1 \cdot E_j^0\right) \tag{2}$$

are arbitrarily satisfied for $E_j^{t-2}, \ldots, E_j^1 \cdot E_j^0$.

One transition matrix can be created with the transition probabilities $p_{j\to k}^{t} = p_{j\to k}^{t+\tau}$ for each moment t, where $t \in [1,336]$. This matrix is denoted by M_t and, for a given moment t, it assumes the following form:

$$M_t = \begin{bmatrix} p_{M\to M}^t & p_{M\to R}^t & p_{M\to C}^t & p_{M\to I}^t \\ p_{R\to M}^t & p_{R\to R}^t & p_{R\to C}^t & p_{R\to I}^t \\ p_{C\to M}^t & p_{C\to R}^t & p_{C\to C}^t & p_{C\to I}^t \\ p_{I\to M}^t & p_{I\to R}^t & p_{I\to C}^t & p_{I\to I}^t \end{bmatrix} \tag{3}$$

All the elements $p_{j\to k}$ of the matrix, being probabilities, are non-negative.

Supposing that a PEV is in the state E_j, the event where, as a result of one trial, the PEV remains in the state E_j or passes to any of the states E_k, where $j \neq k$, is the sure event. Since the events E_k are mutually exclusive, for $k = M, R, C, I$, the following equation holds:

$$P\left[\textstyle\sum_k E_k^t \middle| E_j^{t-1}\right] = \textstyle\sum_k p_{j\to k}^t = 1 \tag{4}$$

Thus, the sum of the terms in each row of the matrix M_t equals one. However, the sum of the terms in a column might be different from one.

Figure 1 presents an overview of the Markov chain developed for this study.

Initial State and State Transitions Probabilities

The Markov chain developed is a composition of two sub-cycles with the duration of one day: one for the week days (repeated five times in a row) and other for the weekend days (repeated twice consecutively).

Therefore, to have the Markov chain completely characterized, it is only needed to define the initial state probabilities, $P(E_k^t)$ for $t = 0$, and the state transition probabilities, $p_{j\to k}^{t} = P\left(E_k^t \middle| E_j^{t-1}\right)$ for $t \in [1,48]$, of these two sub-cycles and then repeat them to compose the full weekly cycle.

The required probabilities were determined by analyzing the results of a statistical study, whose main goal was the characterization of the common traffic patterns in a region in the north of Portugal, covering the city of Porto and other smaller surrounding cities [3]. The values obtained from this study for the initial state probabilities ($t = 0$) were 0.89 for "*parked in a residential area*", 0.04 for "*parked in a commercial area*", 0.02 for "*parked in an industrial area*" and 0.05 for "*in movement*". Regarding the state transition probabilities, the values obtained for the week day sub-cycle, weekend day sub-cycle and full weekly cycle are presented in Figures 1, 2 and 3, respectively.

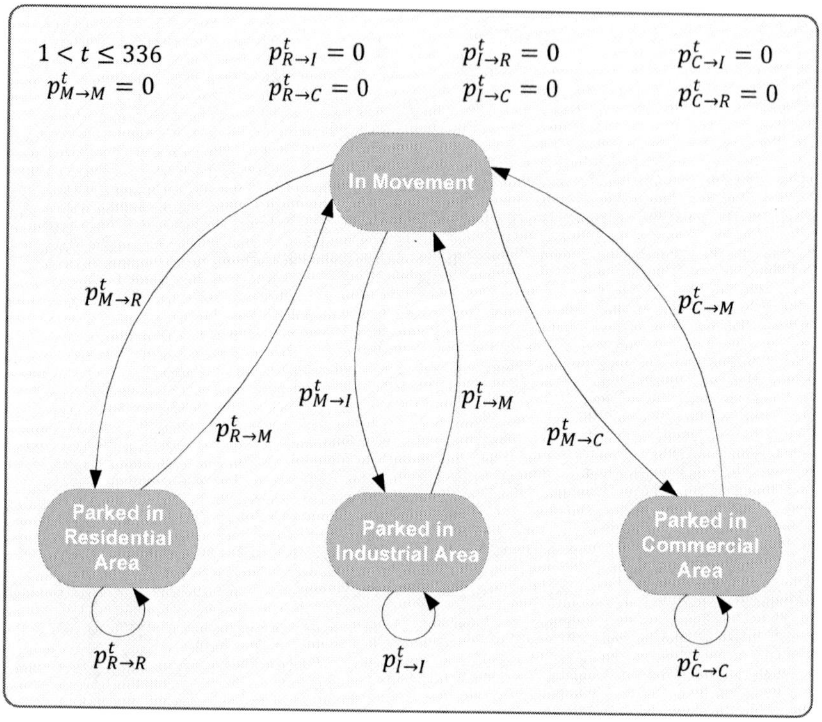

Figure 1. Discrete-state and discrete-time Markov chain.

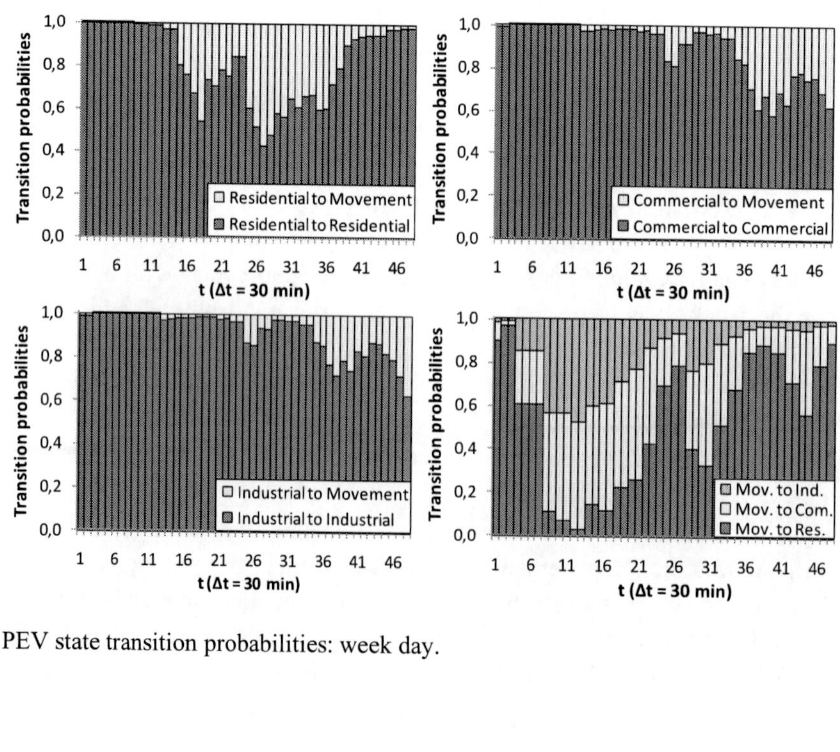

Figure 2. PEV state transition probabilities: week day.

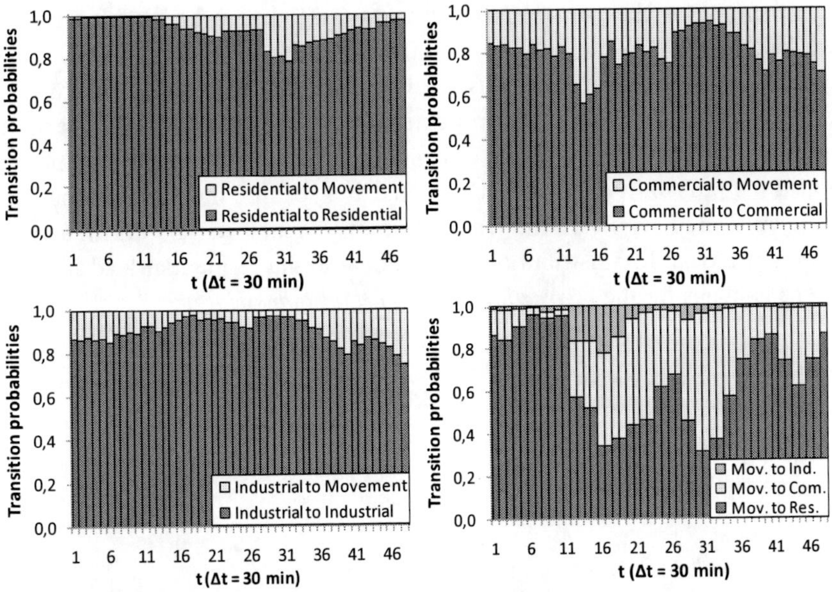

Figure 3. PEV state transition probabilities: weekend day.

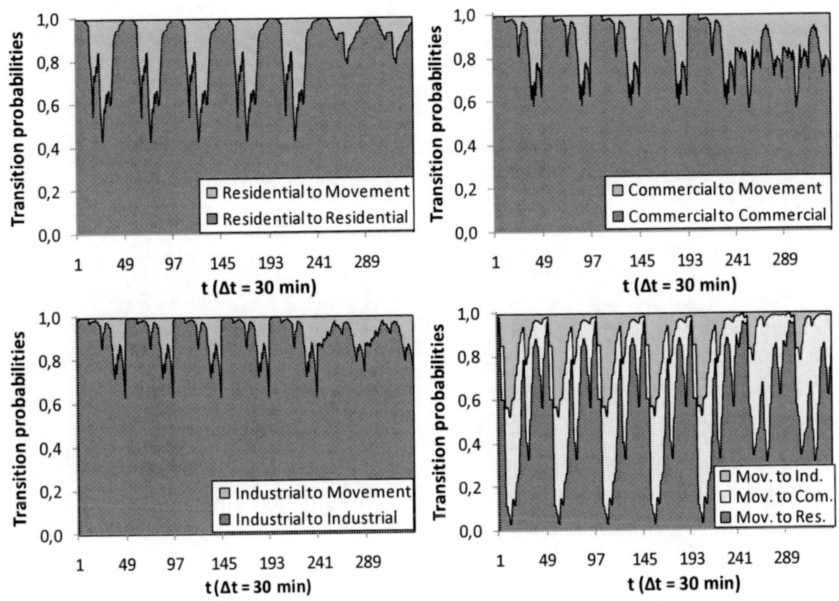

Figure 4. PEV state transition probabilities: full weekly cycle (Monday to Sunday).

Buses Allocation Probabilities for Parked PEV

After defining the PEV states for each time step, it is required to determine the probabilities of PEV in "*parked*" states and to allocate them to the network buses (a bus is a node of the electricity network which can have consumers connected). If a PEV is in the state "*in movement*", there is no need to define its location. However, if it is in a "*parked*" state and connected to the grid for charging purposes, it is crucial to know the PEV location to allocate its load to a specific network bus.

The procedure adopted to solve this issue is to consider the real nature of the loads connected to each network bus. Thus, for the network under study, all the existing loads need to be classified as industrial, commercial or residential loads. After, using equations (5) to (7), it is calculated the probability of a PEV be located at a specific bus. For instance, if a PEV is "*parked in a residential area*" at time instant t, a bus location will be drawn and assigned to it, according to a probability distribution proportional to the residential load installed in each bus. The same happens for the "*parked in a commercial/industrial area*" states.

$$P^R(Bus\ b) = \frac{Load^R_{Bus\ b}}{\sum Load^R} \quad (5)$$

$$P^C(Bus\ b) = \frac{Load^C_{Bus\ b}}{\sum Load^C} \quad (6)$$

$$P^I(Bus\ b) = \frac{Load^I_{Bus\ b}}{\sum Load^I} \quad (7)$$

where:

$P^{R/C/I}(Bus\ b)$ – probability of a PEV be located in bus b, if "*parked in a residential / commercial / industrial area*";

$Load^{R/C/I}_{Bus\ k}$ – residential / commercial / industrial load installed in bus b;

$\sum Load^{R/C/I}$ – network total residential / commercial / industrial load.

Monte Carlo Sampling

The flowchart of the Markov chain Monte Carlo algorithm developed in this work is presented in Figure 5.

The first step of the sampling process is to make the initial characterization of all the PEV. The PEV battery capacity, charging power, energy consumption and initial battery State-of-Charge (SOC) are defined according to truncated Gaussian probability density functions, whose average, standard deviation, maximum and minimum values allowed are presented in Table 1.

Table 1. Gaussian distributions for PEV characterization

	Average	Standard deviation	Maximum value allowed	Minimum value allowed
Battery capacity (kWh)	24.73	17.19	85.00	5.00
Slow charging rated power (kW)	3.54	1.48	10.00	2.00
Energy consumption (kWh/km)	0.18	0.12	0.85	0.09
Initial battery SOC (%)	75.00	25.00	95.00	25.00

While the initial battery SOC values were assumed for the purpose of this work, the values for battery capacity, slow charging rated power and energy consumption were obtained from the information made available by the manufacturers of 42 different PEV. In (Micro-vett) (Mitsubishi) (Peugeot) (Venturi)are presented some of the Internet sites from where PEV characteristics were obtained for this study. The maximum and minimum values allowed, presented in Table 1, were used to confine the values drawn for each PEV within realistic boundaries. As mentioned previously, a given driver behavior was also assigned initially to each PEV. The different behaviors considered in this study were defined according with the findings of a survey made within the framework of the MERGE project (N. Downing, 2010). The results revealed that there are four major types of behaviors regarding PEV charging, as presented in Table 2.

For the purpose of this work, regarding the behaviors modeling and simulation, there was no relevant differences between the drivers that "charge at the end of the day" and those who "charge whenever is convenient and they have time". Therefore, the PEV to which one of these drivers' behaviors was assigned, were assumed to behave equally along the simulations.

Table 2. Drivers' behaviors considered

	Percentage of the responses
PEV charge at the end of the day	33%
PEV charge only when it needs	23%
PEV charge whenever possible	20%
PEV charge whenever is convenient and the driver has time	24%

For the drivers who charge their PEV only when it needs, it was assumed that the minimum battery SOC that triggers the need for charging was 30%.

The second step of the sampling process was to simulate PEV movement along one year. To start with, the PEV states for each time instant were defined according with the previously described Markov chain and presented probabilities. After, for each time instant, a bus location was attributed to parked PEV, as explained in the previous sub-section. For the PEV in movement, a procedure was developed to account their energy consumption and the respective reduction in the battery SOC. First a Gaussian probability density function was used to draw the travelled distances for all the PEV in movement. Therefore, if a PEV was in movement in time instant t and its battery SOC went below a predefined threshold (assumed to be 15%) in time instant $t + 1$, it was considered that the PEV would make a short detour to a fast charging station for recharging purposes. The travelled distance during the detour was obtained using also a Gaussian probability density function, whose parameters are presented in Table 3.

Table 3. Gaussian distributions for travelled distances

	Average	Standard deviation	Maximum value allowed	Minimum value allowed
Travelled distance in common journeys (km)	9.01	4.51	27.03	0.90
Travelled distance to fast charging station (km)	4.51	2.25	13.52	0.45

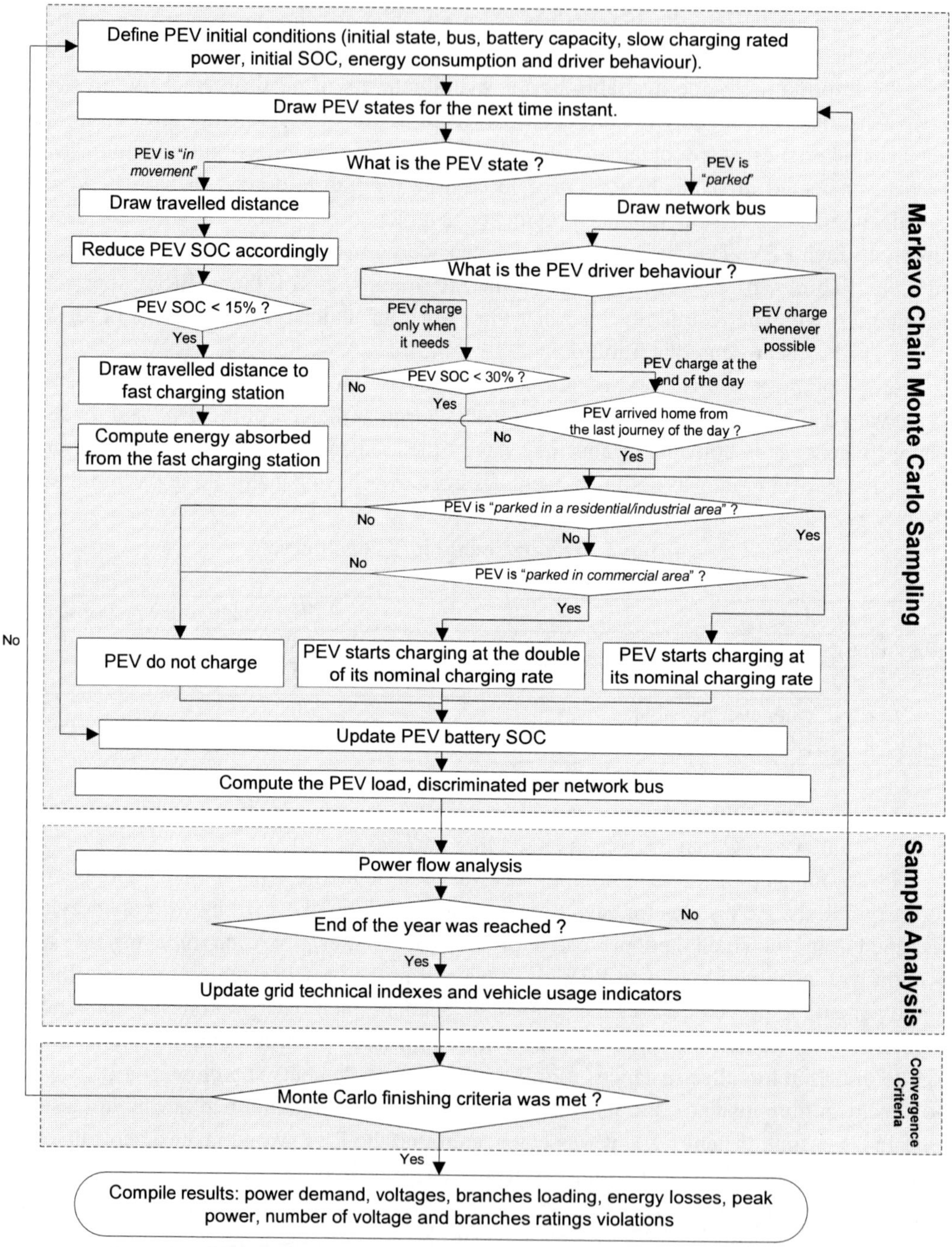

Figure 5. Markov chain Monte Carlo algorithm flowchart.

The fast charging was assumed to be made during 15 minutes with a power of 40 kW(Decreto-Lei n.º 39/2010, Diário da República, 1.ª série — N.º 80 — April 26th, 2010 (Decree-Law in Portuguese), 2010). The average of the Gaussian distribution used to characterize the travelled distance in common journeys was obtained by dividing the average daily mileage in Portugal (35 km) (Azevedo, 2008) by the average number of journeys per day (3.88 journeys). The standard deviation was assumed to be 50% of the average.

The values of the Gaussian function for the travelled distance to the fast charging station, were obtained by assuming that they were 50% of those used in the travelled distance in common journeys distribution. At each time instant, the PEV battery SOC is updated according with the energy spent travelling or accordingly with the energy absorbed in slow charging mode or in fast charging stations. It was assumed that PEV "*parked in a residential area*" and "*parked in an industrial area*" charge at their nominal charging rate while "*parked in a commercial area*" PEV have the capability of charging at the double of their nominal charging rate. With this procedure it is possible to fully characterize the PEV power consumption and, together with the definition of the conventional load of the network under consideration, it is possible to compute the total amount of power required from the network, discriminated per bus and per time instant.

Sample Analysis

The evaluation of the samples is made by running a power flow for each time instant, using the *PSS/E* software, to gather information regarding voltage profiles, power flows in branches and energy losses. During the simulation, the average, maximum and minimum power demand and voltage value is recorded for each bus of the system under study. A similar procedure is adopted for the power flows in the network branches. The day where the highest peak load occurs is also recorded, in order to provide an idea of the worst situation that might occur when a percentage of conventional vehicles are replaced by PEV.

In order to keep track of the most problematic buses and branches within the grid, the number of out of limit voltages and lines overloading occurrences are recorded along the simulations. According to (European Committee for Electrotechnical Standardization – CENELEC), voltages must be kept with the interval 0.90 – 1.10 p.u. during 95% of the time, in a weekly basis. It was considered in this work that a voltage violation occurs when the values are outside the referred interval.

Convergence Criteria

To terminate the Monte Carlo process, two criteria are used: number of iterations and the variances of the aggregated network load of each one of the 336 time instants. The latter means that one variance value is computed for the total network load per time instant t, $t \in [1,336]$. The process is set to perform 4000 iterations (4000 weeks) and check, in the end, if the variation of all the 336 variances in the last 5 iterations is lower than $1e^{-3}$. If at least one of the 336 variances did not meet this convergence criterion, the process is kept running more iterations until all the variances variations are lower than the predefined value.

SIMULATIONS AND RESULTS

In the studies performed, a PEV integration of 25% was assumed, meaning that 1759 PEV exist in the geographical area covered by the network used as case study (described in the next section). Three scenarios were simulated, considering:

- All PEV as dumb charging adherents;
- All PEV as multiple tariff adherents;
- All PEV as smart charging adherents.

For each of the scenarios, 4000 iterations were run with the Monte Carlo simulation method described in the previous section.

Case Study

Figure 6 depicts a rural area used as test case in which the presented MV grid is supplying the electricity consumers and consequently the PEV owners (15 kV grid). It is a large geographical region composed by residential, industrial and commercial areas, and thus allows tracking each PEV while commuting to and from work and to and from leisure activities.

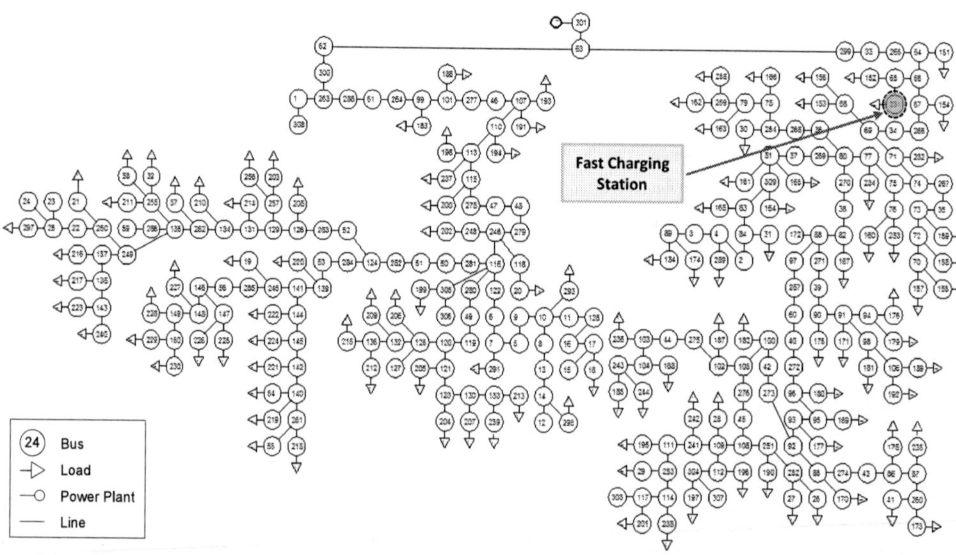

Figure 6. Diagram of the rural area and its electricity distribution network.

There is ca. 13200 inhabitants, 4200 households and a total of 7035 PEV enclosed in the geographical area served by this network (ca. 533 vehicles per 1000 inhabitants). Regarding the PEV charging points available, the following assumptions were made for each charging level:

- Level 1 – 4405 level 1 charging points were assumed to exist in the network, 4200 in the households and 201 in the 41 industries (5 charging points per each industry);
- Level 2 – 135 level 2 charging points were assumed to exist in the network, 5 per each commercial building (there are 24 commercial buildings in the network);
- Level 3 – Only one fast charging station exists, located in a very robust area of the network (bus 231), not prone to violate the electricity grid technical limits. The fast charging station has capability for charging 25 PEV simultaneously.

In order to perform the simulations, a typical weekly load diagram for this network was used. This diagram, depicted in Figure 7, was obtained by aggregating the load diagrams of the different types of consumers within the network.

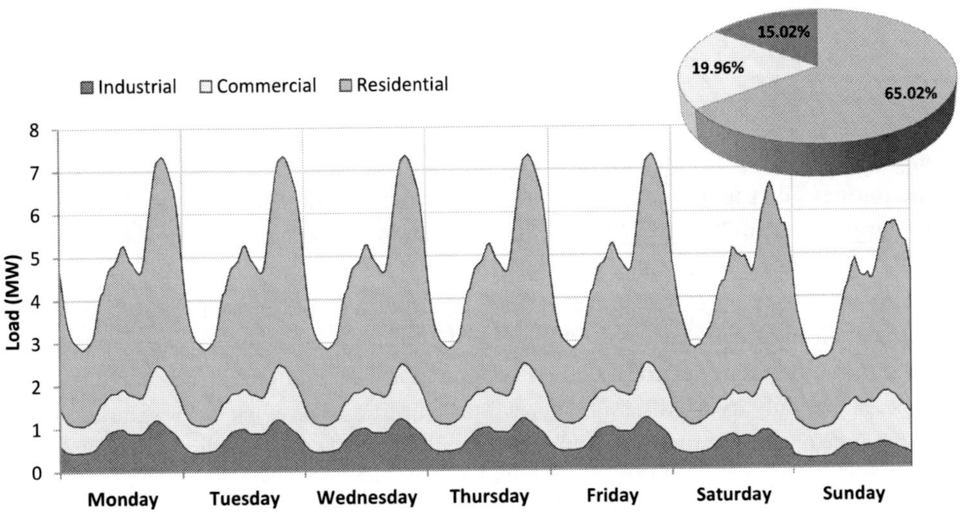

Figure 7. Load diagram of a typical week (the pie chart shows the energy consumption per sector).

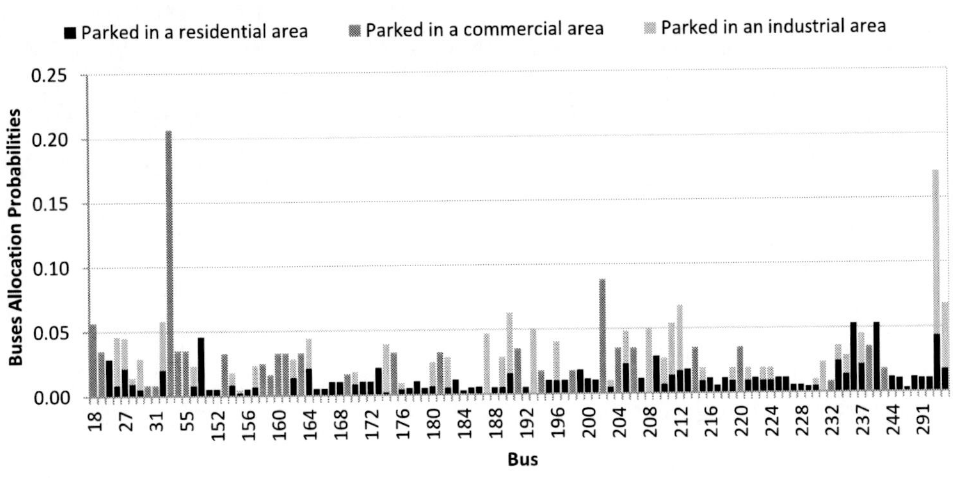

Figure 8. Buses allocation probabilities.

The residential, commercial and industrial consumers' diagrams were combined taking into account the proportion of installed power related with each type of these consumers. As shown in the pie chart of 0, the final diagram has a contribution of 65% of the residential sector, 20% of the commercial and 15% of the industrial, as these are the proportions of installed power related with each type of load within this grid. The night consumption in the commercial sector is essentially due to the activities developed by restaurants and hotels in the geographical area covered by the network.

The peak of conventional load is 7.3 MW, distributed over 115 of the 309 network buses, and the energy consumption during a typical week assumes the value of 789 MWh.

Figure 8 shows the probabilities used to define the grid buses where PEV plug-in for charging when they are "parked in a residential area", "parked in a commercial area" and "parked in an industrial area". The values presented were defined assuming that the probability of a PEV be located at a given bus is proportional to the amount of residential/commercial/industrial load installed, as shown in Equations (5), (6) and (7).

Results

Mobility Patterns

The journeys distribution during a week and a weekend day (Friday and Saturday) for the dumb charging, multiple tariff (22h – 8h) and smart charging scenarios are presented in Figure 9. As it can be observed, the curves for the three charging strategies follow the same trend. This is, in fact, an expected result, as the same assumptions were used to simulate the PEV movement in all the scenarios addressed (the discrete-time, discrete-state Markov chain described in the previous section). Only a week and a weekend day results were presented in Figure 9, since only two different sets of probabilities were used to simulate the PEV movement, one applied to the five week days and the other applied to the two weekend days.

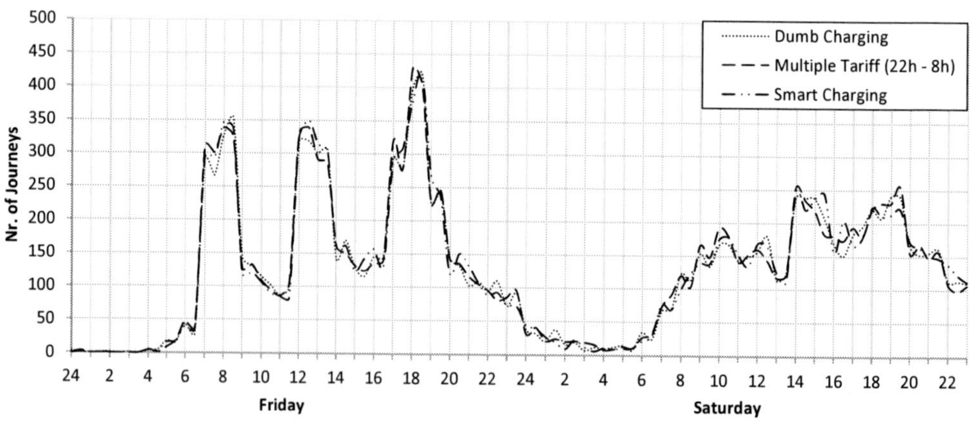

Figure 9. Journeys distribution during a week and a weekend day for the dumb charging, multiple tariff (22h – 8h) and smart charging scenarios.

During the week days, three peaks are clearly noticeable in the figure, two most likely related with household – work commuting (around 8h and 18h), and the third, slightly after noon, probably related with people leaving their working places to have lunch somewhere else. In the weekend days, probably due to the absence of the household – work commuting, the journeys are more distributed during the day.

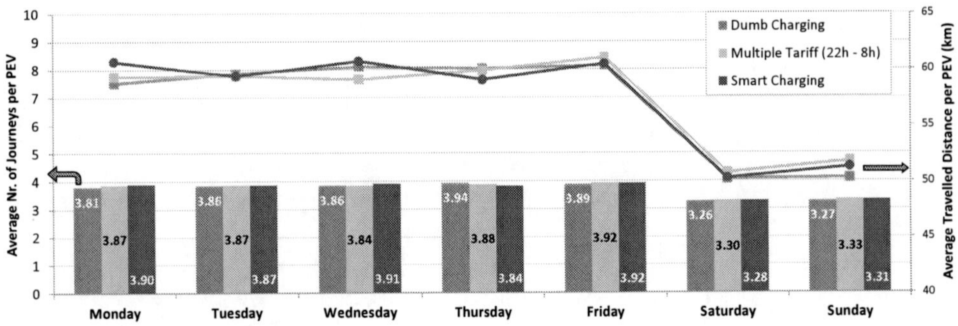

Figure 10. Average nr. of journeys and average travelled distance per PEV for the dumb charging, multiple tariff (22h – 8h) and smart charging scenarios.

It should be noted that one set of results for the PEV journeys are obtained in each of the 4000 iterations run with the Monte Carlo simulation method. The values presented in Figure 9, as well as those presented in the remaining charts of this subsection, are referred to the last iteration performed in the simulation of the scenarios referred in each figure.

Although these values are likely to suffer slight changes from iteration to iteration, their global behaviour is kept almost unchanged along all the simulations performed, as it can be observed by the common trend shown by the three curves in Figure 9.

Figure 10 complements the information provided in Figure 9 regarding the PEV mobility patterns, showing the average number of journeys and average travelled distance per PEV, in each day of the week, for the dumb charging, multiple tariff (22h – 8h) and smart charging scenarios.

As the travelled distance per journey was obtained through a Gaussian probability density function (Table 3), the distribution of the total travelled distance is very similar to the number of journeys distribution.

The total distance travelled by PEV at each time step is presented in Figure 11.

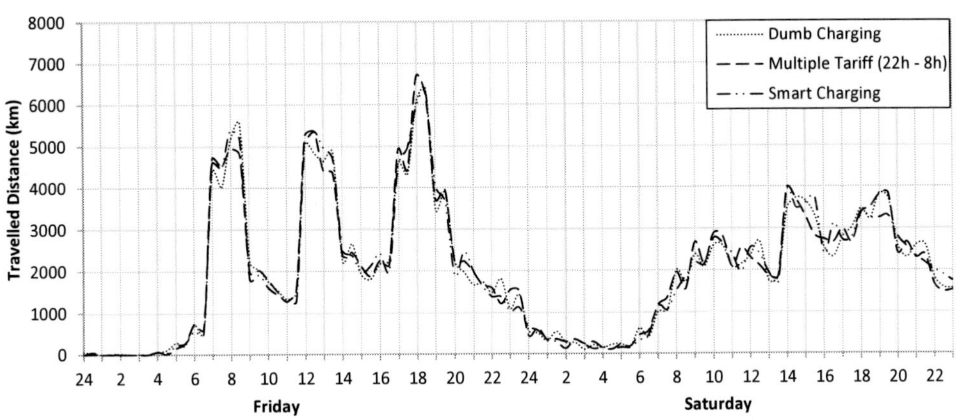

Figure 11. Travelled distances distribution during a week and a weekend day for the dumb charging, multiple tariff (22h – 8h) and smart charging scenarios.

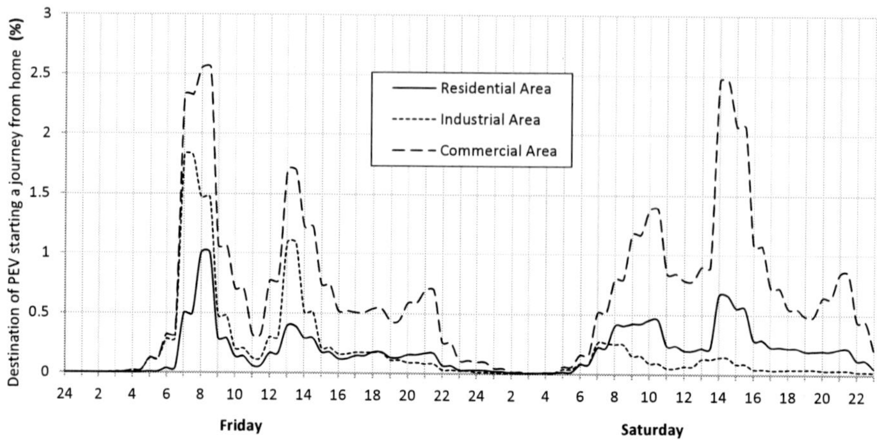

Figure 12. Destination of PEV starting a journey from home.

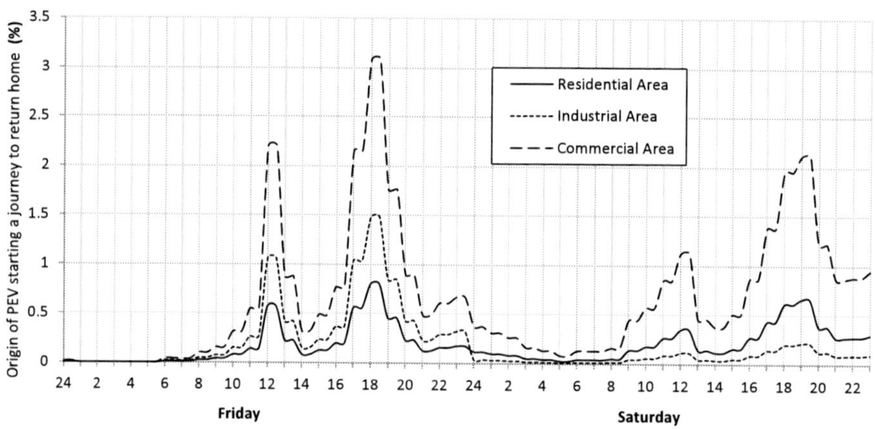

Figure 13. Origin of PEV starting a journey to return home.

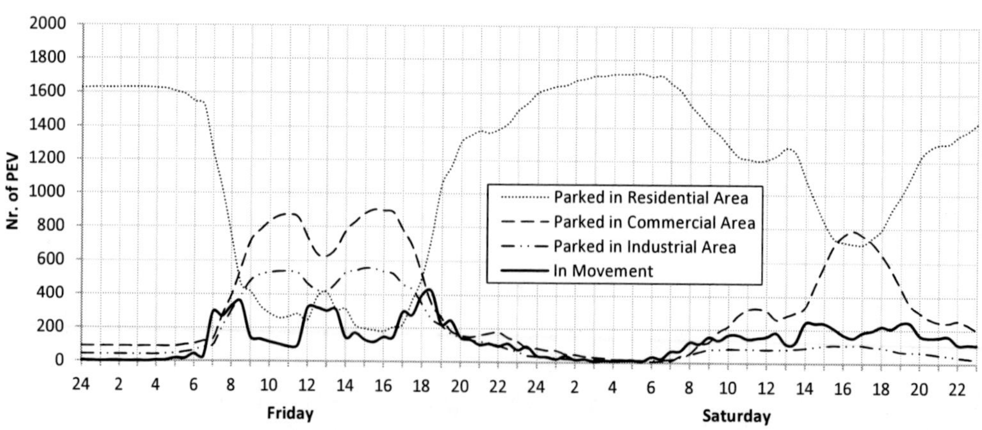

Figure 14. Nr. of PEV in movement and parked (in residential, commercial and industrial areas) during a week and a weekend day in the smart charging scenario.

The following two charts present the destination of PEV starting a journey from home and the origin of PEV starting a journey to return home.

In order to provide some insights about the places where the PEV stay parked during the day, it is presented in Figure 14 the number of PEV parked in residential, commercial and industrial areas, obtained in the last iteration of the simulation performed for the smart charging scenario.

In what regards residential areas, as expected, there is a large number of PEV parked during the night period, both on the week and on the weekend days. During the day, the number of PEV parked in these areas is considerably lower during the week than during the weekend, probably due to the fact of most of the people not working during the weekend.

The results are quite different for commercial and industrial areas, where the number of PEV parked reaches the highest values during the day, both on week and weekend days. Nevertheless, while the number of PEV parked in commercial areas reaches almost the same maximum value during all the days of the week, the number of PEV parked in industrial areas is considerably lower during the weekend than during the week.

Since the fact of a PEV being parked does not means that it is plugged-in to the grid, it is defined, during the simulations, essentially based on the drivers' behaviours, the number of PEV that effectively connect to the grid for charging purposes at each time step.

Figure 15, 16 and 17 show the results obtained for the number of PEV plugged-in to level 1, level 2 and level 3 charging facilities, respectively, obtained in the simulations performed for the dumb charging, multiple tariff and smart charging scenarios. In what regards level 1 charging facilities[1], as it can be observed in Figure 15, in the multiple tariff scenario, the PEV only plug-in for charging during the period when the electricity price is assumed to be lower.

The curve related with the smart charging scenario distinguishes itself from the others only during the night periods, since it was assumed that an incentive's policy exist, where compensations are provided to the PEV owners proportionally to their availability during the periods when it is more profitable to buy energy from the market (assumed to be between 23h and 9h).

Therefore, for the purpose of this work, it was assumed that if a smart charging adherent is parked during the referred period and has the possibility of charging, it will be always plugged-in to the grid.

Concerning level 2 charging facilities[2], as shownin Figure 16, the curve related with the multiple tariff (22h – 8h) scenario it is the only one that clearly distinguishes itself from the others, as the PEV only plug-in for charging during the period when the electricity price is assumed to be lower, similarly to what happened for level 1 charging facilities.

Regarding level 3 charging facilities[3], as shown in Figure 17, the curves for the three scenarios reveal similar behaviours. Yet, it is perceptible that the curve related with the multiple tariff (22h – 8h) scenario is slightly above the others during the majority of the time, revealing that with this charging strategy PEV owners are obliged to charge more often in fast charging stations.

---

[1] It is assumed that PEV plug-in to level 1 charging facilities when their state is "parked in a residential area" or "parked in an industrial area".

[2] It is assumed that PEV plug-in to level 2 charging facilities when their state is "parked in a commercial area".

[3] It is assumed that PEV plug-in to level 3 charging facilities when their state is "in movement" and they go to a fast charging station.

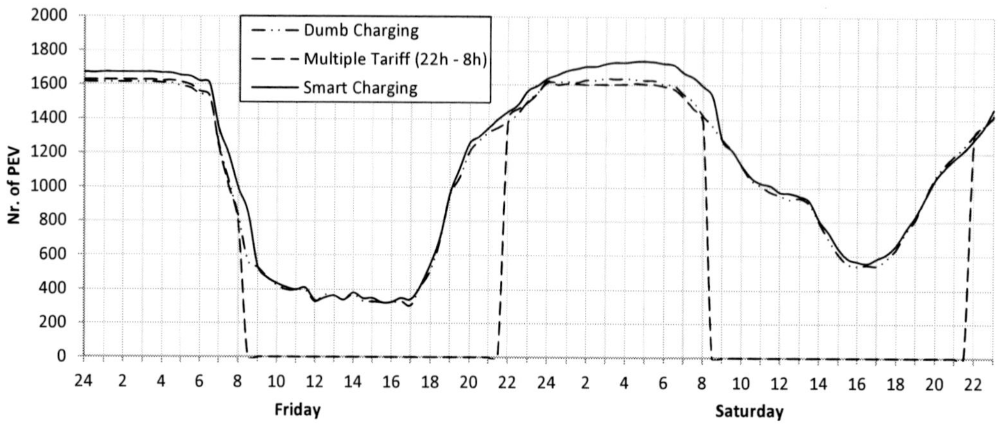

Figure 15. Nr. of PEV plugged-in level 1 charging facilities during a week and a weekend day for the dumb charging, multiple tariff (22h – 8h) and smart charging scenarios.

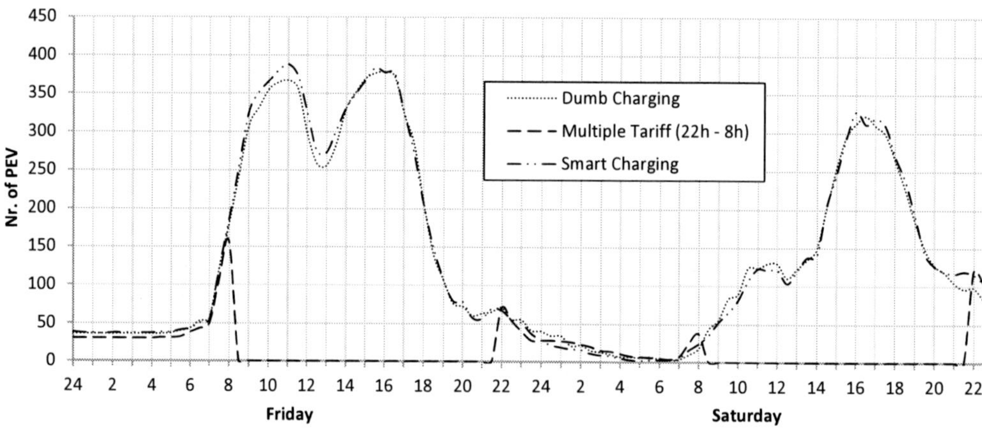

Figure 16. Nr. of PEV plugged-in level 2 charging facilities during a week and a weekend day for the dumb charging, multiple tariff (22h – 8h) and smart charging scenarios.

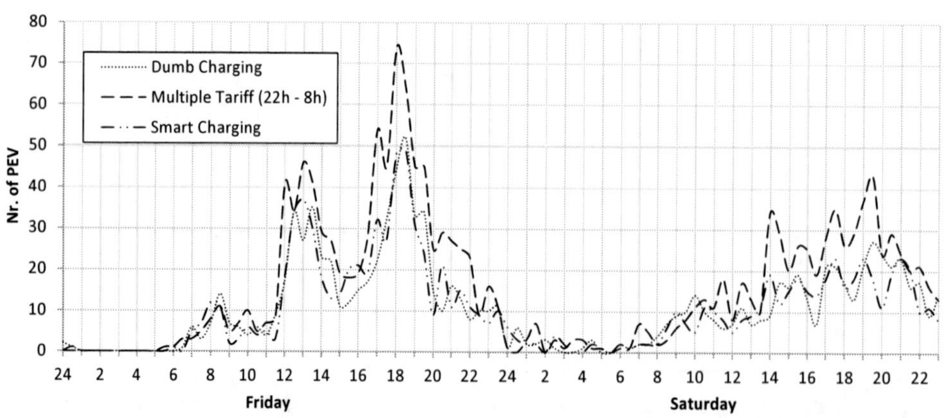

Figure 17. Nr. of PEV plugged-in level 3 charging facilities (fast charging station) during a week and a weekend day for the dumb charging, multiple tariff (22h – 8h) and smart charging scenarios.

This happens due to multiple tariff adherents only charging in level 1 and 2 facilities between 22h and 8h. Therefore, when a PEV makes several journeys outside this period, it might need to recharge its batteries and the only solution is resorting to a fast charging station.

During the week day (Friday), two peaks appear during the day, the first slightly after noon and the second around 18h. As expected, these peaks coincide with the periods when the number of journeys is higher, since PEV only resort to fast charging stations when they are travelling and the energy stored in their batteries is not enough to reach their destinations. The third peak depicted, around 8h, has not correspondence, as during the night the majority of the PEV will fully charge their batteries, not having the need to go to a fast charging station during the first journey of the day.

PEV Energy Consumption

In this subsection, three charts are presented, showing the power absorbed by PEV in level 1, 2 and 3 charging facilities in the dumb charging, multiple tariff (22h – 8h) and smart charging scenarios.

The dashed lines presented in the figures represent the maximum power consumption possible, assuming that all PEV plugged-in at level 1 are capable of absorbing 3 kW (blue line) and all PEV plugged-in at level 2 are capable of absorbing 12 kW (red line). They were included in the charts only to provide an idea of the worst case scenario theoretically possible, that would be having all the plugged-in PEV charging simultaneously. Nevertheless, this scenario is very unlikely to occur, since it disregards that some of the PEV will have their batteries fully charged over time and will not absorb more power from the grid. No dashed line was included in the charts for the maximum power consumption possible at level 3, since it coincides with the line for the power consumption at level 3 (green line). This happens since it was assumed that when a PEV plugs-in in a fast charging station, it will always be charging. Thus, it is impossible to have a PEV connected to a level 3 charging facility without being charging. In the dumb charging scenario, Figure 18, the PEV tend to charge at level 1 mostly at the end of the day, which is the time period when people arrive home from work. The amount of power requested by the PEV during these periods would provoke some violations of the voltage limits of several network nodes and, in order to avoid them, the DSO would have to curtail part of the PEV load (dotted blue line).

When the system enters in the emergency operating state, the DSO tries to solve the problems detected only by curtailing load from PEV charging at level 1. Then, if this measure is not enough, the DSO might also curtail load from PEV charging at level 2 (in commercial areas) and at level 3 (in fast charging stations). However, it should be stressed that in the present case, only load from PEV charging at level 1 was curtailed.

In what regards charging at level 2, as expected, the power absorbed by PEV follows the trend of the number of PEV parked in commercial areas, as it can be observed in Figure 14. Concerning level 3 charging, the power absorbed by PEV follows the trend of the number of PEV plugged-in in fast charging stations, as depicted in Figure 17.

In the multiple tariff scenario, Figure 19, the PEV only charge in level 1 and 2 charging facilities between 22h and 8h, which is the period when the energy prices are assumed to be lower. For this reason, there are a high number of PEV connecting to the grid for charging at 22h and the amount of power requested provokes the violation of the technical limits of several network components. In order to avoid them, as it happened in the dumb charging scenario, the DSO would have to curtail part of the PEV load (dotted blue line).

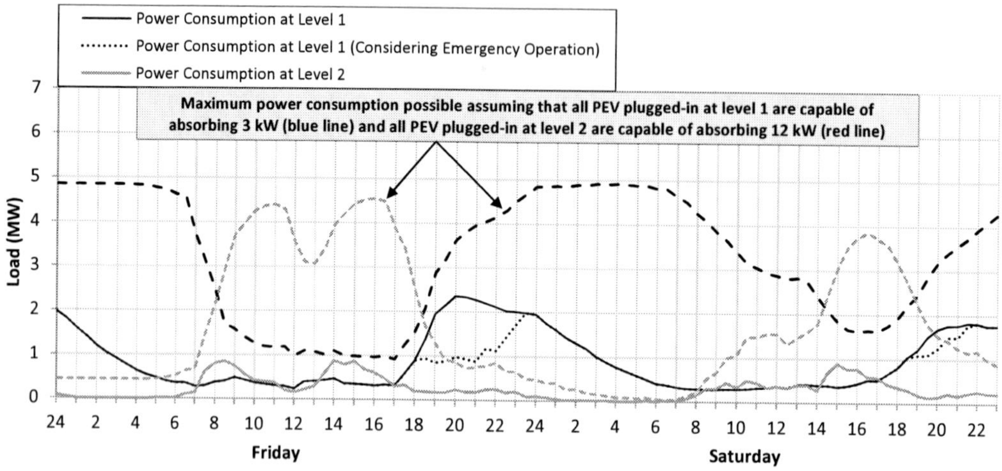

Figure 18. Power consumption by PEV plugged-in in level 1, 2 and 3 charging facilities during a week and a weekend day in the dumb charging scenario.

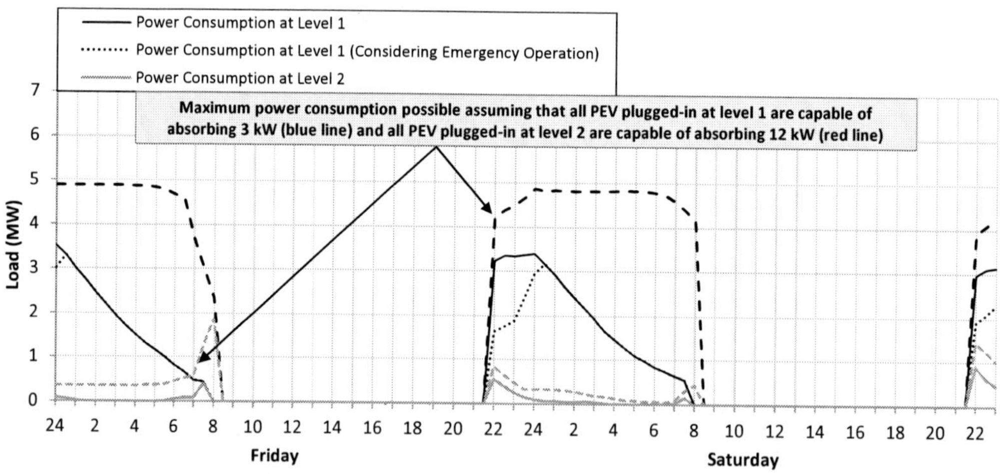

Figure 19. Power consumption by PEV plugged-in in level 1, 2 and 3 charging facilities during a week and a weekend day in the multiple tariff (22h – 8h) scenario.

Similarly to the dumb charging, in level 3 charging, the power absorbed by PEV follows the trend of the number of PEV plugged-in in fast charging stations, as depicted in Figure 17.

The results obtained in the smart charging scenario, Figure 20, are very similar to those obtained with the dumb charging, namely in what concerns level 2 and 3 power consumption. The only relevant differences are related with the power consumption at level 1, where it is clear a shift of the PEV consumption from the 19h – 24h period to the 2h – 7h period.

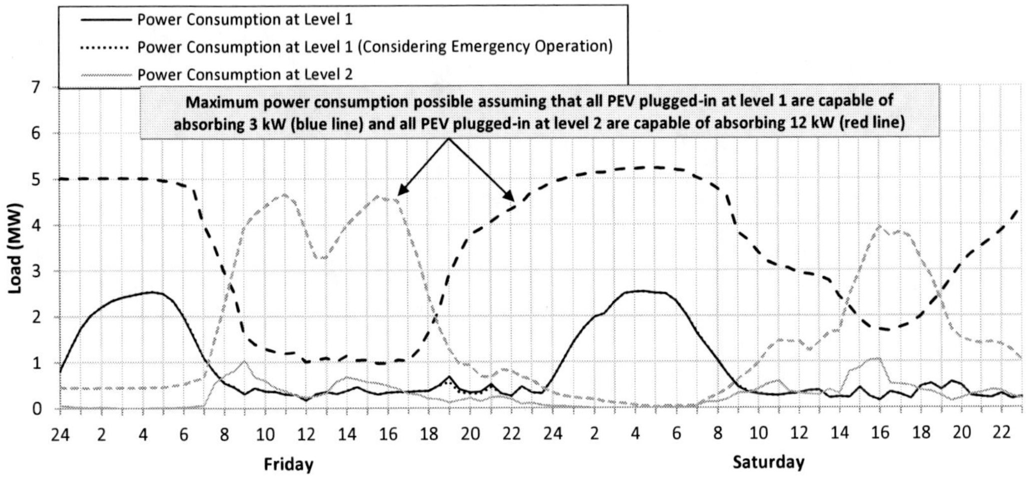

Figure 20. Power consumption by PEV plugged-in level 1, 2 and 3 charging facilities during a week and a weekend day in the smart charging scenario.

Changes in Load Diagrams

Figure 21 shows the load diagrams changes for the three charging strategies addressed.

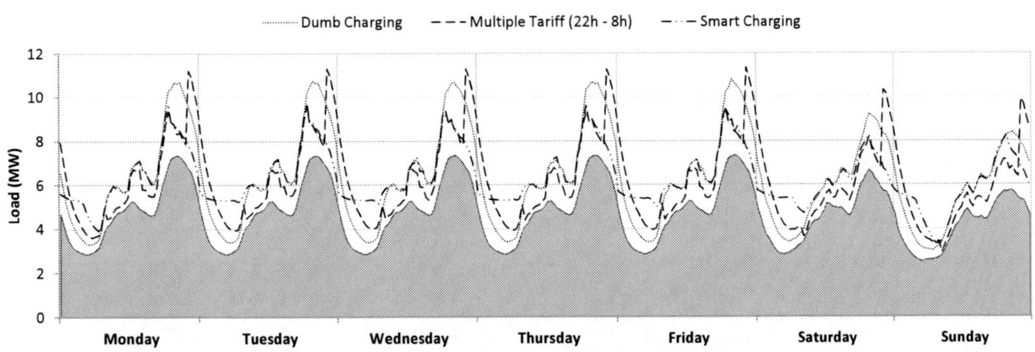

Figure 21. Changes in the load diagram with a PEV integration of 25%.

The changes in the load diagram in the dumb charging scenario are depicted in detail in Figure 22. With this charging strategy, the PEV tend to charge mostly at the end of the day, which is the time period when people arrive home from work.

In the week days, there are two other periods in the day where the PEV load is considerably high. The first occurs in the morning, after the first journey of the day of the majority of the drivers, which is from home to work. This load appears mostly in commercial and industrial areas, as it was assumed that PEV owners have the possibility of charging their PEV at their working places. The second occurs after the lunch period, when there are a considerable number of vehicles in movement that might need to be recharged afterwards.

As shown in Figure 21, the amount of power requested by the PEV in the dumb charging mode provokes a very large increase in the peak load, leading to the violation of the technical limits of several network components. In order to avoid these violations, the DSO would have to curtail 25.66 MWh of the energy demanded by PEV during the week analysed (black areas in Figure 21).

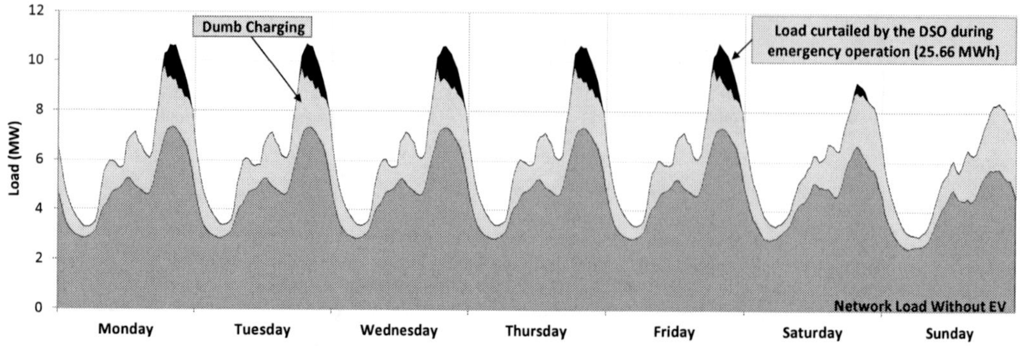

Figure 22. Load diagram in the dumb charging scenario.

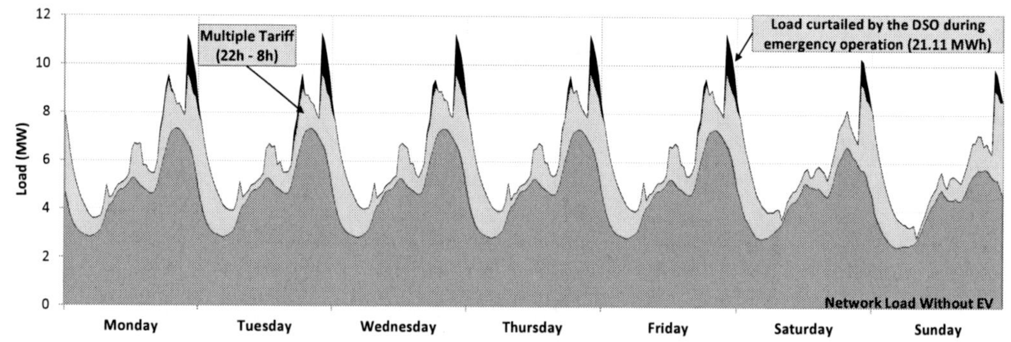

Figure 23. Load diagram in the multiple tariff (22h – 8h) scenario.

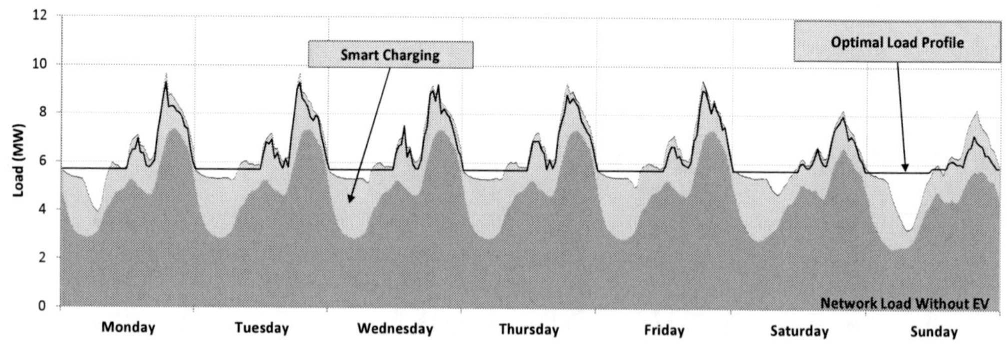

Figure 24. Load diagram in the smart charging scenario.

Figure 23 shows the changes in the load diagram in the multiple tariff (22h – 8h). With this charging strategy, the PEV only charge in level 1 charging facilities between 22h and 8h, which is the period of time when the energy prices are assumed to be lower. For this reason, there are a high number of PEV connecting to the grid for charging at 22h and the amount of power requested provokes the violation of the technical limits of several network components. In order to avoid these violations, the DSO would have to curtail 21.11 MWh of the energy demanded by PEV during the week analysed (black areas in Figure 23).

The remaining PEV consumption, outside the period between 22h and 8h, is due to PEV charging in level 2 facilities (commercial areas) and level 3 facilities (fast charging stations).

In the smart charging, it was assumed that the electricity price has a quadratic dependency of the total consumption and thus the optimal profile that allows minimizing the cost of the energy absorbed by PEV is the one that allows flattening the load diagram as much as possible. This optimal profile is represented by the black line in Figure 24.

Energy Losses and CO_2 Emissions

In Figure 25 it is depicted the average value of the weekly energy losses, obtained in the 4000 iterations performed for each scenario, as well as the percentage of the consumption they represent. The weekly energy losses grow 70% from the scenario without PEV to the dumb charging, 57% to the multiple tariff and 36% to the smart charging scenario.

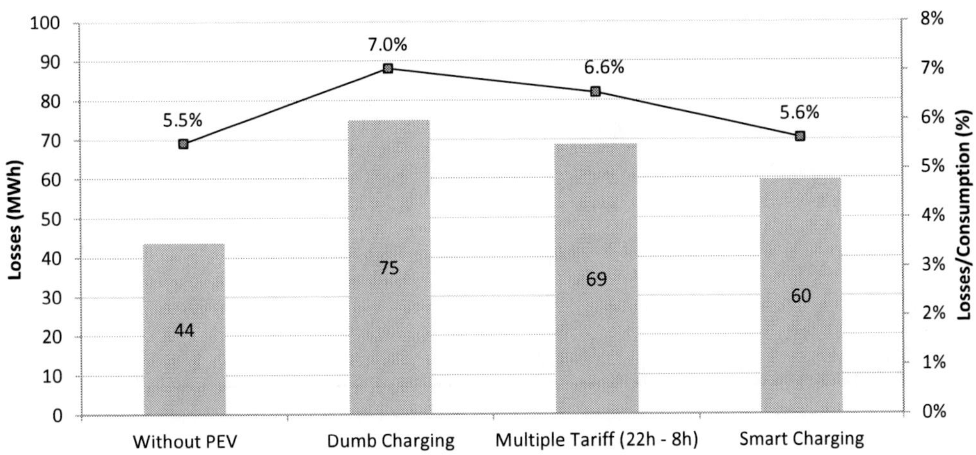

Figure 25. Weekly energy losses in all the scenarios studied.

From the results obtained it was possible to conclude that the multiple tariff and the smart charging, in comparison with the dumb charging, also have a positive impact in the CO_2 emissions associated with the electricity production. When comparing the PEV charging schemes considered, the results obtained show that the multiple tariff can avoid ~2.36 tons of CO_2 emissions per week when compared to the dumb charging. The smart charging, in turn, can avoid ~5.67 tons of CO_2 per week. These figures were obtained considering the CO_2 emissions reference factor for Portugal for 2007, which was 369 kg CO_2/MWh (Covenant of Mayors).

Monte Carlo Convergence and Sample Variance

As referred, the Monte Carlo simulation ends when two criteria are met: when 4000 iterations are performed or when the variations of the 336 variances in the last 5 iterations are lower than 1×10^{-6}. For all the scenarios simulated for this network, the variances variation criterion was met before the algorithm reach iteration 4000. As an example, Figure 27 shows the evolution of one variance variation per scenario studied. As it is shown, the variation rate after iteration 750 is very low, indicating that the algorithm reached the convergence criteria.

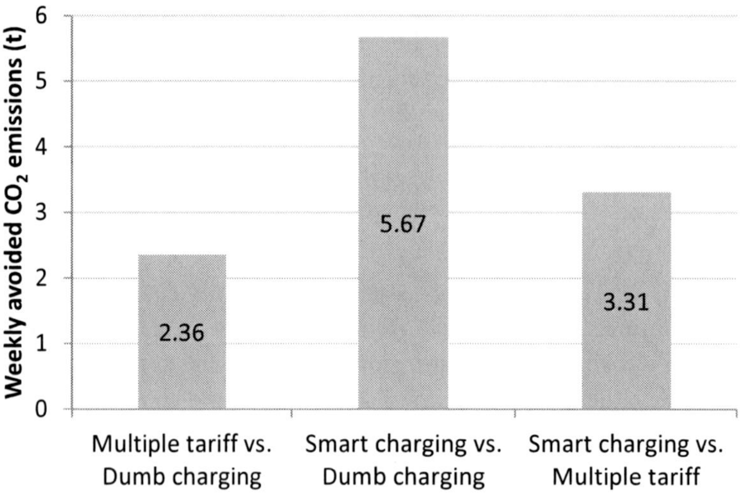

Figure 26. Avoided CO_2 emissions.

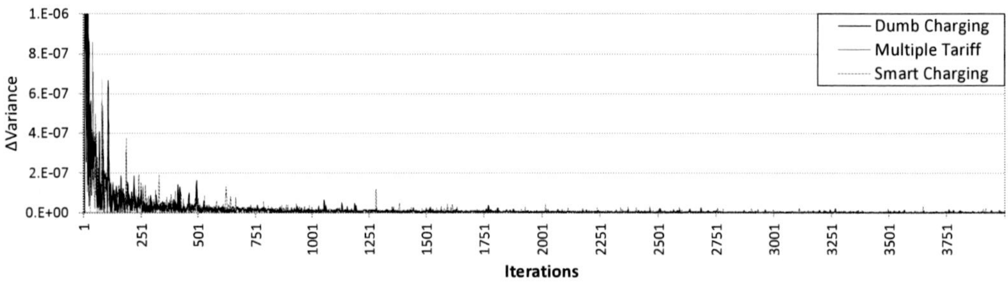

Figure 27. Evolution of the variances variations.

CONCLUSION

In this chapter, a methodology for simulating PEV travelling patterns and energy requirement was proposed and its application to a case study based on real data was presented subsequently. The simulation platform developed in this work proved to be very efficient in performing a realistic evaluation of the commuting patterns and impacts that result from a massive integration of PEV in distribution networks. It is important to notice that the model can be used for both plug-in hybrid and full electric vehicles, as it allows specifying a wide range of characteristics for each vehicle that is considered in the simulation.

The proposed tool uses a stochastic method to simulate the PEV movement, allowing exploring different scenarios in a coordinated way. This methodology grants the possibility of obtaining detailed knowledge on PEV individual daily routes, as well as other global indicators such as the electricity grid impacts provoked by the PEV battery charging.

This tool also allows creating prospective PEV uptake scenarios, which can be used to plan the future of the transportation sector and of the power distribution systems. Additionally, the accurate quantification of the PEV fleet mobility patterns and energy needs can also be used to evaluate the global fleet energy requirements, environmental impacts and recharging/refuelling infrastructure needs.

Concerning the analysed case study, it might be concluded that the network under study is capable of integrating moderate quantities of PEV without the occurrence of line loading and voltage limits violations. However, with 25% of PEV integration the number of voltage violations registered was substantial. These results show that it is impossible to proceed to a 25% replacement rate without making investments in network reinforcements or by adopting smart charging solutions, in order to tackle the low voltage problems identified.

Important findings were also made regarding the energy losses. Their value grows 70% from the scenario without PEV to the dumb charging, 57% to the multiple tariff and 36% to the smart charging. These values show that for higher PEV integration levels, losses are likely to become a very important issue for the system operator. Therefore, the system operator should look for efficient mechanisms to manage PEV charging, in order to avoid wasting large amounts of money in the energy distribution process.

REFERENCES

[1] Commission, European, «Eurostat,» Available: http://ec.europa.eu/eurostat.

[2] S. Bending et al., «Specification for an Enabling Smart Technology,» Deliverable D1.1 of the European Project MERGE, 2010.

[3] J. A. P. Lopes et al., «Integration of Electric Vehicles in the Electric Power System,» *Proceedings of the IEEE,* vol. vol. 99, pp. 168-183, 2011.

[4] EDP Serviço Universal, «Tarifas e horários (in Portuguese),» Available: http://www.edpsu.pt/pt/particulares/tarifasehorarios/.

[5] W. Kempton and J. Tomic, «Vehicle-to-grid power fundamentals: Calculating capacity and net revenue,» *Journal of Power Sources,* vol. 144, pp. 268-279, 2005.

[6] Momber et al., «Identification of Regulatory Issues Regarding Market Design and Network Regulation to Efficiently Integrate Electric Vehicles in Electricity Grids, Deliverable D5.2 of the European Project MERGE,» 2011.

[7] Micro-vett. Available: http://micro-vett.it.

[8] Mitsubishi, Available: http://www.mitsubishi-cars.co.uk/imiev/innovation.

[9] Peugeot, Available: http://www.peugeot.co.uk/vehicles/peugeot-car-range/peugeot-ion/.

[10] Venturi, Available: http://www.venturi.fr.

[11] N. Downing et al., «EU MERGE Project: Mobile Energy Resources in Grids of Electricity," deliverable D1.1, "Identification of Traffic Patterns and Human Behaviours,» 2010.

[12] *Decreto-Lei n.º 39/2010, Diário da República, 1.ª série — N.º 80 — April 26th, 2010 (Decree-Law in Portuguese),* 2010.

[13] C. L. Azevedo, Métodos de estimativa de volumes anuais de tráfego rodoviário - um modelo para Portugal (in Portuguese), Master Thesis in Transports, IST, Universi-dade Técnica de Lisboa, 2008.

[14] European Committee for Electrotechnical Standardization – CENELEC, «EN 50160:2007, Voltage characteristics of elec-tricity supplied by public distribution systems».
[15] Covenant of Mayors, Available: http://www.covenantofmayors.eu.

In: Grid Electrified Vehicles
Editor: Carla Alexandra Monteiro da Silva

ISBN: 978-1-62808-839-7

Chapter 8

ENVIRONMENTAL RATING AND CLASSIFICATION OF ELECTRIC VEHICLES

Carla Silva*, Teresa Batista, Alexandre Lucas and Sara Marques

IDMEC, Instituto Superior Técnico, Universidade Técnica de Lisboa, Lisboa, Portugal

ABSTRACT

The growth of the world economy and the pursuit of higher living standards are at the root of higher energy consumption and more intensive private car use. The fuel economy/CO_2 label is a mandatory or voluntary sticker that is affixed to the car and contains information on the vehicle's fuel consumption/CO_2 emissions. It originated in the U.S. in the 1970s and 1980s and has spread throughout the world, recently as far as India. Car labelling aims to raise awareness about consumers' purchasing decisions and encourage manufacturers to explore more efficient powertrains based on electric propulsion. These powertrains include a pure electric vehicle (EV), a hybrid electric vehicle (HEV), a fuel cell hybrid vehicle (FCHEV), and a plug-in hybrid vehicle (PHEV). So far labelling covers only the car usage stage and omits air pollution indicators such as, HC, CO, NOx and PM. This chapter reviews labelling, scoring methodologies and their EV rankings and attempts to derive a more complete life cycle (LCA) based scoring methodology. The ACEEE Green Score methodology rates Nissan Leaf, Toyota Prius Plug-in and Honda FCX highly, with scores of 60 out of 100. The more complete LCA analysis includes the life cycle of vehicle materials, fuel upstream emissions and infrastructure of energy supply materials and cradle-to-grave emissions with regard to power plants, home, fast and normal chargers, hydrogen stations, refineries and conventional gas stations. In this context renewable power plants report higher energy and carbon intensity than conventional ones. As far as vehicle materials are concerned, thanks to Li-ion or Ni-MH batteries the energy and emissions burden is higher for electric vehicles than for conventional ones.

Keywords: Labelling, ranking, pure electric, plug-in hybrid, fuel cell

---

* Corresponding Author: Email: carla.silva@ist.utl.pt.

INTRODUCTION

The growth of the world economy and the pursuit of higher living standards are at the root of higher energy consumption and more intensive private car use. The latter causes environmental and social damage, not only locally but also on a global scale. Among the worst categories of damage is pollution in terms of soil, water, air and noise. The air emissions resulting from car use can be divided into two main categories, GreenHouse Gas emissions (GHG) that affect climate changes, namely global warming, and criteria pollutants, which affect the air quality and have consequences for the health of the population and ecosystems. Criteria pollutants include: carbon monoxide (CO), hydrocarbons (HC), nitric oxides (NO_x), particulate matter (PM) and sulphur dioxide (SO_2). GHG emissions include: carbon dioxide (CO_2), methane (CH_4) and nitrous oxide (N_2O).

Several steps have been already taken to reduce emissions from vehicle use, and technical progress has led to greater energy efficiency, the tightening of vehicle emission standards and improvements in fuel quality, and the use of catalysts has reduced transportation-related emissions of some pollutants, but not enough to offset rising traffic volumes [1].

In 2010, the CO_2 emissions from the transport sector were about 23% of global CO_2 emission levels and cars represented about 41% of that figure. With the higher levels of transport demand, and depending on the fuel mix, the total CO_2 emissions from the transportation sector are, according to the World Energy Council, expected to grow between 16% (Tollway scenario, which describes a more regulated world where governments decide to intervene in markets to promote technology solutions and infrastructure development that put common interests at the forefront) and 79% (Freeway scenario, which describes a world where pure market forces prevail to create a climate for open global competition) by 2050. Detailed analysis of the predictions for CO_2 emissions specifically from cars forecasts an increase of about 35% between 2010 and 2050 in the Freeway scenario. In the Tollway scenario, CO_2 emissions from cars in 2050 are likely to drop by 46% below 2010 level[2].

The contribution of the road transport sector in the U.S. in 2002 to the total emission of CO was around 55%, almost twice that in Europe (EEA-32) in 2009, which was 29% [3][4]. A significant and continued decreasing trend of CO emissions from the transportation sector in both OECD and non-OECD regions until 2050 is expected (Figure 1).

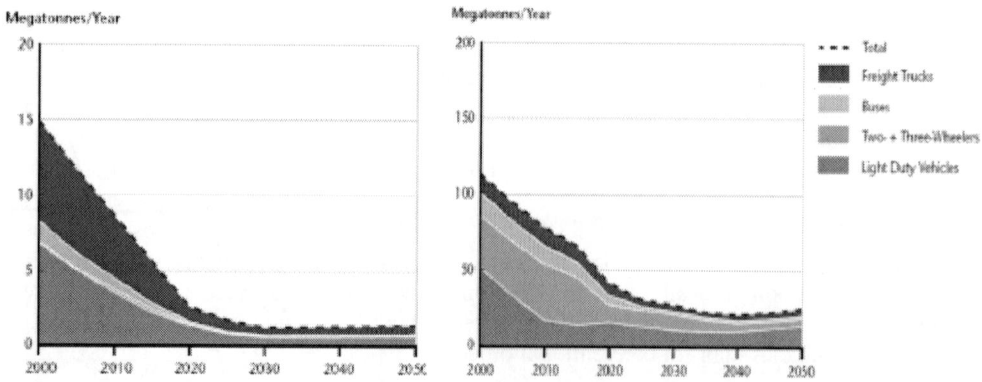

Figure 1. Transport-related carbon monoxide (CO) emissions by mode OECD regions (left), non-OECD regions (right) [5].

The road transport contribution of NOx to the total emissions of this pollutant in Europe in 2009 was around 33%, not significantly different from that in the U.S. in 2002, where the contribution was about 35% [3][4]. Considering the transport-related emissions predicted by the World Business Council for Sustainable Development (WBCSD), it is expected that in both OECD and non-OECD (Figure 2) regions the NOx emissions will drop significantly until 2050; however, for non-OECD they will slightly increase at first [5].

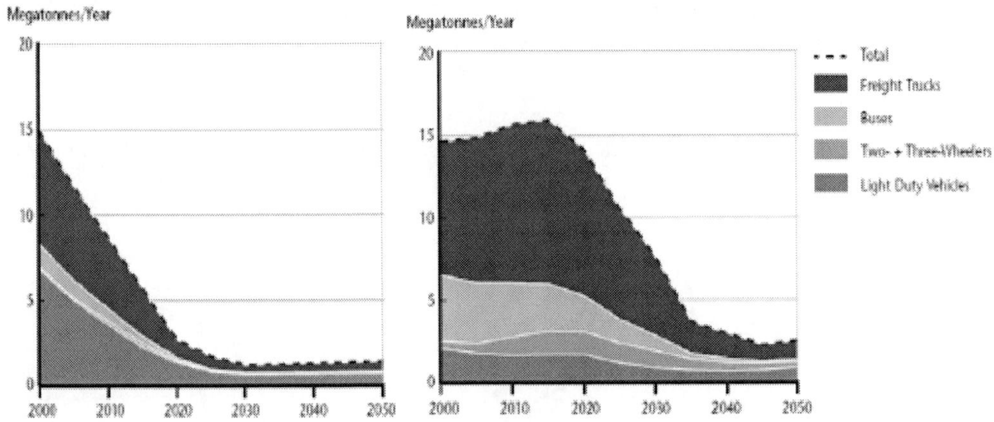

Figure 2. Transport-related nitrogen oxide (NOx) emissions by mode, OECD regions (left), non-OECD regions (right) [5].

The road transport (exhaust) contribution of PM_{10} to the total emissions in Europe (EEA-32) in 2009 was around 7%, higher than in the U.S. in 2002, which was around 3%. Considering the predictions presented in WBCSD for transport-related emissions, the PM_{10} emissions from road transport will decrease in OECD (Figure 3) and non-OECD countries until 2050, but the decrease in non-OECD regions will become significant only in 2020.

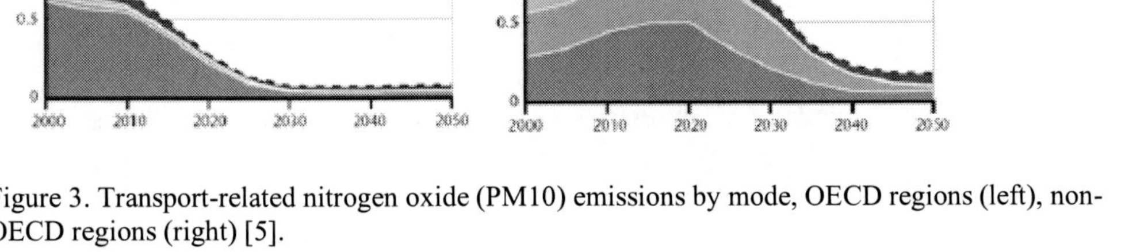

Figure 3. Transport-related nitrogen oxide (PM10) emissions by mode, OECD regions (left), non-OECD regions (right) [5].

The contribution of the road transport sector to the total emission of SO_2 in Europe in 2009 (EEA-32) was less than 1% [3]. The world emissions of SO_2, in 1990 were about 122 million tons and in 2000 the emissions were 20% lower, which was mainly because of the strict controls implemented in Western Europe, but also thanks to economic restructuring in

Central and Eastern Europe and in Russia and the Newly Independent States. Under the baseline assumptions, the world emissions decreased to about 97 million tons in 2010 and will increase again to 105 million tons in 2030, but the implementation of the best available control technology can reduce the emissions to about 20% of the 1990 level [5].

Hydrocarbon emissions can be reported in a variety of ways depending on the end use of the emission estimates and the measurement technique used. The general forms are total hydrocarbons (HC), nonmethane hydrocarbons (NMHC), nonmethane organic gas (NMOG), volatile organic compounds (VOC) and nonmethane volatile organic compounds (NMVOC). The contribution of the road transport sector to the total emission of NMVOC in Europe in 2009 was 13%, whereas in the U.S. in 2002 it was more than twice as much, about 27% [3][4].

These types of pollutants are expected to decrease in both OECD and non-OECD nations (Figure 4).

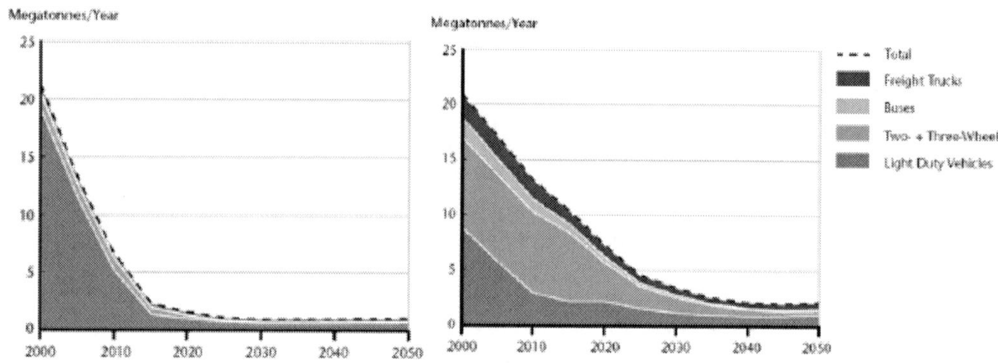

Figure 4. Transport-related VOC emissions by mode OECD regions (left) and non-OECD Regions (right) [5].

Road transport is still one of the main noise sources in urban environments. The total vehicle noise derives from several sources (engine, intake, powertrain, exhaust, tyres, etc). Modern vehicles are 90% quieter than those of 40 years ago. Many of the possible avenues for further noise reduction can actually be achieved by addressing the non-vehicle inputs, such as the road surface, the infrastructure and the tyres. It is the tyres' interaction with the road surface that most determines how much noise the passage of a vehicle makes. Typically, the sound pressure of a car varies in the range of 60 to 74 db(A), which is mainly owed to tyres.

One of the political instruments that can contribute to the decrease of emissions is car labelling. Car labelling aims to raise awareness in terms of consumers' purchasing decisions and to encourage manufacturers to explore more efficient powertrains based on electric propulsion. These powertrains include the pure electric vehicle (EV), hybrid electric vehicle (HEV), fuel cell vehicle (FCHEV) and plug-in hybrid vehicle (PHEV).

Labelling of road vehicles has largely focused on energy consumption. It requires a mandatory or voluntary sticker that is affixed to the car with information designed to convince the customer to buy and the manufacturers to produce more fuel-efficient vehicles. The U.S. was one of the first to use energy label programmes as far back as 1975 and other countries worldwide have also started adopting labels for fuel efficiency (see Figure 5). The

EU legislation requires a label showing fuel economy and CO_2 emissions to be attached to all new cars or displayed nearby at the point of sale.

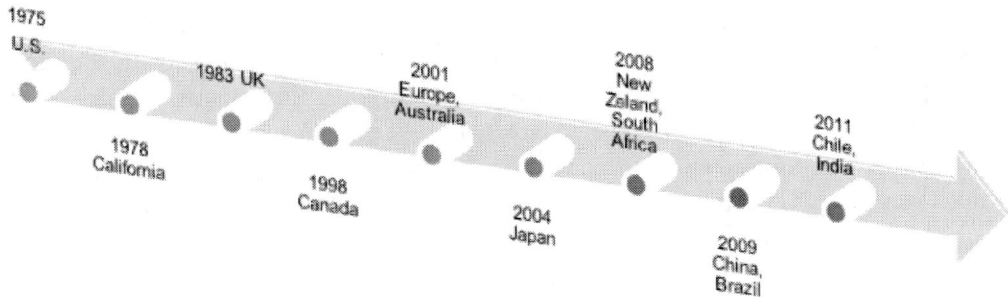

Figure 5. History of labelling implementation.

Environmental ranking methodologies attribute a score to each vehicle, so vehicles can be compared with each other. Usually the environmental indicators relate to greenhouse gas emissions (GHG) and air pollutant emissions (HC, CO, NO_x, PM, SO_x). US rating methodology developed by ACEEE is guided by life cycle analysis (LCA) and scores vehicles from zero to 100 (best). The Belgian Ecoscore methodology adopts an LCA perspective, disregarding the materials of vehicle impact, but giving some importance to the noise impact (scores from zero to 100(best)). The German VCD's current rating system does not purport to be an LCA, and it has been modified to accommodate evolving views about the importance of various automobile impacts. VCD scores from zero to 10 (best). It considers impacts on nature by both CO_2 (vehicle manufacturer information) and pollutants (European Standards), on human health by pollutants, and noise. Australian rates consider only the usage stage, giving zero to 10 (best) points to GHG and air pollutants, disregarding noise. The overall rating is based on the sum of the air pollution and greenhouse ratings (zero-five star (best)). ACEEE scores and Ecoscore account for GHG and air pollutants emitted during fuel production, refining, transport and distribution at fuel stations or charging points. So far, however, they disregard infrastructures construction, maintenance and decommissioning.

This chapter provides information on the current state of the art in environmental classifications and rankings for conventional and alternative vehicles. Six vehicles are used to exemplify the environmental ranking classification with/without energy supply infrastructures construction, maintenance and decommissioning, in the U.S. and Belgium.

STATE OF THE ART

Environmental Classification

Car labelling is typically focused on fuel efficiency, but some examples refer to CO_2 tailpipe emissions (UK example, Figure 6), GHG emissions rating (tailpipe only, see Figure 8) and smog rating (tailpipe only, a result of HC and NO_x release, see Figure 8). Beginning with the 1998 model year, the Smog Index Label began appearing on all new cars sold in California. The label was intended to help consumers compare smog-forming emissions from different vehicles within that model year. After 1 January 2009, manufacturers were required to affix the Environmental Performance (EP) Label. This label provided both a smog score and a global warming score, both from zero to 10, with 10 being the cleanest and five being the average vehicle. In 2011, the U.S. Environmental Protection Agency and the U.S. Department of Transportation adopted a new Fuel Economy and Environment label that is required on all vehicles from the model year 2013. In order to harmonise with the federal labelling programme, this new label replaces the California Environmental Performance Label on most new cars sold. European directive 1999/94/EC requires member states to ensure that relevant information is provided to consumers, including a label showing a car's fuel efficiency and CO_2 emissions. In July 2005, the United Kingdom launched the 'Green Label', which is intended to let the consumer know about the environmental impact that a particular car will have by detailing the amount of CO_2 the car will produce per kilometre, and presenting a scale constituted by CO_2 emission bands that allow the end-user to frame each vehicle CO_2 emission (Figure 6).

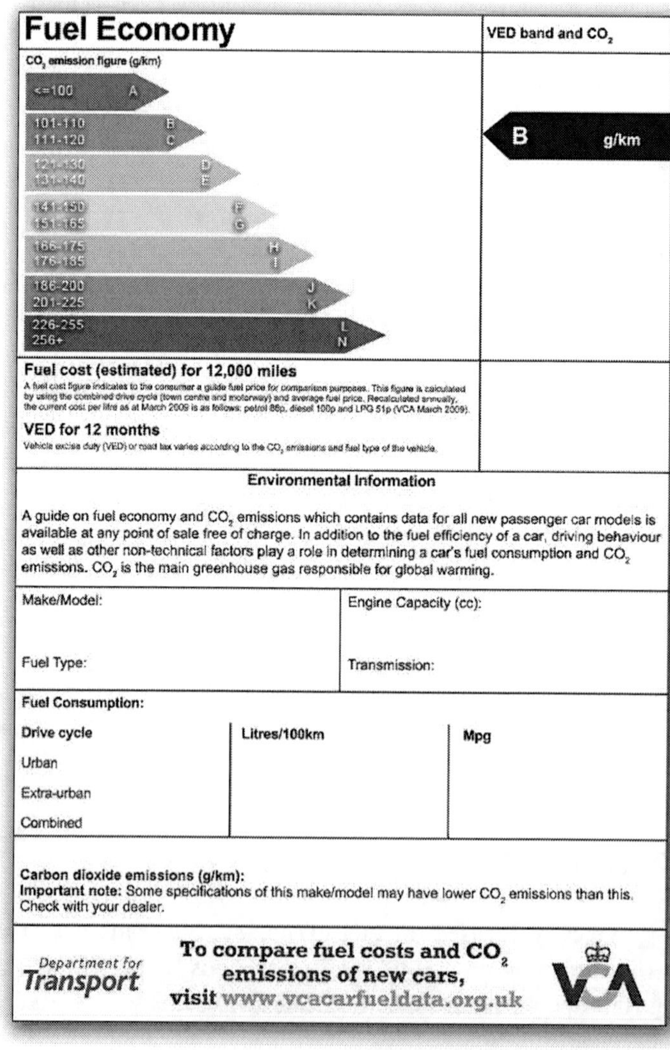

Figure 6. UK label design.

These bands are consistent with the vehicle registration tax (VRT) and annual motor tax for new cars, calculated on the basis of carbon dioxide (CO_2) emissions from vehicles. Seven emission/tax bands apply and VRT and motor tax payable are determined by the relevant band for each vehicle. The banding system is based on the European energy product label, although other member states use a different colour-coded label.

The Brazilian label design is similar to European labels for energy efficiency of electric appliances and Figure 7 shows an example of a flexifuel vehicle with information on urban and extra-urban fuel consumption (100% alcohol or 100% gasoline) (no information on emissions is included).

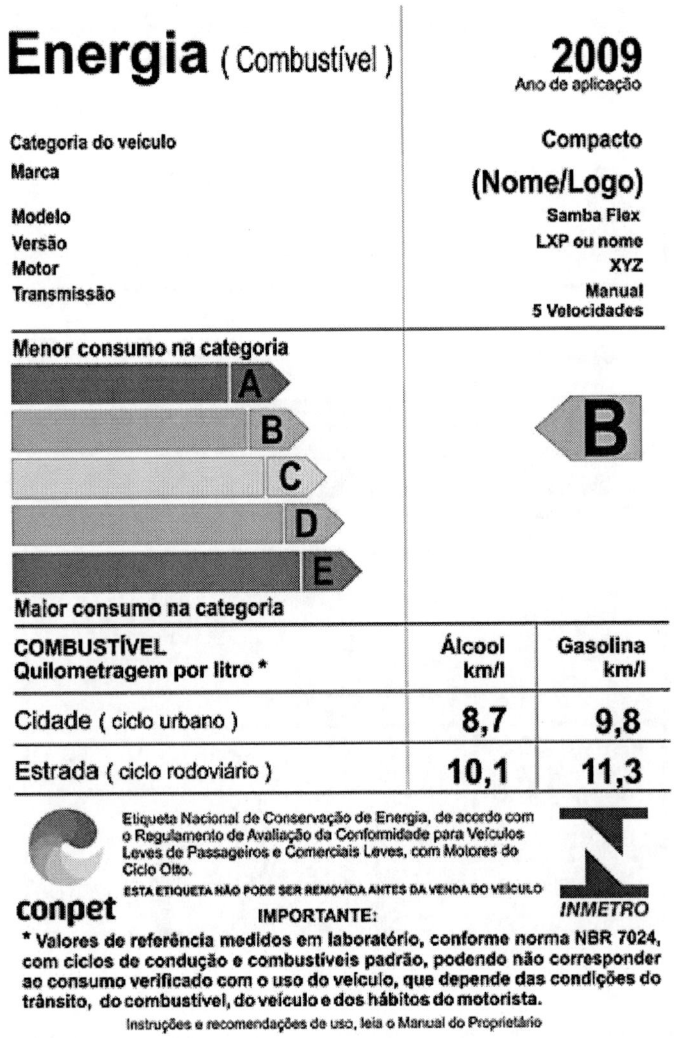

Figure 7. Brazilian label for flexifuel vehicles.

The U.S. national label incorporates the EP Californian label (smog and GHG ratings at tailpipe), and Figure 8 shows label examples. An estimation of annual fuel cost is displayed, using an annual mileage of 15,000 miles and a projected gasoline price based on projections from the U.S. Energy Information Administration for the relevant model year.

Figure 8. U.S. label for pure electric, hybrid plug-in and fuel cell vehicles.

The estimated fuel cost over a five-year period for the vehicle compared with the average new vehicle is also displayed. The label for plug-in hybrid electric vehicles displays fuel economy for both modes of operation. The left-hand fuel economy box on the label displays the fuel economy when the car is operating on electricity. This is described in miles per gallon of gasoline-equivalent (MPGe). This is similar to MPG, but instead of presenting miles per gallon of the vehicle's fuel type, it represents the number of miles the vehicle can go using a quantity of fuel with the same energy content as a gallon of gasoline.

The right-hand fuel economy box displays the fuel economy when the battery is empty and the vehicle is using gasoline. This is described in miles per gallon.

Both of these values represent the combined city and highway value. Combined fuel economy is a weighted average of city and highway MPG values calculated by weighting the city value by 55% and the highway value by 45%. In electricity consumption it also indicates how long it takes to charge a completely empty battery using a 240-volt electrical supply.

Other examples of labels worldwide can be found in [7].

Environmental Rankings

Beyond tailpipe emissions there are emissions related to production of the fuel used to power the vehicle and emissions related to its embodied materials. Current labels do not have environmental information or have incomplete information, e.g., the U.S. label displays smog and GHG (mostly CO_2).

Since 1998, the American Council for an Energy-Efficient Economy (ACEEE) has published an annual consumer-orientated guide, now titled *ACEEE's Green Book®*, which provides environmental performance rating information for the U.S. automotive market. The methodology is based on principles of life cycle assessment (LCA) and environmental economics [8]. The information provided by *ACEEE's Green Book®* is useful for informing consumers, but it also provides information to manufacturers, with potential influence on technology and product planning activities. Life cycle assessment is performed by dividing vehicle impact into three main phases: vehicle in use (tailpipe emissions), also known as tank-to-wheel (TTW), emissions from the fuel supply cycle (upstream emissions), also known as well-to-tank (WTT) and emissions embodied in the vehicle (manufacturing, assembly and disposal), also known as cradle-to-grave. Then all impacts are combined into a single Green Score that allows consumers to compare the environmental performance of automobiles easily. The *ACEEE's Green Book®* defines an environmental damage index (EDX). This index is a sum of damage functions, each based on attributes associated with the life cycle of the vehicle and its fuel.

An evaluation based on environmental economics uses monetised damage functions so that the EDX expresses an estimated life cycle environmental cost of the vehicle. A monetised environmental damage function reduces to:

$$EDX = \sum d_{ij} e_{ij} \tag{1}$$

where i is an index of emission type (air pollutants, including greenhouse gases), j is an index of locations of emissions, d_{ij} is an environmental damage cost (e.g., cents per gram), e_{ij} is the quantity of emissions averaged over a vehicle's operational life (e.g., grams per mile). The damage index so defined represents environmental impacts averaged over vehicle lifetime travel distance and the units can be given in cents per vehicle mile (¢/mi).To facilitate communication and comparison between vehicles an indicator was derived from the EDX to convey rankings in *ACEEE's Green Book®*, that is the Green Score on a higher-is-better scale of zero to 100:

$$\text{Green Score}=a.\frac{e^{-EDX/c}}{(1+\frac{EDX}{c})^{b}}, \text{ with a=100, b=3 and c=8.19¢/mi} \quad (2)$$

The Green Score ranges inversely to EDX and it spreads out the future 'green vehicles' at the expense of less differentiation among current vehicles. For base damage costs, i.e., those representing impacts of pollutants emitted directly from motor vehicles, the geometric means of low and high health cost estimates from [9] were adopted. Then, a reduction factor of five for factories and refineries and a reduction of 10 for electric power plants was applied. Table 1 shows the damage cost estimated for principal air pollutants used in this methodology. For GHG emissions a damage cost of 0.0237$/kg of CO_2-equivalent was assumed.

Table 1. Damage cost estimated for principal air pollutants [10]

	Marginal Cost Estimates for Principal Air Pollutants (2004$/kg)		
Pollutant	Motor Vehicles	Refineries and Factories	Electric Power Plants
CO	0.04	0.008	0.004
HC/ VOC/ NMOG	0.47	0.094	0.047
NO_x	6.24	1.25	0.62
SO_2	29.42	5.88	2.94
PM_{10}	50.09	10.02	5.01

EDX for global warming is calculated by multiplying the GHG life cycle emissions by the corresponding damage cost. Greenhouse gas emissions from each source, in-use emissions, vehicle materials life cycle (embodied) and fuel cycle (upstream) are summed and then multiplied by the CEF (CO_2 Equivalency Factors) (see Table 2).

Table 2. Global warming potential by type [11]

Type	CO_2 Equivalency Factors (CEF)
CO2	1
HC	2
NOx	4
CO	5
CH4	22
N2O	355

In-use or direct tailpipe emissions are taken from fuel consumed in the standard FTP cycle, as in *Table 3.*

Table 3. Fuel consumption dependent emission factors [12]

Pollutant (vehicle standard)	Gasoline (g/gal)	Diesel (g/gal)
SOX	0.19	0.12
CH4	1.26	0.89
N2O	1.67	0.60
CO2	8558	9936

The fuel upstream emissions considered come from Table 4. Embodied emissions are taken from linear correlations with vehicle and battery weight.

Table 4. Upstream emissions [12]

Pollutant	Gasoline (g/gal)	Diesel (g/gge)	Electricity (g/kWh)	Hydrogen NG reforming (g/gal)
NMOG	6.1	1.7	0.014	1.00
CH_4	27.7	26.7	0.023	47.04
CO	8.4	7.7	0.158	7.871
N_2O	0.14	0.11	0.008	0.193
NOx	9.9	9	0.861	15.21
SOx	77	7.3	0.914	6.779
PM_{10}	0.4	0.3	0.075	0.307
CO_2	2308	1669	598	10816

EDX for air quality is calculated by multiplying the values for each pollutant by its corresponding damage cost and finally adding up all partial EDX. Total global warming and air quality EDX are transformed into a score by equation 2.

Ecoscore is an environmental score, created for the Belgium road vehicles fleet and defined for three vehicle categories: light-duty vehicles, heavy-duty vehicles and two-wheelers. The Ecoscore calculation takes into account the most important pollutants emitted by the vehicle during the driving phase (exhaust emissions, GHG, CO, HC, NO_x, particulate matter (PM) and sulphur dioxide (SO_2)) and emissions from the production and distribution phase of the fuel (fuel cycle emissions, WTT). Emissions from the vehicle assembly and from the production of its components, as well as vehicle maintenance and end-of-life, are not taken into account. Ecoscore considers the following categories of damages: climate change, air quality depletion (health impairing effects and effects on ecosystems) and noise pollution. These impact categories are combined through a weighting system to create the Ecoscore Indicator [13].

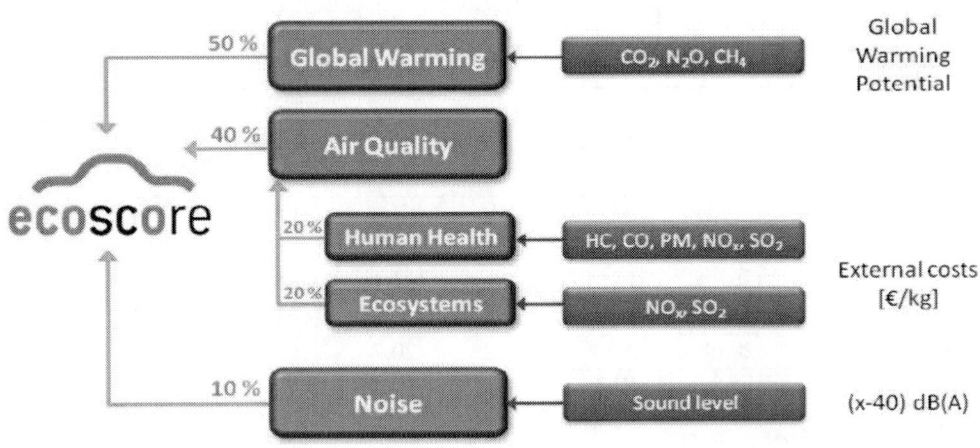

Figure 9. Ecoscore scheme.

A total impact (TI) is calculated by:

$$TI = \sum_i \alpha_i . q_i \text{ with } \sum_i \alpha_i = 1 \quad (3)$$

where TI – total impact of the assessed vehicle, α_i – weighting factor of damage category i , as in Figure 9. The final score is given on a higher-is-better scale of zero to 100:

$$EcoScore = 100. \exp(-0.00357. TI) \quad (4)$$

Each normalised damage (q_i) in category i (global warming, air quality or noise) is calculated by using the values for a reference vehicle:

$$q_i = \frac{Q_i}{Q_{i,ref}} \quad (5)$$

where Q_i – total damage of the assessed vehicle in category I, $Q_{i,ref}$ – total damage of the reference vehicle in category i. The reference vehicle is taken to be a Euro IV, with 120 g/km of tailpipe CO_2 and 70 db(A) noise level.

The damage Q_i for global warming includes direct tailpipe emissions and indirect fuel upstream emissions. Direct CO_2 is based on complete combustion of the consumed fuel (Table 5). CH_4 is assumed to be 0.02 g/km for gasoline and 0.01 g/km for diesel [10]. N_2O is assumed to be 0.005 g/km for gasoline and 0.008 g/km for diesel[12].

Table 5. Fuel characteristics and unregulated direct emissions of CO_2 and SO_2 used in the Ecoscore methodology (direct emissions) [14]

	Energy content (MJ/kg)	Density (g/L)	CO2 (kg /L)	SO2 (g /L)	N2O (g/km)
Petrol	42.7	755	2.21	7.55E-04	Euro I 0.027 Euro II 0.013 ≥Euro III 0.005
Diesel	43.3	850	2.70	8.50E-04	Euro I 0.002 Euro II 0.005 ≥Euro III 0.008

Table 6. Indirect emissions (g/kg for H_2)) petrol, diesel [15]; electricity/ hydrogen [16]

	CO mg/kWh	NMHC mg/kWh	NOx mg/kWh	PM mg/kWh	CO_2 g/kWh	SO_2 mg/kWh	N_2O mg/kWh	CH_4 g/kWh
Petrol	18.4	761.4	151.9	8.6	33.1	236.2	0	6.2
Diesel	16.6	315.4	129.6	3.6	24.5	174.2	0	5.6
Elec Belg	64.51	20.68	271.9	51.25	20.7	185.35	1.46	0.16
H_2	5.7	16.8	12.3	2	10620.6	9.5	0.04	59.8

The global warming potential (GWP) of greenhouse gases is based on the Intergovernmental Panel on Climate Change values from 2001 and a time horizon of 100 years. The damage Q_i for human health and ecosystems consider direct tailpipe emissions from Euro standard limits or measured in NEDC and multiply it by its respective damage cost. A ratio of 25% urban to 75% non-urban averages is used for the calculation of the environmental score. For indirect fuel cycle contributions, Table 6 is used and rural damage costs are taken [13].

Table 7. External costs related to air quality damage [13]

Classification	Weighting	Inventory	Units	Rural
Human Health	HC	€/kg	3	3
	CO_2	€/kg	0.0008	0.0032
	PM10	€/kg	103.49	418.61
	NOx	€/kg	1.152	1.483
	SO_2	€/kg	6.267	14.788
Ecosystems	NOx	€/kg	0.176	0.176
	SO_2	€/kg	0.113	0.113

The damage Q_i for noise,

$$Q_{noise} = E_{noise} - [40db(A)]$$

where E_{noise} is the noise level of the vehicle at the utilisation stage.

The fuel cycle WTT that is used in the two rating methodologies, U.S. and Belgium, is based on emissions from energy flows and disregards infrastructure-embodied materials. Other examples of methodologies could be found in the Internacional Energy Agency, e.g. for Europe (Green Global NCAP Workshop – IEA, Paris, 30th April 2013) and in the Australian Department of Infrastructure and Transport (green vehicle guide).

METHODOLOGY

Energy Supply Infrastructure-Embodied Materials

The GHG contribution to energy supply infrastructures regarding construction, maintenance and decommissioning can be estimated by a proposed model [17][18][19] applied to both electricity and hydrogen from centralised NG steam methane reforming (SMR). Conventional fuel (diesel and gasoline) supply infrastructures were identified as minor contributors, with a maximum of 2.5% [17] of vehicle life cycle and so the same absolute values were considered in both scenarios under study. The model includes two general equations that estimate GHG emissions from infrastructures by inputting information on regional electricity mix, the annual distance driven, supply chain losses and the number of vehicles per station or charger. The evaluation of the number of vehicles served per each charger/station as a function of annual distance driven is considered. Each parameter is presented in gCO_{2eq}/MJ and may be corrected by the loss factor of the downstream stages, so it can reflect a final energy-based unit. The electric embodied material for infrastructure carbon, in gCO_{2eq}/MJ, is shown in Equation (6) and is a function of parameters that refer directly to the supply chain, power plant, transmission grid (TG) and normal (Nc), home (Hc) and fast charger (Fc) impact:

$$LCA_{CO_{2eq}\ Electricity} = \frac{M_c}{(1-L_{\%TG})} + 0.07 + N_c + H_c + F_c \quad (6)$$

The parameter given by variable Mc refers to the carbon intensity of the construction, maintenance and decommissioning of all power plants, according to the region's electric mix. The 0.07 gCO_{2eq}/MJ constant refers to the electricity transmission grid carbon intensity per unit of final energy. Nc, Hc and Fc define the emission life cycle analysis of normal, home and fast chargers respectively. All parameters are further developed in [19].

For hydrogen, carbon intensity in gCO_{2eq}/MJ is shown in Equation (7) and is given by parameters that also refer directly to the supply chain. The first and third ones refer to the NG pipeline infrastructure that supplies the SMR plant and distributes H_2 to the refuelling stations respectively. The second parameter is related to the SMR hydrogen plant, where CF stands for the capacity factor of the plant. The fourth parameter refers to the H_2 refuelling station. This is then divided by the total energy that flows through the refuelling station per year. All parameters are further developed in [19].

$$LCA_{CO_{2eq}H_{2(a)}} = \frac{0.354}{(1-L_{\%H_2Plant})\times(1-L_{\%pline})\times(1-L_{\%Co})} + \frac{0.17/CF}{(1-L_{\%pline})\times(1-L_{\%Co})} + \frac{0.354}{(1-L_{\%Co})} + \frac{1.84E^{+07}}{(MJsold/year)} \quad (7)$$

U.S. and Belgian scenario electricity mix inputs and annual driven distance per vehicle are shown in Table 1. Table 2 presents all inputs required to calculate N, H and F in Equation (1) in the electricity pathway. Service ratios are based on the region's average annual driven distance per vehicle using the model by [19]. Charging behaviour is considered to be the same in both.

Table 10 presents input values considered in the application of Equation (7). Inventory values are presented in energy output-based units and not final energy. Since the ratio variable of vehicles per station is crucial for the estimations, it was assumed that they were in the same proportion as conventional fuel stations in each country. This ratio was obtained by dividing the number of light-duty vehicles in each country by the number of fuel stations [22][23].

Table 8. Input values considered for electricity pathway (electric mix by country)

M1	M2	M3	M4	M5	M6	M7	M8	
Electric Mix %[20]								
Coal	Oil	NGCC	Nuclear	Hydro	Geo	Wind	Solar	km/year [21]
5.2%	2.0%	36.0%	54.2%	1.8%	0.0%	2.2%	0.04%	Belgium-15000
45.6%	3.34%	23.4%	19.1%	6.8%	0.4%	1.4%	0.06%	U.S.-19100

Table 9. Input values considered for electricity pathway (chargers)

N	Int_c Nc=187.0[a]	LTimeNc= 6years[18]	TTW= 0.54	
Scenario\Inputs	%NcSR	SRTotal	AvrcarMJ/Yearryear	%Nc
Belgium	12.0%	1.113	8,100	15%
U.S.	14.9%	1.149	10,314	15%
H	Int_c Hc=42.0[b]	LTimeHc= 15 years[18]		
Scenario\Inputs	%HcSR	SRTotal	AvrcarMJ/year	%Hc
Belgium	87.8%	1.113	8,100	79.5%
U.S.	84.9%	1.149	10,314	79.5%
F	Int_c Fc=3146.9[c]	LTimeFc= 12 years[18]		
Scenario\Inputs	%FcSR	SRTotal	AvrcarMJ/year	%Fh
Belgium	0.3%	1.113	8,100	5.5%
U.S.	0.3%	1.149	10,314	5.5%

a, b, c, refer to carbon intensity of normal, home and fast chargers respectively. TTW is the considered energy in MJ/km for an average EV [18].

Table 10. Input values considered for H_2 pathway

Intpline[a]	3.54E-01	IntH_2Plant[b]	0.17	IntH_2stationc	1.84E+07	TTW=1.25	CFd =0.9
Inputs	L%H2 plant	L%pline	L%Co	km travelled /year	MJ sold/ station/year	Vehicles/ station	
Belgium	21%	2%	16%	15,000	3.19E+07	1,703	
U.S.	21%	2%	16%	19,100	4.71E+07	1,974	

[a] refers to carbon intensity of NG and H_2 pipeline, b refers to carbon intensity of SMR plant, c refers to carbon intensity per year of H_2 refuelling stations respectively. Units are in gCO_{2eq}/MJ; d refers to the capacity factor of the SMR plant. TTW is the considered energy in MJ/km for a FC-HEV[18].

Table 11 presents the results for the two scenarios under study for all technologies assessed. Estimations show minimum, maximum and best estimate values obtained taking into account only model uncertainty values. The electricity pathway presents the highest GHG impact using gCO_{2eq}/MJ of final energy functional unit, whereas the Belgium scenario reports higher impacts of both electricity and hydrogen pathways.

Table 11. Results of energy supply infrastructure GHG emissions for Belgian and US scenarios per pathway

gCO_{2eq}/MJ Min-Max (Best estimate)	Electricity	Hydrogen	Gasoline [18]	Diesel [18]
Belgium	5.1-18.4 (7.3)	0.8-2.5 (1.8)	0.3-0.9 (0.5)	0.3-1.1 (0.7)
U.S.	5.1-16.2 (7.0)	0.6-2.3 (1.6)	0.3-0.9 (0.5)	0.3-1-1 (0.7)

In both scenarios uncertainty regarding the electricity pathway comes mostly from the hydro and nuclear power plants. The model gives a high amplitude for these parameters (compared with its best estimate value) and this results in the high uncertainty shown in Figure 10. In addition to this the hydrogen pathway registers high uncertainty compared with the best estimate value, although this is dependent on the H_2 pipeline uncertainty included in the model.

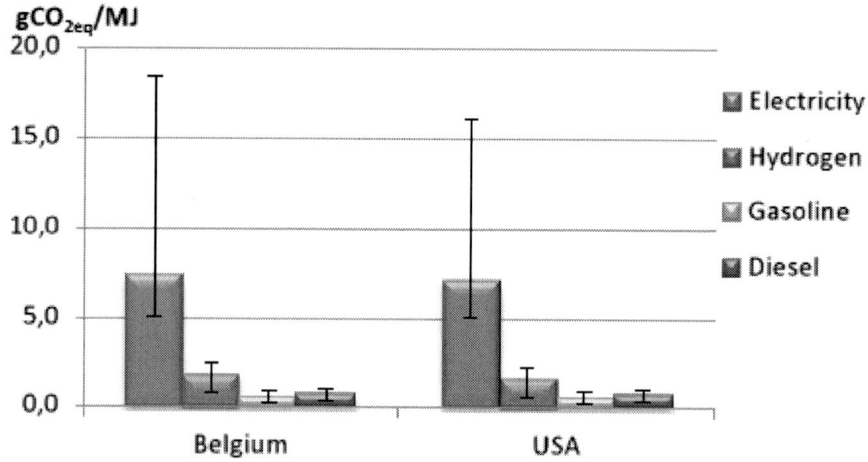

Figure 10. Emissions (a) and energy (b) LCA of energy supply infrastructures for Belgium and US scenarios.

Although they have different variables that could contribute to a higher or lower GHG impact by infrastructure, the estimates of both scenarios are similar. For example, although the US scenario has higher annual driven distance per vehicle than the Belgian one (which would decrease the impact per MJ of final unit), it also has a higher hydro power plant component in its electricity mix. On the other hand, the US scenario has a higher service ratio owed to a higher annual driven distance per vehicle; however, it benefits from a large scale gain, since it provides more energy per charger.

Regarding the H_2 pathway, differences derive from the number of refuelling H_2 stations considered. The model estimates that the number of H_2 stations will be the same as for the conventional ones. In this sense, since Belgium has fewer vehicles per station than the US scenario, it will supply less energy per facility, hence harming the final GHG emission value.

Vehicle Technologies

Six vehicles were used to exemplify the environmental ranking classification with and without energy supply infrastructures construction, maintenance and decommissioning, in the U.S. and Belgium. The vehicle specifications are provided in Table 12.. Standard limit emissions values were taken from the certification class in which the vehicles were qualified Tier 2 (U.S.) or LEVII (California) for the calculation of the Green Score, this last being the one that prevails, as it is more stringent. For the Ecoscore calculation the limit emissions were taken from the VCA database.

Table 12. Specifications of the analysed vehicles and fuel consumption in different standard cycles

Vehicles (Lev II; Euro V)	ICE (kW)	Motor (kW)	Weight (kg)	Battery (kg)	FC (kW)	UDDS	HWEFT	Combined (EPA)	Combined (NEDC)
ICEVDiesel-VW Golf (2.0 TDI Bluemotion)	103[a]	-	1366[a]	-	-	30[g]	42[g]	35[g]	4.3(l/100km)
ICEVPetrol-VW Beetle (2.0 TSI 200PS M6)	147[a]	-	1439[a]	-	-	21[g]	30[g]	25[g]	7.4(l/100km)
HEV-Toyota Prius (1.8 Petrol)	100[b]	60[b]	1500[b]	Ni-MH 41[b]	-	51[g]	48[g]	50[g]	4.1(l/100km)
EV-Nissan Leaf (Electricity)	-	80[c]	1525[c]	Li-Ion 300[c]	-	106 (0.31kWh/mile)[g]	92 (0.37 kWh/mile)[g]	99[g]	0.173(kWh/km)
PHEV-Toyota Prius Plug-in (1.8 Petrol + Electricity)	100[d]	60[e]	1525[e]	Li-Ion 80.[e]	-	CS: 51[g]	CS: 49[g]	CS: 50[a][h] CD:500+0.29(kWh/mile)	2.37(l/100km petrol)[i] [j] +0.092(kWh/km electricity)
FCHEV-Honda FCX clarity (Compressed Hydrogen Gas)	-	100[f]	1625[f]	Li-Ion	100[f]	60 (miles/Kg)[g]	60 (miles/kg)[g]	60 (miles/kg)[g]	0.002(kg H_2/km)

[a] (www.carfolio.com); [b] (www.toyota-tech.eu);[c] (www.nissan.pt); [d] (www.toyota.pt);[e] (www.toyota.com);[f] (http://automobiles.honda.com);[g] (www.fueleconomy.gov); [h] Regarding FTP/CAFE Cycle, for PHEV the fuel economy was calculated by a proportion of 39% of charge-depleting and (CD) 61% of charge-sustaining (CS). This proportion is a result of 11 miles of charge-depleting in the total US cycle considered of 27.96 miles (www.fueleconomy.gov). The combined fuel economy in charge-sustaining (CS) was calculated according to the proportion city/highway advised by EPA. For charge-depleting (CD) the fuel economy was available in gasoline, and electricity. Conversion factor: 1 gallon of gasoline= 33.7Kwh; [i] (http://www.vcd.org); [j] Regarding the NEDC cycle, for PHEV the fuel consumption was calculated by a proportion of 83% of charge-depleting (CD) and 17% of charge-sustaining (CS). This proportion is a result of 25 km miles of charge-depleting in a total cycle of 30km (cycle considered for Europe). The value of fuel consumption of petrol is already an average of the CD and CS, and the electricity consumption only relates to CD.

Noise levels were considered from the specifications of each vehicle. An extra vehicle was considered for the Ecoscore calculations; because this methodology has a normalisation step, it needs a reference vehicle. The reference vehicle chosen was conventional petrol ICEV, and its fuel consumption was calculated according to the conversion factor from the Pew Centre [24] based on CO_2 emissions. The CO_2 emissions had a reference level of 120 g/km; the other emissions were taken from the thresholds of EURO IV.

RESULTS AND DISCUSSION

Results of the application of the US environmental ranking Green Score® methodology to the set of studied vehicle technologies (Table 12) were assessed with and without reference to the energy supply infrastructures. Table 13 shows the EDX and Green Score results for criteria pollution for each phase and total GHG. The methodologies rank highly new alternative technologies of fuel cell, pure electric and plug-in hybrid vehicles.

Table 13. Results of the application of the Green Score methodology

	Air quality			Climate change	Totals	
	In-use	Fuel cycle	Embodied	GHG	EDX	Green Score
FCHEV-Honda FCX clarity	0.00	0.10	0.29	0.68	1.08	60.39
EV-Nissan Leaf	0.00	0.11	0.39	0.70	1.20	57.23
PHEV-Toyota Prius Plug-in	0.11	0.12	0.28	0.70	1.21	57.05
HEV-Toyota Prius	0.15	0.13	0.28	0.74	1.30	54.79
ICEVDiesel-VW Golf	0.10	0.14	0.21	0.96	1.41	52.32
ICEVPetrol-VW Beetle	0.11	0.25	0.22	1.34	1.92	42.02

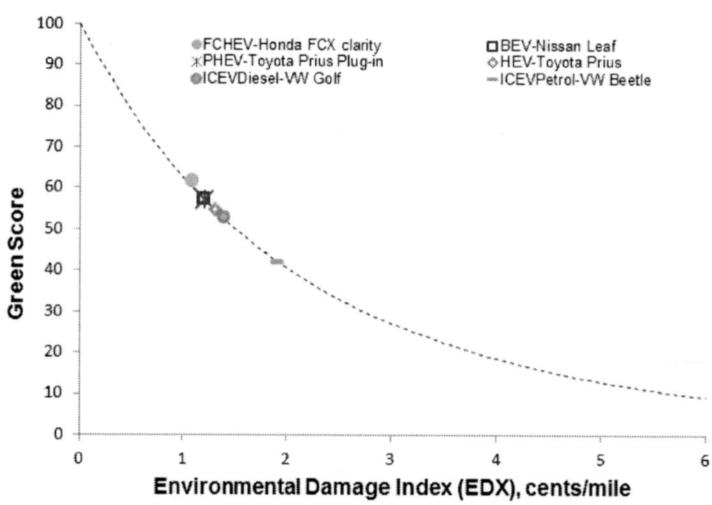

a)

Figure 11 (Continued).

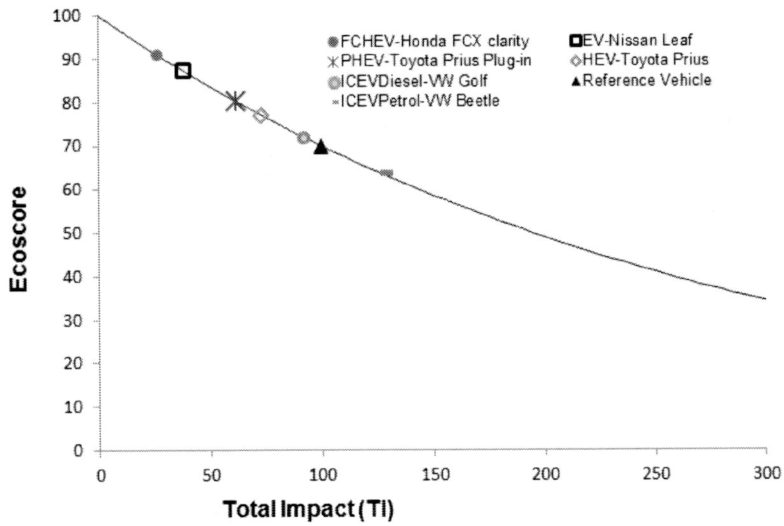

b)

Figure 11. a) Green Score vs. environmental damage cost (EDX) and b) Ecoscore vs. Total Impact (TI).

Figure 11 shows how the vehicles perform in terms of U.S. and Belgian environmental scores. The consideration of infrastructure construction, maintenance and decommissioning has negligible effects in the final ranks, decreasing the scores by less than 2%. The major decrease is in EV-Nissan Leaf, followed by the PHEV-Toyota Prius Plug-in.

Results of the application of the Ecoscore methodology to the set of studied vehicles are as follows. Table 14 shows the results of the application of the methodology: partial impact categories (absolute values) that contribute to the TI and Ecoscore.

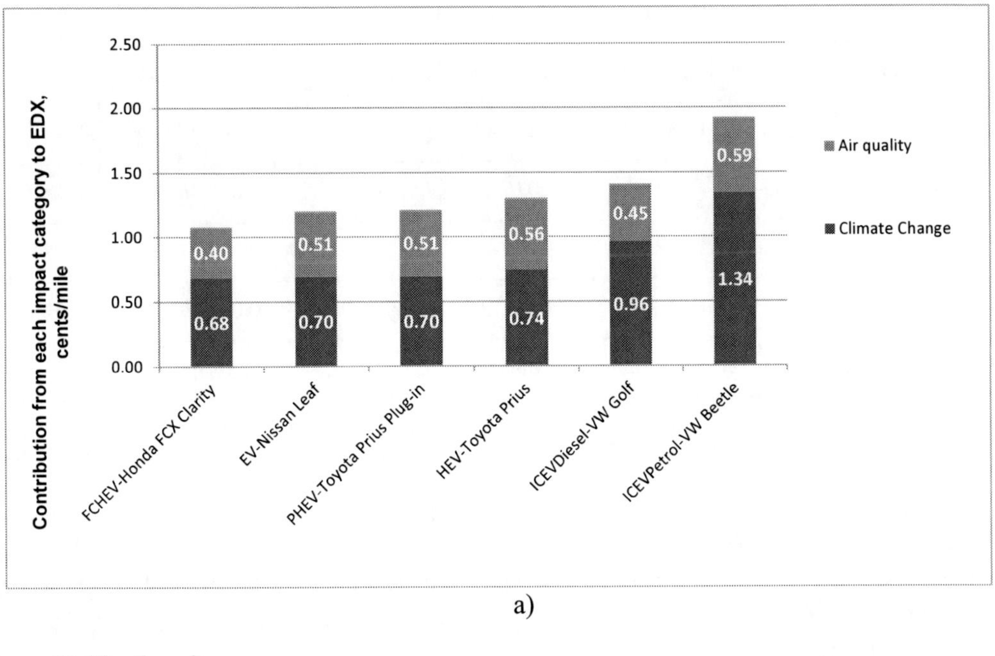

a)

Figure 12 (Continued).

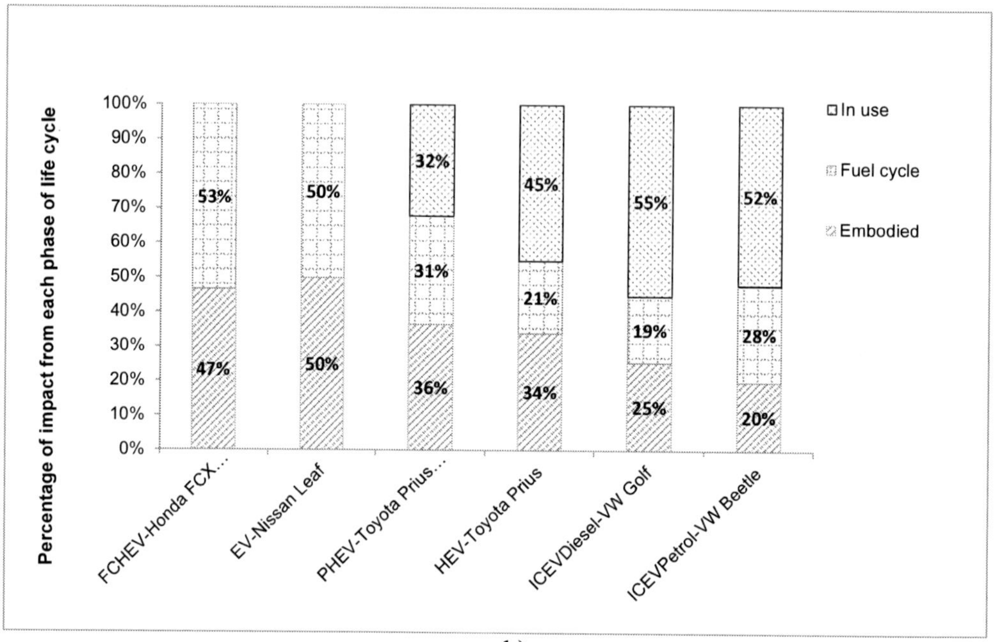

b)

Figure 12. a) EDX breakdown by impact category (disregarding energy supply infrastructures) and b) EDX breakdown by phase of life cycle.

Table 14. Results of the application of the Ecoscore methodology

	Air quality		Climate change	Noise	Totals	
	Human Health	Ecosystem	GHG		TI	Ecoscore
FCHEV-Honda FCX clarity	6.27E-04	6.57E-06	24.38	28.00	26.65	90.92
EV-Nissan Leaf	1.18E-03	1.19E-05	36.58	28.00	37.69	87.41
PHEV-Toyota Prius Plug-in	1.74-03	1.92E-05	80.99	29.00	62.09	80.12
HEV-Toyota Prius	1.96E-03	2.07E-05	105.32	29.00	73.32	76.97
ICEVDiesel-VW Golf	1.62E-03	4.35E-05	129.90	30.00	91.76	72.07
Reference Vehicle	2.64E-03	3.88E-05	137.94	30.00	100.00	70.00
ICEVPetrol-VW Beetle	3.41E-03	4.19E-05	188.52	31.00	126.13	63.74

Figure 11 shows that the positions of the vehicles are the same in both methodologies. The vehicle that scores highest according to Green Score and Ecoscore is FCHEV, followed immediately by EV. The intermediate positions are occupied by the hybrid technologies, PHEV and HEV, in third and fourth place respectively. ICE vehicles occupy the final places, ICEV Petrol being in the last position.

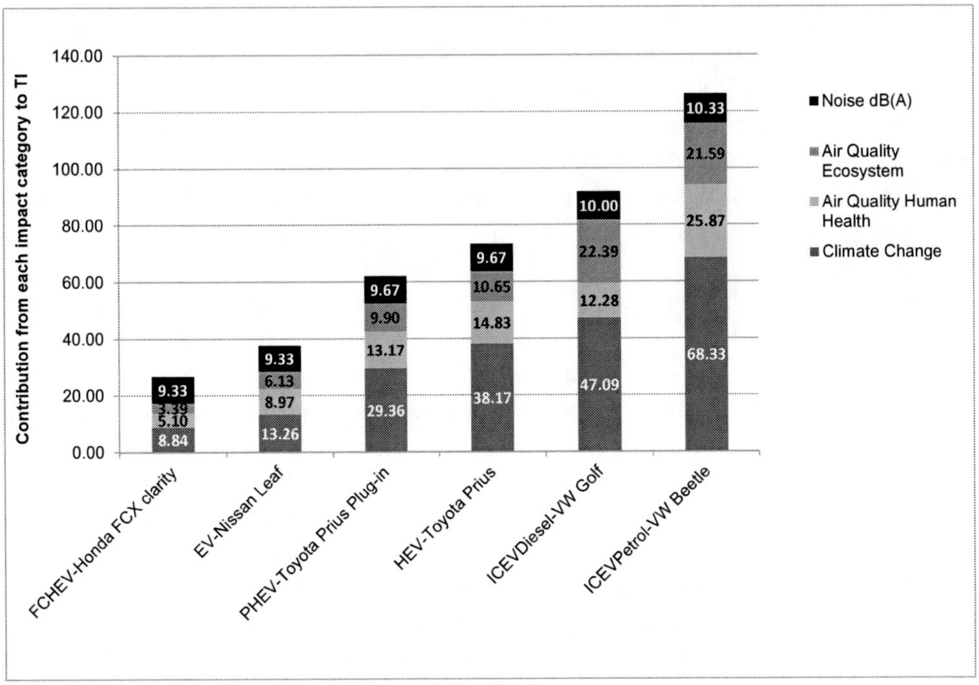

a)

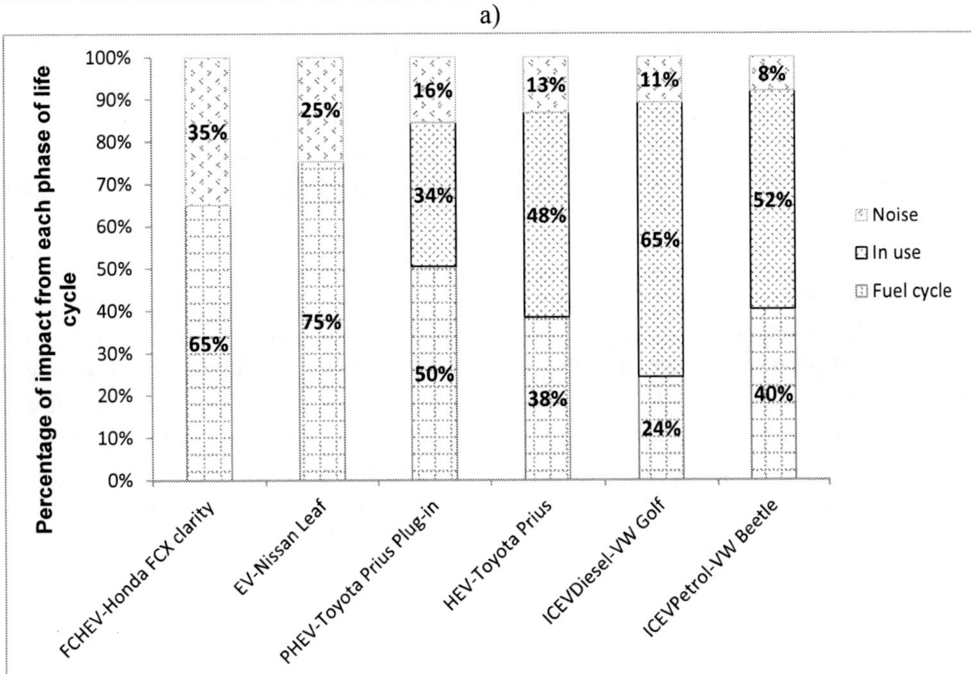

b)

Figure 13. a) Total impact (TI) breakdown by impact category (disregarding energy supply infrastructures) and b) TI breakdown by phase of life cycle.

Analysing the impact categories and the total impact of vehicles, Figure 12 and Figure 13 show that according to both methodologies climate is more damaged than air quality in relation to all the studied technologies. ICEV-Petrol is the vehicle with most impact on climate, mainly because of the CO_2 emitted during the in-use phase. The vehicle that has lowest impact in this category is FCHEV-Honda FCX Clarity. In general, vehicles with ICE have more impact on the climate than other alternative technologies. Regarding air quality, the vehicle with the most impact is ICEV-Petrol due to the HC emitted during the fuel cycle.

If we examine the breakdown of the total impact calculated in both methodologies by phase of the life cycle of the vehicle, it is clear that in general embodied emissions have a large responsibility for the damage caused by partially or fully electric vehicles and are lower in ICE vehicles.

Regarding in-use emissions phase, as expected, FCHEV-Honda FCX Clarity and EV-Nissan Leaf have no damage impact, whereas HEV-Toyota Prius, ICEV Diesel-VW Golf and ICEV Petrol-VW Beetle have their main damage impact.

Noise, only taken into account in the Ecoscore methodology, presents a slight variation in absolute values between technologies. After comparison of the Green Score and Ecoscore methodologies, we conclude that the impact categories and the phases considered in the LCA vary between them. Whereas ACEEE methodology considers embodied, fuel-cycle and in-use emissions, Ecoscore only considers fuel-cycle and in-use emissions.

Concerning impact phases, ACEEE takes into account the air quality pollutants and GHG, whereas Ecoscore also takes noise into account.

A new and more complete methodology based on these two is being developed and is intended for application internationally. This new methodology will include at least three phases of the life cycle, embodied, fuel-cycle and in-use emissions, and use the Green Score methodology. In terms of impact phases it will be based on Ecoscore, considering air quality pollutants, GHG and noise. The weighting system used in the Ecoscore methodology is subjective and will be analysed further on to see its effect on modifying the rankings.

It is easy to see the influence of the usage phase in both methodologies, and it is notable that some technologies do not result in direct emissions although the fuel cycle is dependent on this usage phase through the energy consumed.

Directly connected to this impact is the drive cycle that the methodologies use for this assessment [25], Green Book methodology uses the North American regulation cycle (FTP-75) which represents urban driving and the CAFE cycle for fuel economy and Ecoscore methodology uses the European regulation cycle (NEDC) which represents urban and extra-urban driving profiles. The final deviations in the final score may reach 29% for these methodologies depending on the cycle they use; however, the ranking order for the technologies is retained [25]. Further research should focus on the difference between e real versus standard driving across the different technologies and its impact on vehicle environmental rankings.

ACKNOWLEDGEMENTS

The research was partially supported by the MIT-Portugal project 'Assessment and Development of Integrated Systems for Electric Vehicles' (MIT-Pt/EDAM-SMS/0030/2008), funded by FCT. The author Sara Marques is sponsored by FCT grant SFRH/BD/73416/2010.

NOMENCLATURE

ACEEE	American Council for an Energy-Efficient Economy
CAFE	Corporate Average Fuel Economy
CD	Charge depleting
CS	Charge sustaining
EDX	Environmental Damage Index
EV	(battery) Electrical Vehicle
Fc	Fast Charger
FCHEV	Fuel Cell Hybrid Electrical Vehicle
GHG	GreenHouse Gas emissions
Hc	Home charger
HEV	Hybrid Electric Vehicle
ICE	Internal Combustion Engine
ICEV	Internal Combustion Engine Vehicle
L	Losses
LCA	Life Cycle Analysis
LTime	Lifetime (years)
Mc	Carbon intensity of the construction, maintenance and decommissioning of all power plants
Nc	Normal charger
NEDC	New European Driving Cycle
OECD	Organisation for Economic Co-operation and Development
PHEV	Plug-in Hybrid Electric Vehicle
SMR	Steam Methane Reforming
TG	Transmission grid
TI	Total Impact
TTW	Tank-to-Wheel
VRT	Vehicle Registration Tax
WBCSD	World Business Council for Sustainable Development
WTT	Well-to-Tank

REFERENCES

[1] Baptista, P. (2011). Evaluation of the impacts of the introduction of alternative fuelled vehicles in the road transportation sector. IST, UTL.
[2] World Energy Council. (2012). *Global Transports Scenarios* 2050.
[3] European Environmental Agency [Internet] (www.eea.europa.eu) last access February 2013.
[4] U.S. Department of Transportation, Federal Highway Administration (www.fhwa.dot.gov) last access February 2013.
[5] WBCSD. (2004). Mobility 2030: meeting the challenges to sustainability. The sustainability mobility project - full report 2004.

[6] Cofala, J., Markus, A., Mechler, R. (2005). Scenarios of World Anthropogenic Emissions of Air Pollutants and Methane up to 2030. *International Institute for Applied Systems Analysis*. Laxemburg, Austria.

[7] Mahlia, T.M.I., Tohnoc, S., Tezukac T., (2012). History and current status of the motor vehicle energy labeling and its implementation possibilities in Malaysia. J. Renew. Sust. Energ. Rev.,16, 1828-1844.

[8] Langer, T., & Vaidyanathan, S. (2011). Rating the Environmental Impacts of Motor Vehicles: ACEEE´s Green Book ® Methodology. *American Council for an Energy-Efficient Economy.*

[9] Delucchi, M. A. (2004). The Annualized Social Cost of Motor-Vehicle Use in the United States, 1990-1991: Summary of Theory, Data, Methods, and Results. UCD-ITS-RR)96-3. ITS Davis.

[10] Pohl, H. (2003). Vehicle Environmental Rating. Proceedings of the Fuel Cell, Hybrid and Battery Electric Vehicle Symposium, EVS20. Long Beach, CA.

[11] Delucchi, M. A. (2005). A Multi-Country Analysis of Lifecycle Emissions From Transportation Fuels and Motor Vehicles. ITS-Davis.

[12] Choudhury, R., et al. (2002). GM Well-to-Wheel analysis of Energy Use and Greenhouse Gas Emissions of Advanced Fuel/Vehicle Systems - A European Study. Germany: L-B_ Systemtechnik GmbH.

[13] Timmermans, J.-M., Mathews, J., Mierlo, J., & Lataire, P. (2006). Environmental rating of vehicles with different fuels and drive trains: a univocal and applicable methodology. EJTIR,6, no.4, 313-334.

[14] Boureima, F., et al. (2009). Well-to-Wheel analysis of new vehicle technologies of future vehicle markets. ETEC.

[15] Joumard, R. (1999). Methods of estimation of atmospheric emissions from transport:Final Report, LTE 9901 report,. France: Institut National de Recherche sur les Transports et leur Sécurité.

[16] Prieur, A., et al. (2009). Well-to-Tank technology pathways and carbon balance. Roads 2 hycom project.

[17] Lucas, A., Silva, C., Neto, R. (2013). Life cycle analysis of energy supply infrastructure for conventional and electric vehicles. *Energy Policy*;41:537-547.

[18] Lucas, A., Neto, R., Silva, C. (2012). Impact of energy supply infrastructure in life cycle analysis of hydrogen and electric systems applied to the Portuguese transportation sector. *International Journal of Hydrogen Energy*, 37(15):10973-0985.

[19] Lucas, A., Neto, R., Silva, C. (2013). Energy supply infrastructure LCA model for electric and hydrogen transportation systems. *Energy*, Vol. 56, Pages 70–80.

[20] Itten R., Frischknecht, R., Stucki, M. (2012). Life Cycle Inventories of Electricity Mixes and Grid. ESU-services Ltd. Uster, Switzerland 229 pp.

[21] Metschies, G. (2003). Harmonisation of Excise Taxes. Exercises for Fuel and vehicle taxation in African countries. Workshop 11-13 June 2003 Pretoria/South Africa 15 pp.

[22] Datamonitor.com [Internet] http://www.datamonitor.com. Service station retailing in Belgium 2012. [updated 2013 Jan; cited 2013Fev 25].

[23] Census.gov [Internet] http://www.census.gov. United States Census Bureau 2011. Industry Statistics Sampler, NAICS 447 Gasoline Stations 2007. [updated 2011 May; cited 2013 Fev 25] .

[24] An, F., & Sauer, A. (2004). Comparison for the Pew Center on Global Climate Change. Pew Center on Global Climate Change.

[25] Marques, S., Batista, T., Silva, C. (2013). Eco-rating methodologies for private cars: driving cycle influence. EWGT2013 – 16th Meeting of the EURO Working Group on Transportation. Porto: Procedia - Social and Behavioral Sciences.

In: Grid Electrified Vehicles
Editor: Carla Alexandra Monteiro da Silva

ISBN: 978-1-62808-839-7

Chapter 9

REAL-WORLD ACTIVITY, ENERGY USE, AND EMISSIONS OF A PLUG-IN HYBRID ELECTRIC VEHICLE

H. Christopher Frey* and Brandon M. Graver
North Carolina State University, Raleigh, North Carolina, US

ABSTRACT

Plug-in hybrid electric vehicles (PHEVs) are gaining market share. PHEVs have more complex operational characteristics than conventional vehicles. A method for quantifying real-world PHEV activity, energy use, and emissions is demonstrated based on in-use measurement of a selected vehicle. Vehicle activity was measured using an electronic control unit (ECU) data logger and global position system (GPS) receivers with barometric altimeters. Vehicle emissions are quantified based on in-use measurements using a portable emissions measurement system. The relationship between fuel use and emission rates versus power demand is quantified based on Vehicle Specific Power (VSP), which is a function of vehicle speed, acceleration, and road grade. Road grade is inferred from the GPS/altimeter data for every 0.1-mile segment of road driven by the vehicle. A vehicle fuel use and emissions rate model based on VSP is calibrated using data measured in and near Raleigh, NC. Incremental fuel use and emissions associated with an engine cold start are quantified. Several approaches for estimating fuel economy are compared. Activity data from a 300-mile long distance trip is used to illustrate the operating modes of a PHEV. Direct tailpipe emissions and indirect grid emissions are estimated and compared. The proportion of total emissions associated with cold start is investigated for short and long distance trips. The implications of the real-world data for quantification of PHEV cold starts, fuel economy, and grid-based energy use and emissions are discussed.

Keywords: Vehicle, energy, emissions, gasoline, electricity, driving cycles, road grade

* Corresponding author. Department of Civil, Construction, and Environmental Engineering, North Carolina State University, Raleigh, NC 27695-7908 USA. E-mail: frey@ncsu.edu.

INTRODUCTION

A growing number of manufacturers are offering hybrid, plug-in hybrid, and electric vehicles. In particular, plug-in hybrid electric vehicles (PHEVs) are gaining market share, although their sales are still small compared to conventional gasoline vehicles in the U.S. Examples of PHEVs now on the market include the Toyota Prius Plug-In and Chevrolet Volt. According to Edmunds.com, new PHEV models expected on the market in the next two years include the Audi A3, Chevrolet Cruze, Ford C-Max Energi and Fusion Energi, Honda Accord, Hyundai Sonata, Mercedes-Benz S-Class, Mitsubishi Outlander, and Porsche Panamera and 918 Spyder. Thus, there are likely to be a growing number and wide range of body styles of PHEVs from U.S., European, and Japanese manufacturers.

PHEVs typically include an internal combustion engine (ICE), a traction battery (TB), and an electric motor-generator (MG). PHEVs receive power from fuel, typically gasoline, and from grid electricity used to recharge the traction battery. PHEVs can meet power demand associated with vehicle speed, acceleration, and road grade in a variety of ways depending on specific details of their power trains. For example, in the Toyota Prius family of hybrid electric vehicles (HEVs) and PHEVs, the ICE can provide power to the MG to help recharge the TB, or the TB can be discharged to power the MG. Furthermore, the MG can be used as a generator during braking, a process referred to as regenerative braking. Hybrids are sometimes classified as "series," in which only the MG is connected to the drive wheels, or "parallel," in which either the ICE or the MG can provide power to the drive wheels.

PHEVs have multiple operating modes that depend on how much electrical charge in the TB remains from grid-based electricity previously used to recharge the battery, and whether the engine is on or off [1]. Charge depleting (CD) mode refers to operation while a TB containing electrical power derived from the electrical grid is discharged to provide power to the MG. Once the grid-based electricity is discharged from the TB, the TB can still be used as a power reservoir during charge sustaining (CS) mode. In CS mode, the TB can be periodically recharged via regenerative braking or based on power output of the ICE, depending on the type of hybridization. However, typcally, the TB is not recharged to as high a level of state of charge (SOC) as is possible when connecting directly to the grid via a shorepower electricity source. Another level of complexity with PHEVs and HEVs is that the ICE can cycle on or off depending on overall power demand and availability of stored energy from the TB.

The emergence of PHEVs on the commercial market has motivated a new approach to quantify vehicle fuel economy [2]. In the U.S., PHEVs have two fuel economy ratings: gasoline only and electricity+gasoline. The gasoline rating is a combined city/highway estimate based on a weighted combination of chassis dynamometer tests with standardized driving cycles. The electricity+gasoline rating is also derived from dynamometer measurements on the premise that the energy content of electricity used by the car is best represented by the energy equivalent of the electricity itself. Thus, the amount of primary energy input required to generate a given amount of electricity at a power plant is not taken into account.

Because there are relatively few originally manufactured PHEVs in use currently, there have been few studies of such vehicles based on real-world operations. Idaho National Laboratory (INL) conducted real-world energy consumption measurements of over 180 retroffited PHEVs [3]. Graver and Frey (2012) conducted a detailed assessment of the real-world activity, energy use, and emissions of a retrofitted 2006 Toyota Prius [4]. Choi et al. (2011) similarly conducted field measurements on a prototype PHEV school bus [5]. However, there is a need for real-world data regarding the activity, energy use, and emissions of production PHEVs.

In this chapter, a method for quantifying the real-world activity, energy use, and emissions of an example PHEV is illustrated based on measurements of a 2013 Toyota Prius Plug-In made during real-world driving. The resulting data are used to illustrate the key characteristics of PHEV operation, including CD and CS mode, engine on and off operations, and the effect of vehicle operations on emission-related factors such as catalyst temperature. A comparison is made of two approaches for estimating real-world fuel economy, illustrating substantially different insights obtained. The combination of direct tailpipe emissions and indirect emissions from power generation is estimated.

METHODS

Measurements of the real-world activity, energy use, and emissions were made of a 2013 Toyota Prius operated on specified study routes in the Raleigh, NC and Research Triangle Park, NC area. The same vehicle was driven on a long-distance trip from Raleigh, NC to Glen Burnie, MD.

Instrumentation for In-Use Measurements

Data were logged from the electronic control unit (ECU) of the vehicle using an on-board diagnostic (OBD) scantool. The ECU broadcasts a number of parameters at greater than 1 Hz frequency during vehicle operation. Specific parameters that were logged include ambient air temperature (^oC), barometric pressure (kPa), intake air temperature (^oC), engine coolant temperature (^oC), catalyst temperature (^oC), manifold absolute pressure (kPa), revolutions per minute (RPM), vehicle speed (miles/hour, or mph), mass air flow (g/s), battery charge remaining (%), and fuel flow rate (l/hr).

Vehicle position and elevation were measured using seven Garmin 76CSx global positioning system (GPS) receivers with barometric altimeters. All of these receivers were positioned on the front dashboard of the vehicle. The primary purpose of using these receivers is to estimate road grade at all locations along the route. Road grade is the ratio of change in elevation versus distance.

The OBD data logger and GPS receivers were used for all routes described here, including a 303 mile trip from Raleigh, NC to Glen Burnie, MD that took place on January 27, 2013. On January 22, 2013, measurements were made on local routes in the Raleigh, NC and Research Triangle Park, NC area for the purpose of developing a modal emission rate model for the vehicle. Thus, for these latter measurements, an Axion model Portable Emissions Measurement System (PEMS) manufactured by Clean Air Technologies International, Inc., was used. This PEMS has two five-gas analyzers in parallel that simultaneously measure exhaust gas concentrations of carbon dioxide (CO_2), carbon monoxide (CO), hydrocarbons (HC), nitric oxide (NO), and oxygen (O_2). CO_2, CO, and HC are measured using nondispersive infrared (NDIR) and NO and O_2 are measured using electrochemical detection. By convention, the mass of NO is reported as equivalent NO_2. The gas analyzers are calibrated using a standard BAR-97 "high" gas mixture for span calibration of gasoline vehicle exhaust measurements. During data collection, the gas analyzers periodically conduct a "zero" calibration using ambient air as a reference to prevent drift. The precision and accuracy of the PEMS in comparison to a chassis dynamometer vehicle emissions lab was quantified by Battelle, as reported in detail elsewhere [6].

Routes

Data were collected on two sets of routes. On January 22, 2013, OBD, GPS, and PEMS data were collected in the Raleigh, NC and Research Triangle Park, NC area for the purpose of developing a model of the vehicle fuel use and emission rates as a function of power demand. The routes used are the same as those originally developed by Frey, Zhang, and Rouphail [7]. They inlude three alternative routes (Routes A, B, and C) between North Carolina State University and North Raleigh, and two alternative routes (Routes 1 and 3) between North Raleigh and Research Triangle Park, NC. Routes A, B, and C have roundtrip distances of 20.4, 23.3, and 22.6 miles, respectively. Route A is comprised of feeder/collector streets, minor arterials, and major arterials. Route B is similarly comprised but also includes a segment on an interstate highway. Route C has a higher proportion of distance on interstate highway. Route 1, with 32.6 miles roundtrip, is approximately 90% interstate highway, whereas Route 3, with 35.2 miles roundtrip, is mostly major arterials with no highways. The total distance driven for data collection is approximately 135 miles.

On January 27, 2013, OBD and GPS data were collected while the vehicle was driven 303 miles from Raleigh, NC to Glen Burnie MD. The travel path was mostly comprised of I-85 and I-95, with some of the distances driven on rural arterials and urban arterials.

Data Synchronization and Quality Assurance

Data from multiple instruments were time aligned, and errors in the raw data were either corrected or removed, prior to data analysis. For time alignment of OBD and GPS data, vehicle speed was used as indicator variables for both data sets. Vehicle speed is reported by the OBD, and was estimated from GPS data from changes in vehicle position. The time trace of vehicle speed from GPS was compared to that from OBD, and the time delay between them was adjusted until changes in speed were matched. For synchronizing PEMS data to OBD data, engine RPM is typically used as the indicator variable for the OBD data and volume percent CO_2 is used as the indicator for the PEMS data. Engine starts and shutdowns have concurrent changes in both of these variables.

Once the data were synchronized, the combined OBD, GPS, and PEMS data for the January 22, 2013 data collection were evaluated for common errors, such as out-of-range concentrations. Where feasible, such errors were corrected. When errors could not be corrected, data were omitted from the final quality assured data file. Details of the quality assurance procedures are given elsewhere [8].

Road Grade

For each GPS receiver with barometric altimeter, road grade was estimated based on cumulative distance traveled along a route. For example, for the January 27, 2013 data, seven GPS receivers were used. For the data file from each receiver, the route was divided into 0.1-mile segments. For each segment, linear regression was conducted for elevation versus cumulative distance. The slope of the linear regression model is the estimated grade. For each second of the datafile, an average estimate of road grade was made using the average of the seven individual GPS receiver-based grade estimates. The precision of the road grade estimate was inferred based on the 95% confidence interval. Further details of this method are given elsewhere.[9].

Vehicle Specific Power (VSP) Modes

Vehicle energy use is typically proportional to positive power demand on the vehicle power train. Based on Jiménez-Palacios, the power demand on a vehicle can be estimated in terms of Vehicle Specific Power (VSP) [10]. For a light duty vehicle, VSP is estimated as:

$$VSP = v \times \{1.1 \times a + 9.81 \times \sin(\operatorname{atan}(r) + 0.132\} + 3.02 \times 10^{-4} \times v^3$$

where,

v	=	vehicle speed (m/s)
a	=	vehicle acceleration (m/s^2)
r	=	road grade (slope)
VSP	=	Vehicle Specific Power (kW/ton)

The four terms in the VSP equation represent changes in kinetic energy, changes in potential energy, rolling resistance, and aerodynamic drag, respectively. If vehicle speed is zero, there is no load on the engine. For a vehicle cruising at a constant speed of 60 mph with no acceleration or grade, VSP = 9.4 kW/ton. However, if the vehicle at that speed accelerates by 1 mph/s, then VSP increases to 22.6 kW/ton. A vehicle traveling at a constant speed of 60 mph on a 4% positive grade would have VSP of 19.9 kW/ton; in contrast, at the same speed with a grade of -3% VSP would be only 1.5 kW/ton. Thus, vehicle power demand can be very sensitive to each of speed, acceleration, and grade. Power demand less than zero implies no load on the engine. Negative power demand can occur as a result of negative acceleration, negative road grade, or combinations of both.

Frey et al. (2002) developed a VSP-based modal modeling approach for light duty vehicle fuel use and emission rates [11]. Based on statistical analysis of a sample of gasoline-fueled vehicles, a set of 14 VSP modes was defined. These include two modes for negative values of VSP, one mode that includes idling, and additional modes for successively larger positive values of VSP, as defined in Table 1. The VSP mode definitions given here are used to quantify fuel use and emission rates for the test vehicle.

Because PHEVs can operate in several operating modes, including CS, CD, engine on, and engine off, the measured data are stratified into these operating modes in combination with the use of the VSP modal approach. Furthermore, cold start is also evaluated as a separate operating mode.

Table 1. Vehicle Specific Power (VSP) Mode Definitions based on Frey et al (2002) [11]

VSP Mode	VSP (kW/ton)	VSP Mode	VSP (kW/ton)
1	VSP < - 2	8	13 ≤ VSP < 16
2	-2 ≤ VSP < 0	9	16 ≤ VSP < 19
3	0 ≤ VSP < 1	10	19 ≤ VSP < 23
4	1 ≤ VSP < 4	11	23 ≤ VSP < 28
5	4 ≤ VSP < 7	12	28 ≤ VSP < 33
6	7 ≤ VSP < 10	13	33 ≤ VSP < 39
7	10 ≤ VSP < 13	14	39 ≤ VSP

Cold Start

Cold start is typically quantified based on comparisons of warm start "Bag 3" of the Federal Test Procedure (FTP) standardized dynamometer test cycle to a cold start "Bag 1" (with the same standard speed profile) to estimate "excess" mass emissions [12]. Hot stabilized tailpipe emissions (after the car is warmed up) have been decreasing over time [13]. Because cold start emissions occur before the catalytic converter is sufficiently warm, these emissions remain relatively high, and thus may be contributing a larger share to total tailpipe emissions for a trip. According to the U.S. EPA, a cold start must be preceded by a soak time (no engine utilization) of ≥12 hours [12]. Therefore, we conducted a cold start measurement based on 12 hours or more of soak time. The incremental emissions associated with a cold start are estimated based on the difference between the average emission rate during the cold start versus the average emission rate during hot stabilized vehicle operation. For simplicity, the measurements and comparisons of cold and hot stabilized rates are based on engine operation for vehicle speed of zero.

RESULTS

Results are given for road grade, hot stabilized emission rates measured in the Raleigh/RTP area, cold start emissions, long distance trip activity, and energy use and emissions during the long distance trip. The sensitivity of fuel economy to trip length is discussed. The sensitivity of trip total emissions to cold start emissions is also discussed.

Road Grade

Road grade was quantified for driving from Raleigh, NC to Glen Burnie, MD over a distance of 303 miles. As illustrated in Figure 1, the average grade varied from -7.8 percent to +7.7 percent, with a 95-percent frequency range of -3.4 to +3.8 percent and a mean value of

0.2%. An example of road grades for a segment of the route, from Norwood Road at Macon Road in North Raleigh to Creedmoor Road (NC 50) just north of Royce Bridge Road is shown in Figure 2. This example represents a hilly area with grades ranging from approximately ±8%. The roads along this portion of the route are rural two lane, two way arterials with speed limits of 45 to 55 mph. The precision of the estimated grade was ±0.5% or better for 57.7 percent of the over 3,000 0.1-mile segments of the route, and was within ±1.0% for 82.1% of the segments. Road grade was similarly quantified for the Raleigh/RTP, NC routes used for local data activity, energy use, and emissions rates collection.

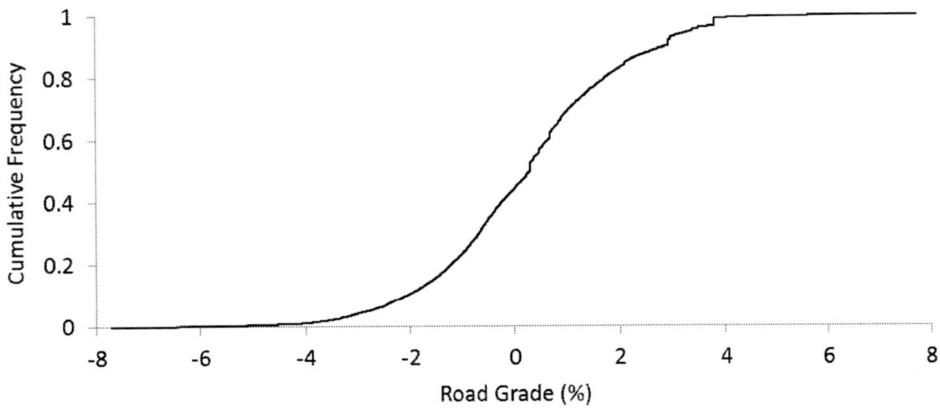

Figure 1. Cumulative distribution function (CDF) of variability in segment average road grade for n=3,030 segments of 0.1-mile segment length between Raleigh, NC and Glen Burnie, MD based on data from GPS receivers with barometric altimeters.

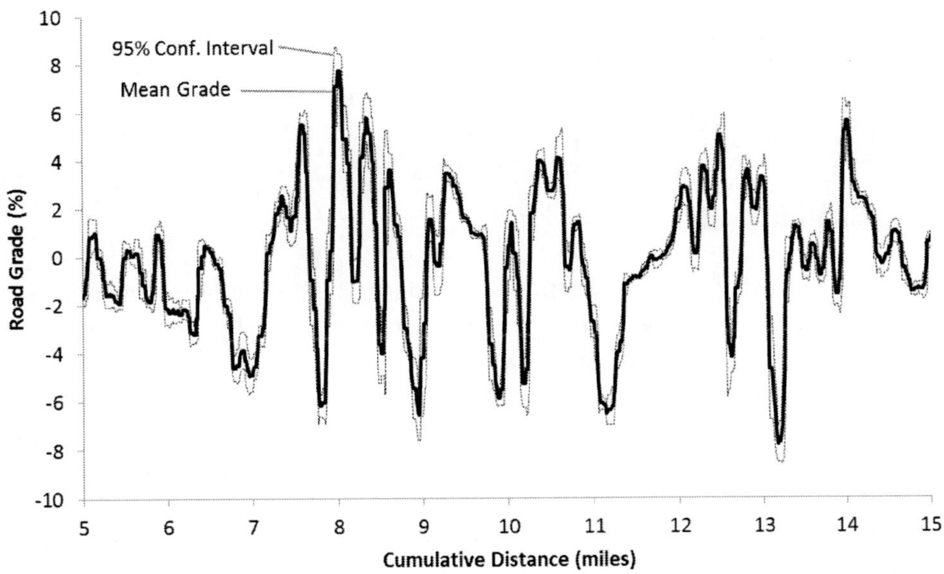

Figure 2. Illustrative 10-mile section of 0.1-mile segment length average grade and 95% confidence interval versus cumulative distance for 303-mile trip from Raleigh, NC to Glen Burnie, MD from Norwood Road at Macon Road in North Raleigh to Creedmoor Road (NC 50) approximately 3.7 miles south of Creedmoor, NC.

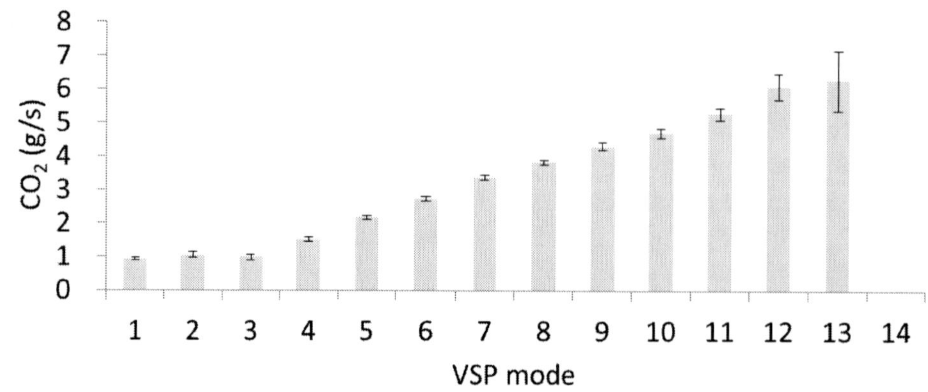

(a) Carbon Dioxide (CO_2)

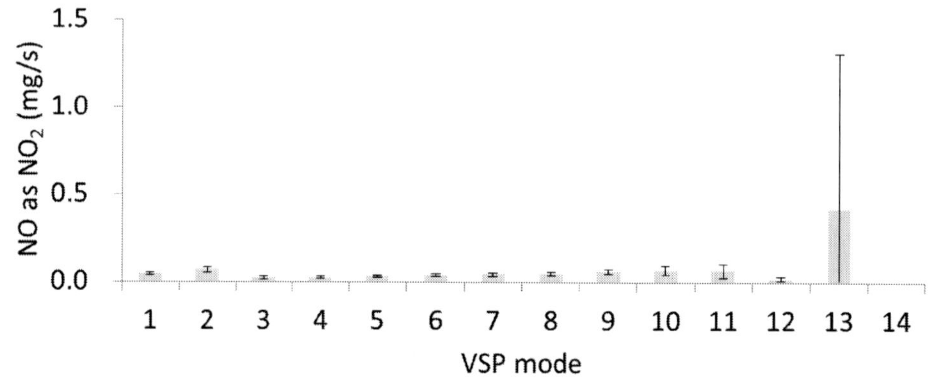

(b) Nitrogen Oxides (NO_x)

Figure 3. Modal Average Emissions Rates of CO_2 and NO_x Versus Vehicle Specific Power (VSP) Mode for 2013 Toyota Prius Plug-In Measured in Raleigh and Research Triangle Park, NC Area on January 22, 2013. (Engine on for 6,402 seconds. No observations in Mode 14.) Measured NO is reported as equivalent mass of NO_2.

Hot Stabilized Emissions

Hot stabilized exhaust gas concentrations were measured in the Raleigh/RTP, NC area and are the basis for estimating VSP modal emission rates for CO_2, CO, HC, and NO. The exhaust concentrations varied depending on VSP mode. For modes 4 to 14, which represent positive VSP, the modal average CO_2 concentration ranged from 12.2 to 14.4 volume percent. Hydrocarbon and carbon monoxide concentrations were below the gas analyzer detection limit. The nitric oxide concentration was below 2 ppm except for Mode 13, which averaged 11 ppm.

Mass per time emission rates were estimated for each pollutant for each mode. Examples of modal average emission rates, and their 95% confidence intervals, are given in Figure 3 for CO_2 and NO_x. These data are based on 6,402 seconds for which the engine was on. The lowest CO_2 emission rate is at VSP Mode 3, which includes idle. The CO_2 emission rates for VSP Modes 1 and 2, which are negative values of VSP, represent situations with no load on the power train. The emission rate increases monotonically from Mode 3 to Mode 13. For this particular day of data collection, there was no vehicle operation at VSP greater than 39 kW/ton, and thus there are no Mode 14 data. The range of the confidence intervals increases, particularly for VSP Modes 12 and 13, because the sample size is relatively small for these modes, at only 44 and 11 seconds, respectively. The NO_x emission rate is 0.070 mg/sec on average for all modes except Mode 13, for which the average rate is 0.42 mg/sec. However, because of the small sample size and large range of variation in emission rate among the small number of seconds of data in this mode, the 95% confidence interval on the mean is wide. Because the CO and HC exhaust concentrations are typically below the gas analyzer detection limits, there is not a clear trend in modal emission rates for these two pollutants.

The EPA certification test emission results for the 2013 Prius Plug-In are reported as 50 mg CO/mile and 6 g NO_x/mile or lower depending on the test cycle, with cold start emissions of 470 g CO/mile.

Cold Start Emissions

A cold start after a 12-hour engine soak period was measured during the morning of January 22, 2013 during the first engine start of the day. The vehicle activity and emissions for the cold start period are shown in Figure 4. The first engine start occurs approximately 380 seconds into the first trip made with the vehicle. After an initial brief period of engine RPM of approximately 2,500 RPM, the engine operates at approximately 1,500 RPM for approximately one minute, and then decreases. The vehicle speed is zero during the first five minutes of the cold start. During this period, the catalyst temperature increases from approximately ambient temperature to 400 °C. The engine coolant temperature also increases. Most of the cold start emissions take place within two minutes of the first engine start. There is a period of about one minute of excess CO_2 emissions associated with the elevated engine idle RPM. There are large but brief peaks in emissions of each of CO, HC, and NO during the first minute of operation. There appear to be secondary, but much smaller, peaks for both CO and HC.

The cold start increment for CO_2 is 70.3 grams, based on an average CO_2 emission rate of 2.18 g/sec during the first 50 seconds after the cold start compared to an average rate of 0.77 g/sec thereafter, while the vehicle speed is zero. The cold start increments for CO, HC, and NO are estimated as 460 mg, 220 mg, and 7.8 mg based on the average emission rate during the first two minutes after the cold start compared to thereafter, when vehicle speed is zero.

Long Distance Trip Activity

Details of the second-by-second vehicle activity for the 303-mile trip from Raleigh, NC to Glen Burnie, MD are given in Figure 5 for battery SOC, catalyst temperature, coolant temperature, speed, engine RPM, and fuel flow rate. These data are reported by the vehicle ECU. The traction battery SOC decreased during the first 2,000 seconds of data collection, reaching a steady state value of approximately 23%. The 2013 Prius Plug-In has an "EV" mode, during which grid power stored in the traction battery is available for use. However, this is not a purely "charge depletion" mode, since there are time periods during EV mode when some recharging of the traction battery occurs, particularly because of regenerative braking. For example, the battery power increased from 70.6% to 71.0% during a brief period of deceleration from 18 mph to 10 mph during EV mode. However, such episodes are infrequent and brief compared to the general trend of depletion of traction battery charge. Therefore, the first 2,000 seconds of data collection are categorized as charge depleting mode for the purpose of further analysis.

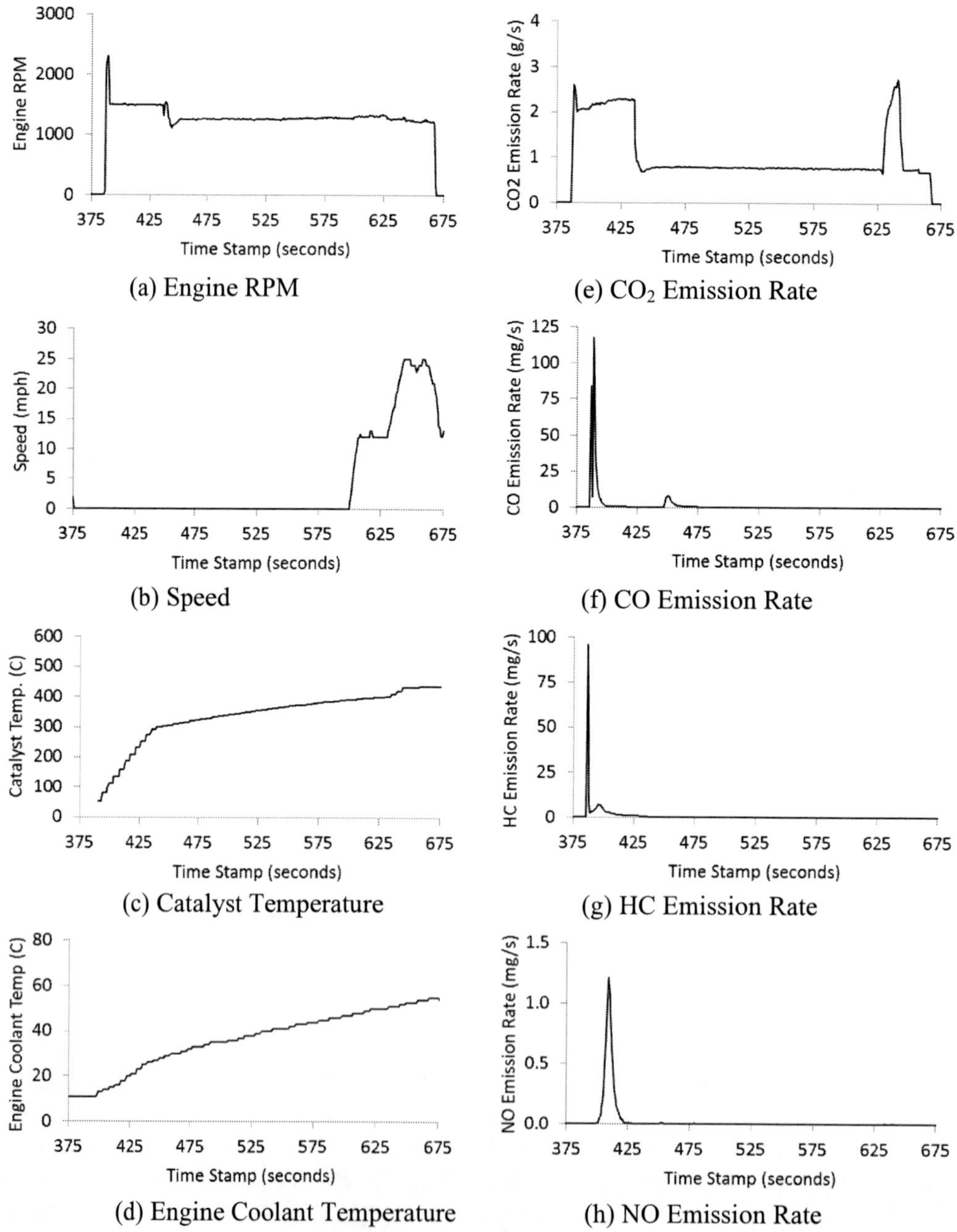

Figure 4. Second-by-Second Activity and Energy Use Data for 2013 Toyota Prius Plug-In During Cold Start on January 22, 2013 in Raleigh/RTP, NC Area.

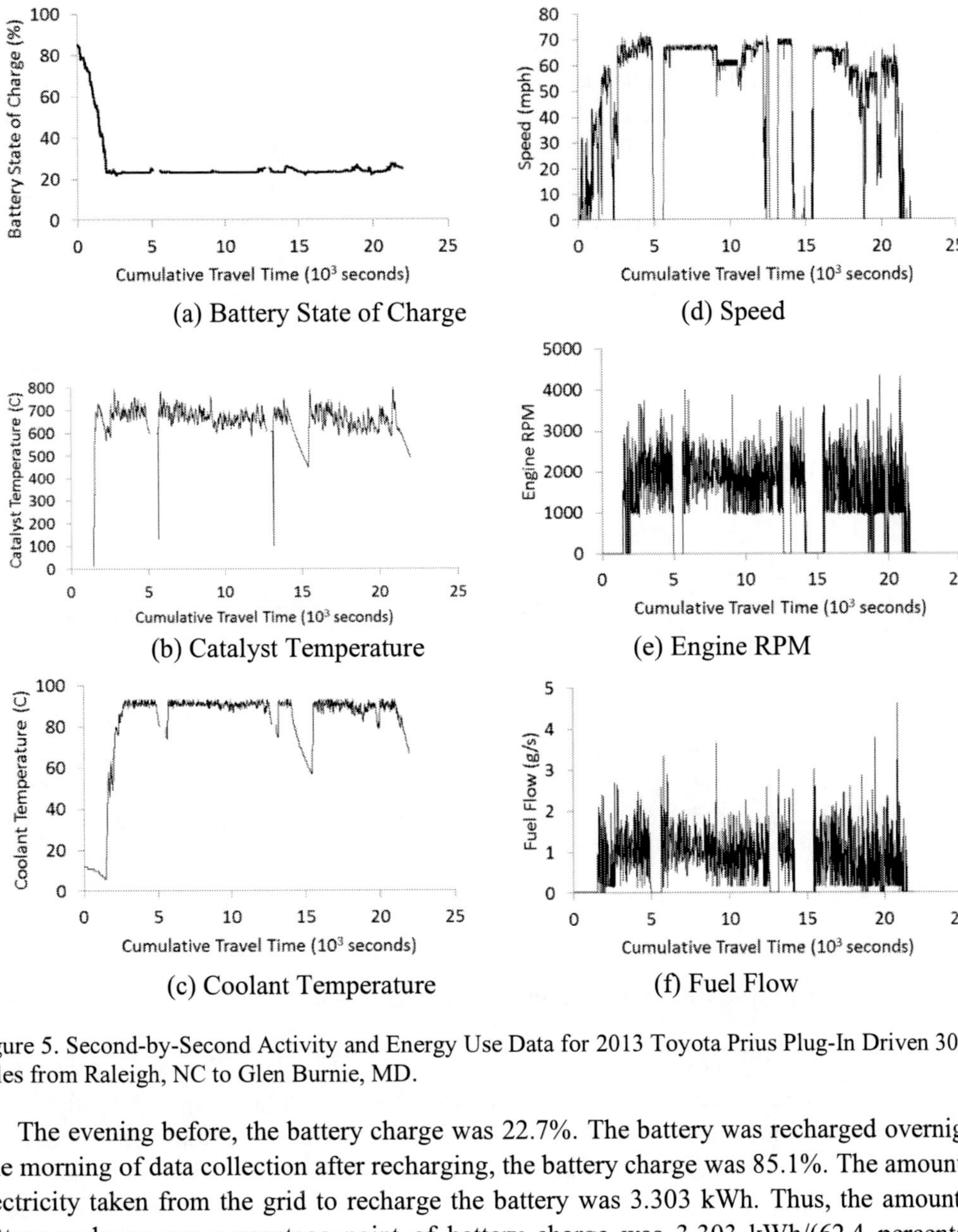

Figure 5. Second-by-Second Activity and Energy Use Data for 2013 Toyota Prius Plug-In Driven 303 miles from Raleigh, NC to Glen Burnie, MD.

The evening before, the battery charge was 22.7%. The battery was recharged overnight. The morning of data collection after recharging, the battery charge was 85.1%. The amount of electricity taken from the grid to recharge the battery was 3.303 kWh. Thus, the amount of battery recharge per percentage point of battery charge was 3.303 kWh/(62.4 percentage points) = 0.0529 kWh/%point.

During CD mode, there were 4 engine starts and shutdowns, with an average engine off duration of 426 seconds per period of engine off. In contrast, during charge sustaining mode, there were 30 engine shutdowns, with an average engine off duration of 137 seconds.

The first engine start did not occur until nearly 1,500 seconds into the trip. At that time, the ECU reported a catalyst temperature of 13 °C, which was approximately the ambient temperature at that time. Furthermore, the engine coolant temperature was reported at 6 °C at that same time. Thus, the engine effectively underwent a cold start during the first engine start. At the first engine start, the engine operated at approximately 1,500 RPM for the first minute, which is typical of the warmup pattern of conventional gasoline engines and similar to the cold start observed on January 22 for the same vehicle. Once the engine was warmed up, the ECU-reported catalyst temperature fluctuated from approximately 650 °C to 700 °C, and the coolant temperature maintained a narrow range typically between 88 °C to 93 °C.

There are three clearly discernible episodes during which both catalyst and coolant temperature dropped as a result of extended stops. At 5100 seconds into the trip, a 437-second rest stop was made during which the vehicle was shut off. Just prior to this stop, the vehicle was operated at reduced speed and low engine load. During this period, the engine shut off. Thus, there was some reduction in coolant and catalyst temperature before the vehicle was powered off. During the time period with no power, the coolant temperature dropped from 80 °C to 74 °C and the catalyst temperature dropped from 598 °C to 133 °C. Thus, this event lead to a partial cold start. Another rest stop was made at 12,738 seconds into the trip for a period of 286 seconds, during which the coolant temperature dropped from 81 °C to 75 °C and the catalyst temperature dropped from 608 °C to 96 °C. A third rest stop occurred at 14,221 seconds into the trip for 1167 seconds, during which the coolant temperature dropped from 81 °C to 57 °C and the catalyst temperature dropped from 690 °C to 470 °C. During the third stop, the vehicle was kept "on" and, thus, the ECU reported continuous data. During the first two stops, the vehicle power was turned off, and, thus, the ECU did not report data during those stops. The decrease in catalyst temperature during the three rest stops may have been substantial enough to result in a short period of excess emissions upon subsequent restart of the engine.

As evident from the speed versus time trace, the vehicle was mostly operated on interstate highways during the trip, with speeds of typically 60 to 70 mph. On major arterial roads, such as those leaving Raleigh or entering Glen Burnie, MD, the speeds were typically 50 to 60 mph. There was also driving on some feeder/collector streets, minor arterial roads, and ramps to and from highways. The engine RPM typically ranged from 1000 to 2500. Engine RPM of less than 500 was used as a criteria for inferring that the engine was off.

Long Distance Trip Energy Use and Emissions

The grid electricity power consumption is estimated based on the rate of change of battery SOC. Charge depletion occurred, with some minor exceptions, during the first 2000 minutes of operation, during which the vehicle was driven 15.0 miles and consumed approximately 0.097 gallons of fuel. This yields a raw fuel economy of 155 miles per gallon of fuel for these first 15.0 miles. However, the concurrent 62 percentage point net change in SOC implies that 3.28 kWh was consumed from the grid. According to Graver and Frey [4], one kWh of electricity is equivalent to 0.33 L of gasoline, taking into account U.S. national average power plant efficiency. Thus, the electricity consumed is equivalent to 1.08 liters (0.2859 gallons) of gasoline. Therefore, the overall equivalent fuel economy is only 39.2 miles per gallon (MPG). In contrast to reality, U.S. fuel economy ratings for plug-in vehicles are based only on the energy content of electricity delivered to the vehicle in terms of MPG_e, not the primary energy content consumed to produce the electricity delivered to the vehicle. Thus, during the 15.0 miles of driving in EV mode, the vehicle had 77.5 MPG_e.

For the entire 303-mile trip, the vehicle consumed 5.84 gallons of gasoline and had a net change in SOC of 60.4 percentage points, for net electricity consumption of 3.20 kWh. If the primary energy input to power plants needed to produce the electricity is taken into account, the fuel economy is 49.5 MPG. If only the energy content of the electricity delivered to the vehicle is taken into account, the overall fuel economy is 51.1 MPG.

In CS mode, the average rate of engine fuel use (when the engine is on) versus VSP mode increases for Modes 3 to 14, as shown in Figure 6, indicating generally increasing fuel use rate with increasing power demand. However, unlike conventional gasoline fueled vehicles for which average modal fuel use is actually monotonically increasing from Mode 3 to 14, for this vehicle the average modal fuel use rates for Modes 10 and 11 are lower than for Mode 9. This exception to the monotonic trend is likely a result of a much lower rate of battery discharge in Mode 9 compared to Modes 10 and 11 when the engine is on, as indicated in Figure 7. Therefore, the electrical portion of the hybrid system appears to be providing more

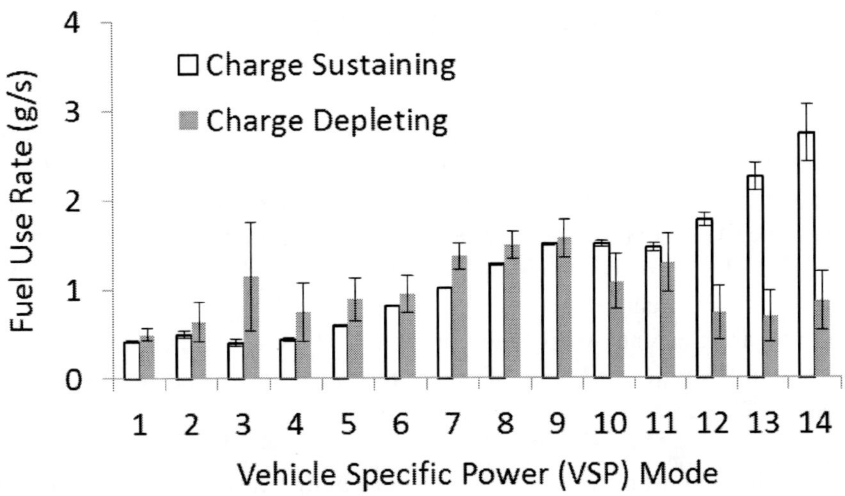

Figure 6. Engine fuel use rate versus Vehicle Specific Power (VSP) mode for charge depleting (n=1,958 seconds) and charge sustaining (n=16,346 seconds) modes for 2013 Prius Plug-In driven 303 miles from Raleigh, NC to Glen Burnie, MD.

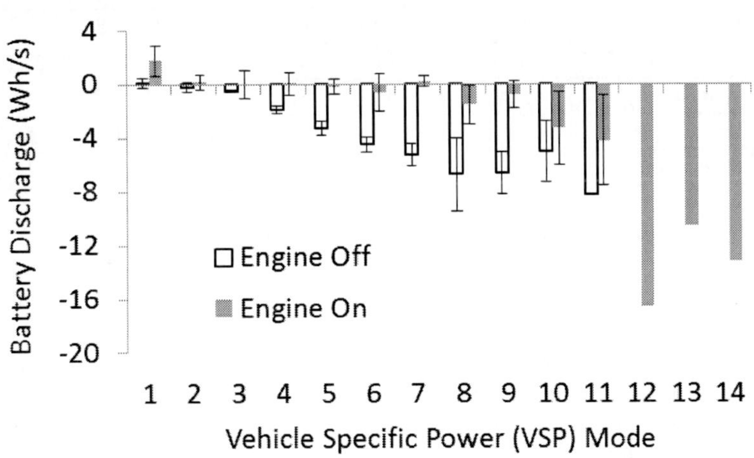

Figure 7. Battery discharge rate versus Vehicle Specific Power (VSP) mode for charge depleting mode for engine off (n=1703 seconds) and engine on (n=255 seconds) modes for 2013 Prius Plug-In driven 303 miles from Raleigh, NC to Glen Burnie, MD.

power assist in modes 10 and 11 than in Mode 9, thereby reducing power demand on the engine itself.

For CD mode, the average fuel use rate is much lower at Modes 12 through 14 than for intermediate power demand modes such as Modes 7 through 11. The concurrent average rate of battery discharge for VSP Modes 12 through 14, ranging from approximately 8 to 16 Wh/s during CD mode when the engine is off, is much higher than for any other VSP modes. Thus, as expected during CD mode, the total power demand is met in part by the traction battery and electrical motor and in part by the engine. Furthermore, these results are consistent with advice provided in the 2013 Prius Plug-In user manual to the effect that electrical power consumption is substantially higher at high speeds compared to lower speeds. Thus, drivers who wish to extend the range of EV driving mode are likely to be more successful in doing so if driving speed is kept below 35 or 40 mph. During data collection, the vehicle was operated in EV mode at speeds of up to 60 mph. The rate of battery discharge rate generally increased with increasing speed.

Based on VSP modal emission rates measured locally in the Raleigh/RTP area and the number of seconds of engine-on operation in each mode on the Raleigh, NC to Glen Burnie, MD route, the estimated tailpipe emission rates are 2.5 mg NO_x/mile, 10 mg HC/mile, and 13 mg CO/mile. The CO_2 tailpipe emission rate averages 170 g/mile.

The trip total fuel use and emissions for the vehicle and for grid-based electricity are shown in Table 2. The vehicle fuel use and emission totals reported here are based on the number of seconds of engine on operation in each VSP mode multiplied by the VSP modal average rate from field measurements of the same vehicle in the Raleigh, NC and Research Triangle Park, NC area.

Table 2. Estimated total fuel use and emissions and estimated mileage-based rates from vehicle internal combustion engine and electricity from grid for 2013 Toyota Prius Plug-In driven 303 miles from Raleigh, NC to Glen Burnie, MD

Source or Description	Units	Fuel	NO_x	HC	CO	CO_2	SO_2
Vehicle Engine [a]	grams	16,200	0.7	3.1	3.8	51,300	0.0
Grid (Power Generation)[b]	grams	800	3.1	0.0	0.3	2,200	10.1
Total (Direct + Indirect)	grams	17,000	3.8	3.1	4.2	53,500	10.1
Total, per mile	g/mile	56.1	0.013	0.010	0.014	177	0.033
Fraction from electricity		0.05	0.81	0.01	0.08	0.04	1.00

[a] Direct fuel use based on mass reported consumed by electronic control unit. Emission rates for NO_x, HC, and CO estimated based on VSP modal emission rates measured in Raleigh/RTP, NC region weighted by number of seconds of engine on operation in each mode for the Raleigh, NC to Glen Burnie, MD trip. CO_2 emission rate estimated based on 86.4 volume percent carbon in gasoline. SO_2 emissions (0.003 grams) estimated assuming 30 ppm by weight of sulfur in gasoline.

[b] Estimated based on electricity consumed to recharge traction battery, national average power plant heating value, typical gasoline heating value (for equivalent fuel mass), and SERC region average power pool emission rates (668 g CO_2/kWh, 939 mg NO_x/kWh, 3.05 g SO_2/kWh, 103 mg CO/kWh, and 10 mg HC/kWh) [4].

The grid fuel use is estimated based on the amount of electricity delivered to the vehicle from the grid to recharge the traction battery taking into account the U.S. national average thermal power plant heat rate (ratio of power plant energy input to electricity output) and a typical electrical transmission efficiency [4]. The grid emission rates are based on power pool average emission rates for the southeastern U.S. as specified in the footnote to Table 2. The overall fuel use rate of 56.1 grams of fuel equivalent per mile equates to approximately 50 MPG for the entire 303-mile trip. Although the contribution of the grid to fuel use, CO_2 emissions, CO emissions, and HC emissions are relatively low at 8 percent or less of the totals for the vehicle and the grid combined, the grid contributes the vast majority of total NO_x emissions and nearly all of the SO_2 emissions.

For the 303-mile long trip, the contribution of cold start incremental emissions to the trip total is conservatively assuming that there were four cold starts during the trip, including the actual cold start early in the trip and three partial cold starts after rest stops during the trip. On this basis, the contribution of cold starts to total emissions is approximately 0.8% for NO_x, 22% for HC, 30% for CO, and 0.5% for CO_2. For just one cold start for the 303-mile trip, cold starts would contribute 6% and 10% to HC and CO trip emissions, respectively. However, for a shorter trip, the contribution of cold starts to trip total emissions would be much higher. For example, the 303-mile trip includes a segment from Raleigh, NC to Creedmoor, NC, a distance of 19.1 miles that took 2321 seconds. For this shorter trip, a cold start would contribute approximately 70 percent of the trip HC emissions and 50 percent of the trip CO emissions, taking into account vehicle tailpipe and grid emissions. The 19.1-mile trip has an average gasoline only fuel economy of 108 MPG and a gasoline and electricity overall fuel economy of 41 MPG, if power plant efficiency is taken into account.

CONCLUSION

A set of methods for quantifying the real-world activity, energy use, and emissions of a production PHEV are demonstrated in this chapter. Vehicle activity is quantified in terms of 1 Hz vehicle speed, acceleration, road grade, and parameters such as battery SOC, engine coolant temperature, catalyst temperature and others. Vehicle energy use is quantified in terms of 1 Hz rate of gasoline fuel use by the ICE and rate of electrical discharge of the TB. Emissions are quantified in terms of 1 Hz mass per time tailpipe emission rates of CO_2, CO, HC, and NO. Fuel use, battery discharge rate, and emission rates were found to be a function of vehicle power demand quantified in terms of vehicle specific power, which is a function of speed, acceleration, and road grade. Road grade was estimated based on GPS receivers with barometric altimeters for each 0.1-mile segment along each route driving by the vehicle. All of these data were collected during real-world operation and thus represent real world driving conditions.

The hot stabilized tailpipe emission rates, which occur after the catalyst and engine have reached a steady-state operating temperature, were found to be low for CO, HC, and NO_x and comparable to or less than certification emission test results for the same year, make, and model of vehicle. However, the first engine start, if the engine and catalyst were at ambient temperature, was found to produce a substantial increment of cold start emissions. Furthermore, the cold start duration was approximately one to three minutes. Periods of no engine activity while the vehicle is not in motion, even if the power to the vehicle is on, can lead to a decrease in catalyst temperature. Thus, it is possible that there may be partial cold starts that are part of an ongoing trip, particularly for long distance trips that involve rest stops. However, the engine catalyst temperature appears to remain high during periods of engine off operation while the vehicle is in motion, as long as engine restarts occur within a few minutes of the previous engine shutdown.

Cold starts can contribute a large percentage of trip emissions, especially for short trips. For example, for a 19-mile trip, a cold start is estimated to contribute half or more of the trip emissions of CO and HC. However, for a long distance trip of several hundred miles, the contribution of the cold start becomes smaller.

Grid-based primary energy use can contribute a large portion of emissions, especially for NO_x and SO_2, based on the typical electric generation power mix in the southeastern U.S. Although not quantified here, it is also expected that primary energy use for electricity to recharge the TB will also lead to a large share of total particulate matter emissions for a gasoline-based PHEV. The actual grid-based emissions are likely to depend on time of day of recharging. For example, electricity demand for recharging overnight, as was the case in these measurements, might likely be met by baseload power generation capacity. For the southeastern U.S., coal-fired and nuclear power plants are typically used for baseload power generation. In contrast, the marginal electricity demand associated with recharging during the day might be met with intermediate or peaking capacity, which could be natural gas-fired combined cycle or simple cycle gas turbines, respectively. Depending on the region of the U.S., hydroelectric power, and to a much lesser extent wind and solar power, may be a portion of the total electricity generation mix. Fossil fuel-fired electric power plants typically have low CO and HC emission rates. Thus, in some states or regions of the U.S., the primary energy emissions of CO_2, NO_x, SO_2, and PM could be higher (or lower) if there is a larger (or smaller) share of coal-based capacity.

For the measured 2013 Toyota Prius Plug-In, the EV range is sensitive to vehicle speed and regenerative braking. EV mode lasted for 15.0 miles for the long-distance trip from Raleigh, NC to Glen Burnie, MD. There were competing factors that influenced this range. One was that it was necessary to operate on a route that included roads with speed limits of 45 to 55 mph, and given prevailing traffic it was necessary to maintain such speeds. The TB electricity discharge rate is substantially higher at high speed compared to lower speeds of less than 40 mph. Furthermore, the portion of the route during EV operation included hills with positive grades as high as 7 percent, which impose significant power demand for hill climbing. However, conversely, when traveling on the down-slope side of each hill, there was a significant opportunity for regenerative braking. When the engine came on, it was used to power the vehicle up the positive slope of the hill. However, upon reaching the crest of the hill, the accelerator position was reduced, leading to an engine shutdown followed by regenerative braking on the down-slope of the hill.

Fuel economy estimates now in use in the U.S. for PHEVs are misleading because they do not account for primary energy consumed to generate electricity. Thus, PHEV operators have a misleading sense of high fuel economy. The current fuel economy standard used in the U.S. for PHEVs should be revised to take into account the energy content of primary energy used to generate electricity. Furthermore, because fuel economy can be sensitive to travel distance between recharging, fuel economy should be reported based on standardized travel distances such as 10, 20, 40, and 100 miles. In the case of the 2013 Prius Plug-In, a distance of 10 miles could be driven entirely in EV mode. However, for longer distances, the proportion of time of engine operation and the proportion of total power demand that must be met by the engine will increase.

ACKNOWLEDGMENTS

This material is based upon work supported by the National Science Foundation under Grant No. CBET- 0853766. Any opinions, findings, and conclusions or recommendations expressed in this material are those of the authors and do not necessarily reflect the views of the National Science Foundation.

REFERENCES

[1] Bradley, T. H.; Frank, A. A. Design (2009), "Demonstrations and Sustainability Impact Assessments for Plug-in Hybrid Electric Vehicles," *Renewable Sustainable Energy Rev.* 13(1):104-117; DOI 10.1016/j.rser.2007.05.003.

[2] SAE (1999), Recommended Practice for Measuring the Exhaust Emissions and Fuel Economy of Hybrid-Electric Vehicles; SAE J1711; Society of Automotive Engineers, Inc.: Warrendale, PA, 1999.

[3] INL. North American PHEV Demonstration Fleet Summary Report: Hymotion Prius (V2Green data logger); U.S. Department of Energy, Idaho National Laboratory: Idaho Falls, ID, 2010.

[4] Graver, B. M, H. C. Frey, and H. W. Choi (2011), "In-Use Measurement of the Activity, Energy Use, and Emissions of a Plug-in Hybrid Electric Vehicle," *Environmental Science and Technology*, 45(20):9044-9051.

[5] Choi, H. W., and H. C. Frey (2010), "Method for In-Use Measurement and Evaluation of the Activity, Fuel Use, Electricity Use, and Emissions of a Plug-in Hybrid Diesel-Electric School Bus," *Environmental Science and Technology*, 44(9):3601-3607.

[6] Battelle (2003). Environmental Technology Verification Report: Clean Air Technologies International, Inc. REMOTE On-Board Emissions Monitor; prepared for the U.S. Environmental Protection Agency by Battelle: Columbus, OH, June.

[7] Frey, H. C., K. Zhang, and N. M. Rouphail (2008), "Fuel Use and Emissions Comparisons for Alternative Routes, Time of Day, Road Grade, and Vehicles Based on In-Use Measurements," *Environmental Science and Technology*, 42(7):2483–2489.

[8] Sandhu, G. S., and H. C. Frey (2013), "Effects of Errors on Vehicle Emission Rates from Portable Emissions Measurement Systems," *Transportation Research Record*, in press.

[9] Yazdani, B., H. C. Frey, and G. S. Sandhu (2013), "In-Vehicle Road Grade Measurement Using Stand-Alone GPS with Barometric Altimeter," *ASCE Journal of Transportation Engineering*, 139(6):605-611.

[10] Jiménez-Palacios, José Luis (1999), Understanding and quantifying motor vehicle emissions with vehicle specific power and TILDAS remote sensing, PhD Dissertation, Massachusetts Institute of Technology, Cambridge, MA.

[11] Frey, H. C., A. Unal, J. Chen, S. Li, and C. Xuan, Methodology for Developing Modal Emission Rates for EPA's Multi-Scale Motor Vehicle and Equipment Emission Estimation System, EPA420-R-02-027, Prepared by North Carolina State University for the Office of Transportation and Air Quality, U.S. Environmental Protection Agency, Ann Arbor, MI, October 2002.

[12] EPA (2001), "Soak Length Activity Factors for Start Emissions," EPA 420-R-01-011, Office of Transportation and Air Quality, U.S. Environmental Protection Agency, Ann Arbor, MI.

[13] Bishop, G. A., and D. H. Stedman (2008), "A Decade of On-Road Emissions Measurements," *Environmental Science and Technology*, 42(5):1651-1656.

In: Grid Electrified Vehicles
Editor: Carla Alexandra Monteiro da Silva

ISBN: 978-1-62808-839-7

Chapter 10

REAL-LIFE COMPARISON BETWEEN DIESEL AND ELECTRIC CAR ENERGY CONSUMPTION

***J. Martins*[1,*], *F. P. Brito*[1], *D. Pedrosa*[2], *V. Monteiro*[2] *and João L. Afonso*[2]**

[1]Dept. Mechanical Eng., University of Minho, Minho, Portugal
[2]Centro Algoritmi, University of Minho, Minho, Portugal

ABSTRACT

Vehicle electrification is one of the strategies with higher potential for increasing the efficiency of vehicle powertrains, reducing the dependency on dwindling fossil fuel sources and meeting stringent emissions targets set by policy makers. Despite all the theoretical assessments and manufacturer's claimed efficiency and emissions records of current vehicles, there is a lack of data concerning real life comparisons of Electric Vehicles (EV) against Internal Combustion Engine (ICE) cars.

A test program comparing the energy consumption of an EV and a diesel powered (ICE) car was carried out.

Both short (at levelled ground and 6% up-hill) and long distance tests were performed for several fixed vehicle speeds. Measurements enabling the assessment of average energy consumption, required power and energy suplied were performed for both vehicles.

Results indicate that in terms of vehicle use (Tank-to-Wheel perspective) the electric powertrain is significantly more energy efficient than the Diesel powertrain, although the difference between the two is less pronounced for higher power events.

Keywords: Electric Vehicle, Energy Efficiency, Energy Consumption

* Corresponding author: J. Martins. Campus de Azurém, 4800-058 Guimarães, Portugal. +351 253 510 235; E-mail: jmartins@dem.uminho.pt.

INTRODUCTION

Energy and Environmental Concerns and Policies

The expansion of car traffic in the last decades has been arising serious environmental concerns over the increase of greenhouse gas (GHG) emissions, urban air pollution and dependence on dwindling fossil fuel reserves.

The oil crises of the seventies, along with the concerns over the volatility of oil prices and the security of fuel supply from politically unstable countries has further spurred the quest for increased energy efficiency and low emission alternatives to conventional fossil fuel-based powertrains [1, 2].

Several national and international agreements and policies have been put forth in order to place even more stringent limits on fuel consumption and/or CO2 emissions of vehicles. Such was the pioneering Kyoto Protocol at the global level or the Corporate Average Fuel Economy (CAFE) regulations in the US [3].

In the European Union voluntary emissions agreements celebrated between the European Commission and institutions like the European Automobile Manufacturers Association (ACEA) were later followed by mandatory regulations on emission performance standards for new passenger cars [4].

Increasing the Energy Efficiency of Conventional Powertrains

A first approach for decreasing vehicle emissions is by increasing the efficiency of the Internal Combustion Engine (ICE). Engine downsizing and turbo-charging are strategies that have proved to be effective in this respect [5].

Also the strategy of overexpansion studied by the group, especially if combined with variable compression ratio and variable valve timing, has a good efficiency potential, as efficiencies higher than those typical of ICEs may be obtained with a gasoline engine over a broad portion of the engine map [6, 7].

There are also the strategies which increase the efficiency of the powertrain by harvesting some of the thermal energy normally wasted through the tailpipe of vehicles and converting it into electricity to reduce alternator use. This can be done namely with Organic Rankine Cycle (ORC) turbines or Seebeck effect thermoelectric generators. The authors has been working in some of these promising technologies [8, 9] but they still need to be further developed before they become mature for vehicle use.

Despite the promising advances in fuel economy technologies it is expected that the automotive market will grow steadily in the next decades, mainly because of increasing standards of living in rapidly developing economies, such as South Eastern Asia and South America. So, between 2000 and 2050 it is expected that the population will grow 1.7 times, but the number of cars is expected to increase even more (3.6 times) [10].

So, it seems that the sole increase of conventional powertrain efficiency will not be sufficient by itself to meet the current medium to long term emission and efficiency goals. Alternatives not relying exclusively on the combustion of fossil fuels need to be considered.

Improving Efficiency through Vehicle Electrification

An increasingly higher degree of vehicle electrification (also called vehicle hybridization) is one of the options which seems logical to take in order to meet the aforementioned targets and diversify energy sources. Electrified vehicles are more efficient than ICE ones namely under urban traffic since they have no idling losses, no unefficient clutching at starts (they have good low end torque) and they can recover a portion of the braking energy through regenerative braking [1]. In fact, hybridization has been quite an effective way of some sports vehicle brands to meet emission standards without sacrificing the performance figures of their vehicles. For instance, the soon-to–be-sold flagship model of Porsche, the 918 Spyder, will be a plug-in hybrid capable of a combined power of 652 kW (887 hp), attaining 0-100 km/h in under 2.8 s. However, its New European Driving Cycle (NEDC) consumption has been rated at only 3.3 L/100 km with CO2 emissions of just 79 g/km [11].

Well-To-Wheel Approaches

Of course, the aforementioned emission and consumption figures require further analysis, since they only reflect the gasoline's consumption and emissions during vehicle usage. The electric energy charged from the power grid into the vehicle's batteries will also have associated emissions at the production, transportation and distribution stage. Even if only the emissions during vehicle usage are the basis for automotive industry mandatory goal compliance, a Well-to-Wheel (WTW) energy efficiency and emissions computation should be used if the full record of primary energy consumption and CO2 emissions of vehicle Life Cycle must to be quantified [12]. The WTW analysis may be divided into Well-to-Tank (WTT) and Tank-to-Wheel (TTW) phases.

Well-to-Tank Phase

The energy spent to extract, refine and supply gasoline and diesel fuels in Europe is around 0.08 and 0.10 MJ/MJ, respectively, which yields WTT energy efficiencies of 93 and 91%, respectively [13].

In EVs the WTT acronym is used analogically of course, meaning the path from the primary energy source down to the vehicle's energy storage system. While the energy conversion efficiency of hydroelectric dams can surpass 90%, the efficiency of the most advanced combined cycle plant (gas + steam turbine) in the world hardly surpasses 60% (typical efficiencies for new facilities is 55%), while modern coal fired steam turbine plants fall within an average 44% efficiency [13]. There is also the efficiency of electric energy transportation, around 92% [14].

Regions with clean electricity production, namely from renewable sources (as in the case of Portugal), will potentiate the global emissions cut obtained with vehicle electrification [15]. Of course, these values would not be as intense in regions which display poor electrical production efficiency or which heavily rely on carbon-intense power sources [13].

In the case of Portugal's electricity production in 2012 (provider EDP Universal, market leader) the emissions linked with electricity production averaged 229 g/kWh (40% from wind energy). During that year the monthly average roughly varied between 100 and 300 g/kWh, with the lowest record having been obtained for peak wind energy production and vice versa [16]. In comparison, the EU-mix calculations made with 1999 data (before heavy investment in low carbon technologies gained momentum) held overall energy conversion efficiencies around 35% and emissions of 430 g/kWh of electricity produced [13]. The EU study found that only emissions in excess of 850 – 900 g/kWh would cause the EV to be worse than an ICE car. Also, WTW GHG emission savings of up to 58% would be possible to obtain with an EV instead of an ICE car for the EU-mix emission conditions [13].

Tank-to-Wheel Phase

Despite the importance of the complete WTW analysis for a global evaluation of the merits and drawbacks of vehicle electrification, the scope of the present work has been reduced to the TTW measurements of the energy consumption related to vehicle use, that is, the energy which has to be supplied to the vehicle by the fuel pump / electric charging station in order to complete a given driving schedule, or the average power needed to cruise at specified driving conditions of speed and road slope. Other studies have focused on the global WTW analysis of plug-in vehicles [12, 13, 15, 17, 18].

The efficiency comparison between an ICE car and an EV is not straightforward. In each vehicle type a chain of processes of energy supply and conversion take place until the energy supplied to the vehicle is effectively used to move the vehicle. Each one of these processes has a given energy efficiency which is not always easy to quantify as it depends on various factors, some of which are very difficult to quantify.

Supply and Charging Efficiency

ICE cars have nearly 100% energy supply efficiency since almost no fuel is lost during refueling (especially in the case of diesel which is non-volatile), nor does the fuel lose any energy content in the process. The same is not true for EV, since there are energy losses in the process of recharging the batteries, mainly due to the charger and to the internal resistance of the battery cells, around 89% and 94%, respectively [14]. But surely these values will vary significantly according to the charging mode (slow / fast) and the battery chemistry [14].

Energy Conversion at the Battery

In an EV the energy conversion efficiency of batteries under discharge depends on factors such as the battery State-of-Charge (SOC), the discharge rate, the battery State-of-Health (SOH) or the battery temperature. This temperature is affected by the ambient temperature and also increases during both the charging and discharging processes, even within nominal conditions of voltage and current [19]. However, the problem of the battery temperature increase is more critical during the batteries discharging process, mainly because greater discharge rates might occur and considering that this process is strongly dependent of the driver behavior [20].

The high temperatures on lithium batteries cause degradation during their lifetime, and the cold temperatures reduce their power and energy output, limiting the performance and the EV driving range autonomy [19]. In order to ensure that the batteries can operate in a broad range of discharging rates and external temperatures, it is essential to use thermal management systems for the batteries, to increase the batteries efficiency, safety and lifetime [19].

Electric Motor and Controller

Regarding the EV powertrain, due to the technological advances in power electronics and electric motors powertrain efficiency figures are now generally above 89%. This comprises the combined efficiencies of the power controller and the motor [14]. Like with the ICE car, the efficiency of the electric motors will be mainly affected by the position in the torque and speed map. High speeds will result in high mechanical losses, while high torques will require high currents, enhancing the losses due to Joule effect [22].

ICE Engine

ICE cars typically display much lower energy efficiency than electric motors when converting the incoming energy (chemical energy contained in the fuel / battery) into traction mechanical energy through the motor / engine. The efficiency of thermal engines is limited to a theoretical maximum (the Carnot efficiency, around 60%) which is much lower than typical electric motor efficiency (which has no theoretical limit). The engine efficiency is very dependent on the region of the engine map is used during driving. Typically, efficiency will top near full load conditions (accelerator pedal fully pressed) and moderate to medium engine speeds. For these conditions a gasoline engine will display efficiencies around 33%, while diesel engines often surpass 40% [21]. However, average efficiencies will fall significantly, especially under urban driving or high performance driving. Some sources cite an average efficiency ranging from 12 to 20% [18]. Specific emissions of Diesel fuel are around 264 gCO_2/kWh, with the volumetric energy content of Diesel (the so-called Lower Heating Value) being 36 MJ/L or 9975 Wh/L. The better the efficiency of the vehicle, the lower will be the CO_2 emissions for each km driven. Then, the emissions for a given trip will be calculated simply by multiplying the specific emissions of Diesel fuel (in gCO_2/kWh) by the energy spent during that trip (in Wh) [21].

Transmission

Most Electric cars have a fixed transmission relation, since electric motors have high torque even at zero speeds. ICE cars, on the contrary, need a multi-speed transmission. Nevertheless, the use of a multi speed gearbox in EVs might be useful to reduce energy loss due to high motor speed and to obtain a torque range which is more flexible. For instance, a small motor should use high reduction ratios to enable an acceptable starting torque.

However, this would reduce the vehicle's top speed. Overall transmission efficiency is normally quite high, around 98% [21].

Approach of This Chapter

There are several factors which affect the overall efficiency of vehicles. Official consumption figures are often made for unreal driving schedules such as the NEDC. Also, OEMs do not frequently disclose a lot of important information that is needed for suitably evaluating the consumption performance of vehicles. Therefore, this chapter tries to address this lack of knowledge concerning real world driving. Two similar car models with Diesel and full electric powertrain have been carefully monitored during charging and for a set of driving conditions / schedules. Consumption figures have been compared and discussed.

Vehicles

There are already some electrified vehicles available in the market. Parallel hybrids such as the Toyota Prius or Series Hybrids (also called Extended Range Electric Vehicles, EREVs) such as the Opel Ampera (Chevrolet Volt in the US). However, usually the name "electric vehicle" is used for the full Electric Vehicle (EV) and in this category there are already some mass produced cars such as the Nissan Leaf which already sold over 60,000 vehicles worldwide, or the sporty and exclusive Tesla Roadster. Particularly, one OEM (Renault) has a full range of electric cars called ZE (Zero Emmission), ranging from the small Twizy to the commercial Kangoo. In the segment of family cars there is also the Zoe and the Fluence ZE. This latter model was the one chosen for this work, having been kindly provided by Renault Portugal, which also supplied a diesel (ICE) car with similar characteristics (Renault Megane).

The characteristics of both the Electric and ICE cars are presented in Table 1. The EV is longer than the ICE car, so as to compensate the extra space needed for batteries in the boot space. The philosophy of Renault is somehow different from the other manufacturers who distribute the batteries throughout the car in order to optimize space. Renault has interchangeable batteries, which means that the batteries have to be packed in a "box" that should be easily replaced by a system like the one proposed by the Company BetterPlace. The batteries are placed between the rear seats and the boot, "robbing" a significant boot space.

Comparing both cars, the electric has a much higher weight (23%) as a result of the added weight of the batteries and has less power (10%). Therefore it is not surprising that it is a slower vehicle, both in maximum speed and in acceleration, as seen in Table 1. But there is another reason for these differences. The EV has only one gear while the ICE car has a 6-speed gear-box. Although the torque of an ICE is much lower than the torque of an electric motor with similar power, having gears allows it to accelerate faster (in 1st and 2nd gears) and to reach higher speeds (in 5th or 6th).

But the largest difference between the vehicles is in terms of the energy that can be stored in each of them, which is 30 times higher on the ICE car, but leading only to 7.4 times higher range (an indication of the lower efficiency of the Diesel powertrain).

Table 1. Characteristics of the EV and ICE Car

Characteristics	EV	ICE (diesel) Car
Vehicle mass (kg)	1605	1307
Dimensions (mm)	4748x2041x1458	4559x1804x1469
Engine Type	Synchronous Motor Rotor Winding	Turbocharged Diesel common-rail
Power (kW)	70	77
Gear Box	Direct drive (1 speed)	6 speed
Storage Energy Type	Lithium-ion battery	Diesel fuel
Stored Energy (kWh)	22	701.4
Range (km)	185	1363 (combined)
Maximum Speed (km/h)	135	190
Acceleration 0-100 km/h (s)	13.7	11.9
Rated Fuel consumption (L/100 km)	-	4.5
Rated Carbon dioxide emissions (g/km)	-	120

The stored energy and ultimately the range is still the Achilles heel of EVs and the Renault Fluence ZE is no exception. In fact, we never saw a predicted range of more than 110 km, and that happened during the slow trips (80 and 90 km/h).

TESTS

Road Tests

Three types of energy consumption tests were performed, all of them at constant speed. One test typology consisted of horizontal short distance trips (2 km) carried out for a broad range of fixed speeds (40 – 120 km/h). Another test typology was a long distance round trip (around 120 km total) made mainly on motorways, with intermediate recharging (prior to the return trip) and made for a more limited constant speed range (80 – 120 km/h). A last test typology was extracted from the long distance trips, consisting on the 6% up-hill sections of these trips, analyzed separately. Not all the tests were possible to perform with both vehicle types.

The speed was kept constant by activating the cruise control of the vehicles. Measurements of average power supplied to the motor controller (kW), as well as total energy supplied to the powertrain for the whole test (kWh/100 km) were recorded for the EV.

Average power for each test was calculated not only from the instantaneous power measurements but also from the total trip energy consumption. This parameter was converted into average trip power. Therefore, the average power plots were obtained from an average of the power and energy measurements in order to decrease the error.

For the ICE car the cumulative fuel consumption (L/100 km) was recorded for each test. To allow the comparison of the ICE and the electric powertrains, the power and energy supplied to the ICE powertrain were calculated from the fuel consumption.

This was made by calculating the thermal energy (power) associated with the mass (mass flow rate) of fuel supplied using the Lower Heating Value (43 MJ/kg 0.835 kg/L) of the diesel fuel. Therefore, the energy (power) used in this work for the ICE car corresponds to the energy (power) released by the fuel (only a fraction will effectively be converted into useful work or power). In order to control the fuel consumption of the vehicle, its tank was fully filled with fuel before and after each round trip.

Battery Charging Monitoring

The battery charging process was monitored with a FLUKE 435 Power Quality Analyzer. This equipment was programmed to register every 1 minute the main parameters related to the battery charging process, such as the RMS values of the power grid voltage and current, the absorbed active power and the energy consumed during the battery charging process. To perform a complete analysis of the battery charging process other parameters were also recorded such as power grid voltage and current harmonics, frequency, and total power factor.

Figure 1 shows the outline of the setup used to monitor the battery charging process. This figure also presents different parts inside the vehicle, as the on-board battery charger, the interconnection box and the batteries. As shown, the battery charger converts the power grid voltage from 230 V (AC) to 400 V (DC), and the interconnection box makes the connections between the battery charger and the batteries. Figure 2 shows the setup used to monitor the battery charging process. In this figure the FLUKE 435 Power Quality Analyzer, the FLUKE current probe and the Electric Vehicle Supply Equipment (EVSE) of the Renault Fluence ZE can be seen.

To charge the batteries the EV was plugged to the power grid using a Type 1 (SAE J1772-2009) connection of the international standard IEC 62196. When charging the vehicle from domestic electric sockets, it is necessary to use an EVSE cable because of the charging mode 3 used in this international standard. This mode is used in single-phase systems with maximum allowed voltage of 250 V (AC) and maximum allowed current of 32 A (80 A in the United States).

Figure 3 shows the results during a random battery charging process. Namely, Figure 3 (a) displays the power grid voltage and current, while the measured powers (active, apparent and reactive powers), the power factor and the RMS values of current and voltage are shown in Figure 3 (b).

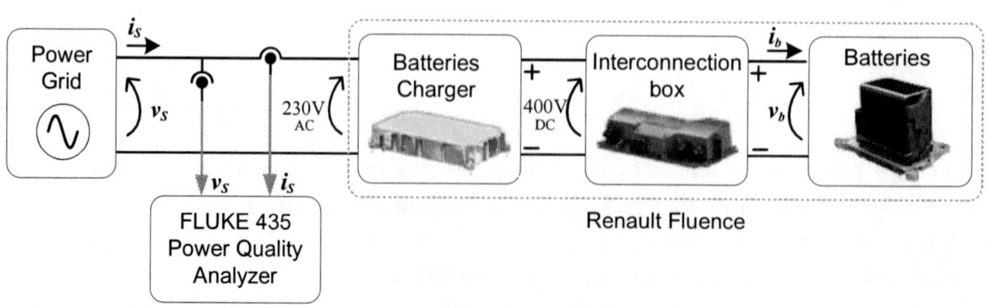

Figure 1. Outline of the setup used to monitor the battery charging process.

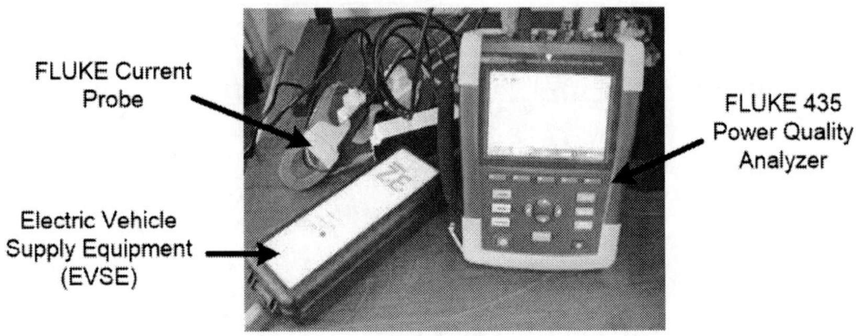

Figure 2. Setup used for monitoring the battery charging process.

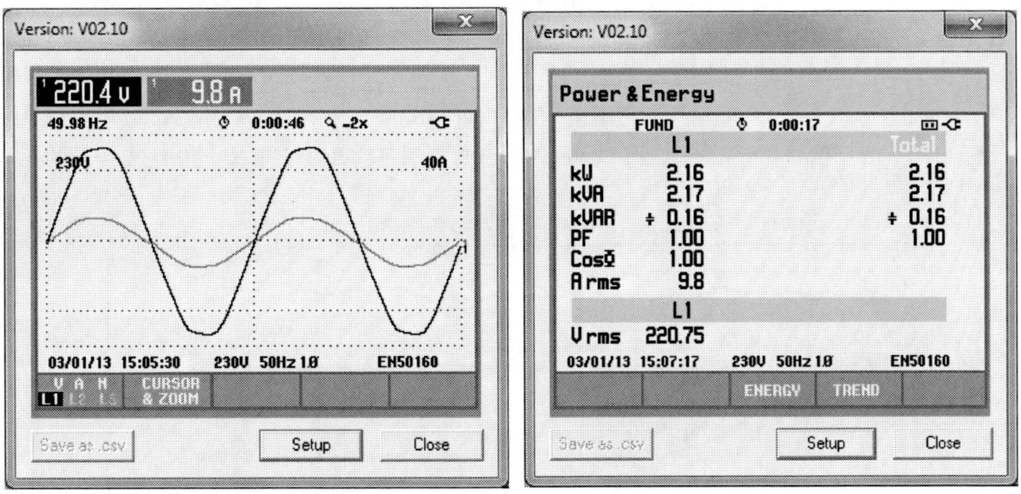

a b

Figure 3. Sample results for battery charging process: (a) Power grid voltage and current; (b) Measured power, power factor and RMS values.

Figure 4 presents the spectral analysis and the Total Harmonic Distortion (THD) of the power grid voltage (a) and consumed current (b) for a typical test. During the battery charging process the power factor was unitary and the consumed current presented a small distortion with a THD of 5%. The power grid voltage presented a THD of 2.4% as a result of the other non-linear loads connected to the power grid.

In this context, considering the predictable proliferation of EVs [23], it is extremely important to regulate the battery charging process [24, 25] in order to preserve the power quality in the power grids [26-28]. The battery charging system of the Renault Fluence ZE accomplishes this objective, operating with unitary power factor and almost sinusoidal current consumption, although it only allows operation in unidirectional mode, denominated as Grid-to-Vehicle (G2V). In the future, smart grids will allow energy to flow in the opposite direction, allowing the concept of Vehicle-to-Grid (V2G), in which part of the energy stored in the batteries can be sent back to the power grid, helping to improve the power grid stability at certain periods [29-32].

The battery charging was performed in two distinct places, in a private home and at a university. Figure 5 presents the power grid voltage and consumed current profiles obtained during a typical test.

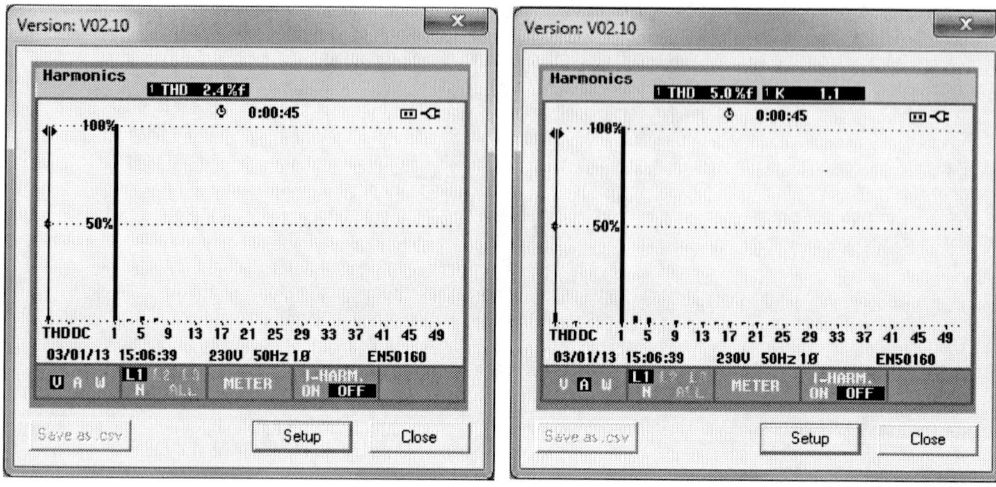

a b

Figure 4. Spectral analysis and THD during a typical battery charging process: (a) Power grid voltage; (b) Consumed current.

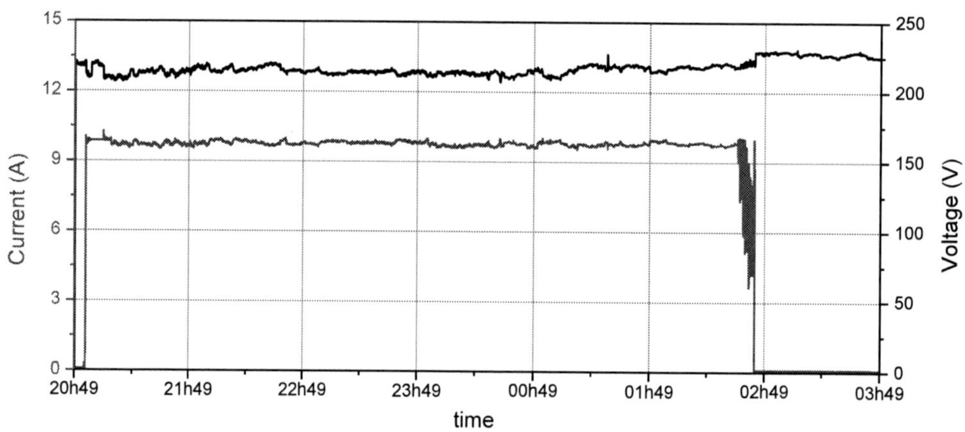

Figure 5. Results recorded during a battery charging process at home.

It can be seen that due to the battery charging process, the power grid voltage in the house was affected and fell slightly but its value was kept within the 10% admissible fluctuation, as defined by the European standard EN 50160.

Despite the voltage fall to a constant value, the current was kept almost constant at 10 A at nearly all the time, meaning that the batteries are charged with constant power. At the end of the battery charging process, during about 10 minutes, the consumed current is non-constant.

During the battery charging process the power factor was also kept constant at with unitary value. No significant deviation from the nominal frequency was ever registered, which attests the good stability of the power grid.

The recorded energy values represent the energy that was supplied by the power grid to the EV, but do not represent the energy effectively stored in the batteries, as the charger and the batteries have a given energy efficiency.

Typically, the battery charger efficiency is specified by the manufacturer and usually is defined as a function of the output power, which can change during the battery charging process. On the other hand, the battery charging efficiency is strongly influenced by several parameters, mainly their temperature, the ambient temperature, and the schedule of the charging, i.e., if the charging process is performed immediately after the discharging process or after a time delay. Typically [33], the efficiency of the battery charger and of the batteries is about 94% and 85%, respectively, so, only about 80% of the energy provided is effectively stored in the batteries [33]. Similar values for the overall charging efficiency are also reported by other authors [34], but based on slightly different efficiencies of battery charger and batteries (both around 90%).

RESULTS

The energy and average power results presented may concern either used or charged energy/power. In the case of the EV, the used energy/power refers to the energy or the average power coming from the batteries, which was used to accomplish a given route (this includes consumption due to auxiliaries such as air conditioning). The charged energy refers to the total energy supplied by the electric grid to recharge the vehicle's batteries back to 100% SOC after a given route. On the ICE car the charging efficiency is 100% so there is no difference between the used and the charged (supplied) energy/power. They refer to the thermal energy/power supplied by the fuel to the engine.

Short Distance, Constant Speed Tests

The tests at constant speed were performed on a horizontal stretch of dual carriageway with an extension of 1 km. They were done in both directions and repeated several times each. Speeds ranging from 40 to 110 km (to 120 km/h in the case of the EV), in increments of 10 km/h, were tested with both cars. The four results (sometimes 6 or 8 for each speed) were then averaged and plotted against vehicle speed. Graphs for used power of the EV and ICE car can be seen in Figure 6 (a) and (b), respectively. Both lines increase monotonically from 40 to 110 km/h. It can seen that the power curves fit quadratic trend lines. This agrees well with the fact that aerodynamic drag, lubricated contact drag (and in general viscous drag) also increase quadratically with speed. Comparing Figure 6 Erro! A origem da referência não foi encontrada. (a) and (b) it can be seen that the ICE (diesel engine) car needs roughly twice the power of the EV to perform the same task (note the factor of 2 between the axis' scales). Surely, the comparison is tricky, as the energy consumptions for producing and supplying electricity / Diesel fuels are not considered here. As already noted in the introduction, that would require a Well-to-Wheel (WTW) Life Cycle Analysis instead of the Tank-to-Wheel (TTW) approach here used. The WTW is out of the scope of this paper but has been used by the group for other works [17]. It is possible to see the same information but in terms of energy consumption in kWh/100 km (Figure 7) for both vehicles. Please note that the increment of energy consumption with speed in each curve is now much lower than in the previous graphs, as the speed is not included in the energy as it is in the power calculation.

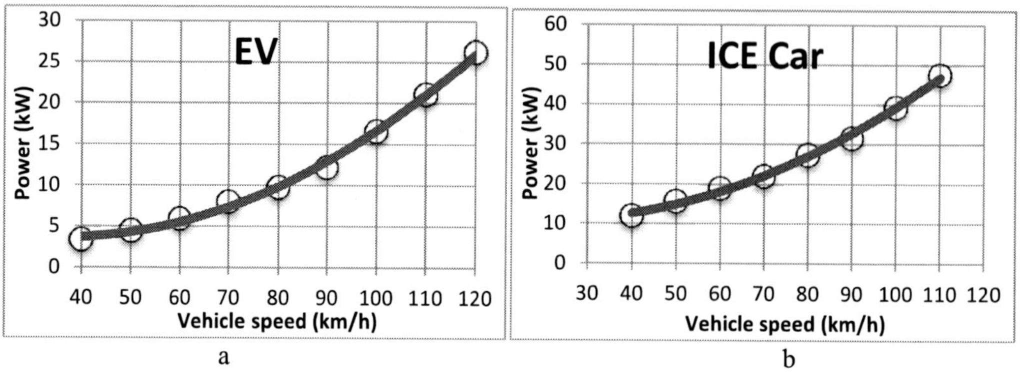

Figure 6. Average power consumption for short tests at constant speed with (a) EV and (b) with ICE car.

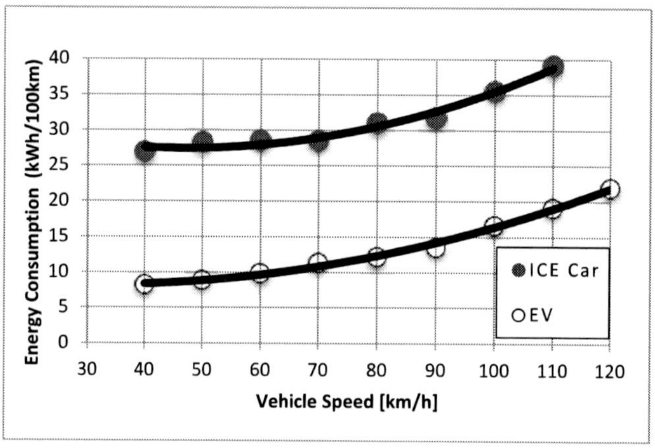

Figure 7. Energy consumption for short tests at constant speed with EV and ICE car.

Figure 8 shows the normalized power consumption using as reference the power for the 110 km/h (maximum tested speed) test. It can be seen that for the EV the ratio of required power between lower (40 km/h) and higher (at 110 km/h) speeds is 0.16, whereas in the ICE car this ratio is 0.25 (Figure 8). This shows that the electric motor is more apt for running at lower speeds with fewer losses. One of the problems with the ICE car is that below 70 km/h 6th gear could no longer be used. This affects mechanical losses as a higher engine / wheel rotational speed ratio will be present below 70 km.

When comparing the energy consumption on both vehicles, the Diesel consumption was always at least 2-fold higher than the consumption of the EV (Figure 9). However, the ratio increased as the speed decreased, showing the reduced capability of the ICE to cope with low speeds efficiently.

Long Trip, Constant Speed Tests

The long trip tests were performed between the cities of Guimaraes and Vila Nova de Gaia (Figure 10) both ways, with a total of more than 120 km. The overall distance was run on dual carriageways, with most of it being done in motorways.

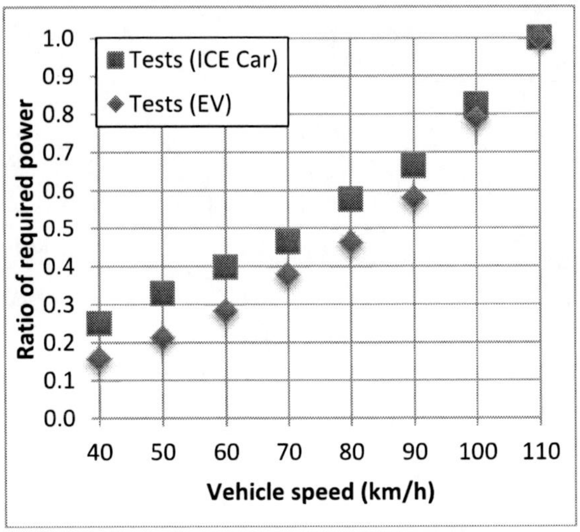

Figure 8. Power required for maintaining speed, as a relation to 110 km/h.

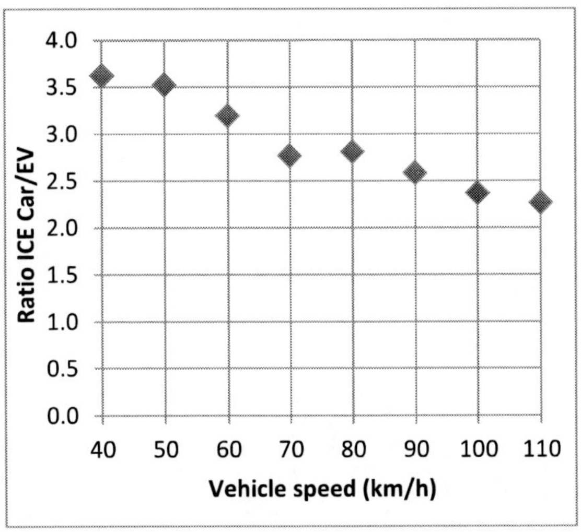

Figure 9. Ratio of energy consumption between ICE car and EV.

However, it was not possible to maintain the speed setpoint for all the distance (Figure 11), because of motorways interchanges (km 27 and 95), toll booths (km 7, 52, 68 and 113) and in the approaches to the cities (start, middle and end). But, as it can be seen in Figure 11, most of the trip was carried out at the speed setpoint. When in the motorway, the cruise control of the car was set to the required speed for all the distance. The small "wrinkles" seen in the trip are a result of the GPS reading, as the cars were controlled (by their cruise control) within ±2 km/h of the setpoint.

An histogram of the speed data points can be seen in Figure 12. Almost 80% of the overall distance was travelled at speeds between 97.5 and 102.5 km/h, showing that most of the trip was in the vicinity of 100 km/h.

The road is somehow hilly, with altitudes ranging from 50 to 250 m (Figure 13). The figure shows the return trip (Guimaraes-Gaia-Guimaraes), which, as it can be seen, is symmetrical.

Figure 10. Road trip (return).

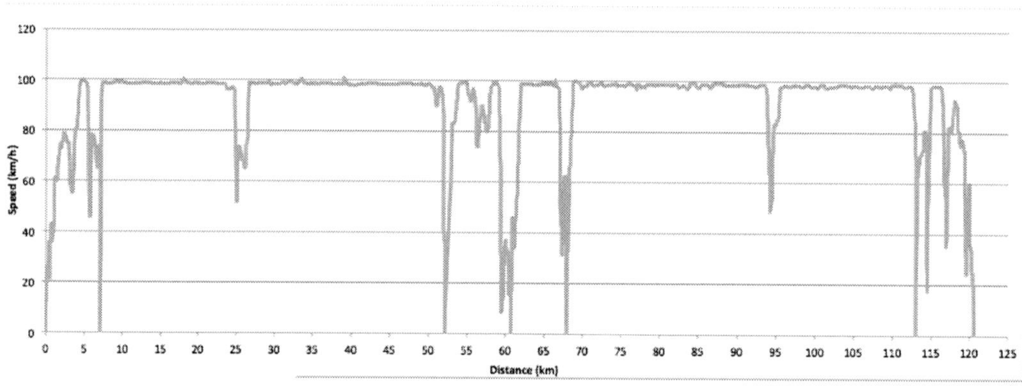

Figure 11. Speed of the round road trip against distance (100 km/h).

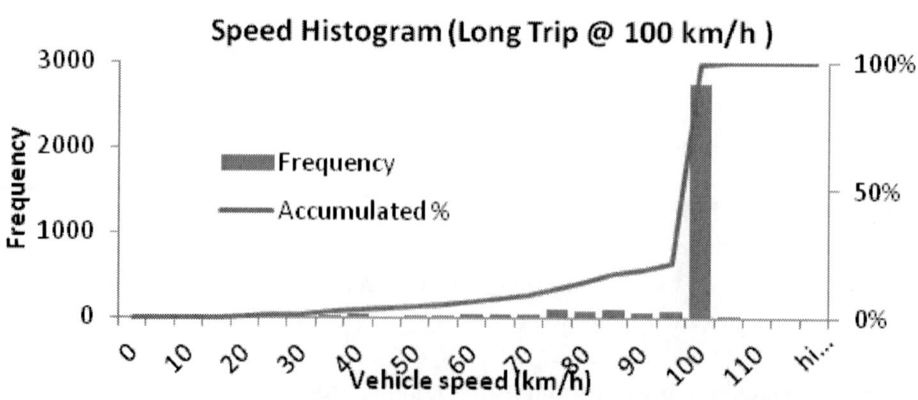

Figure 12. Histogram for speed of the round road trip against distance (100 km/h).

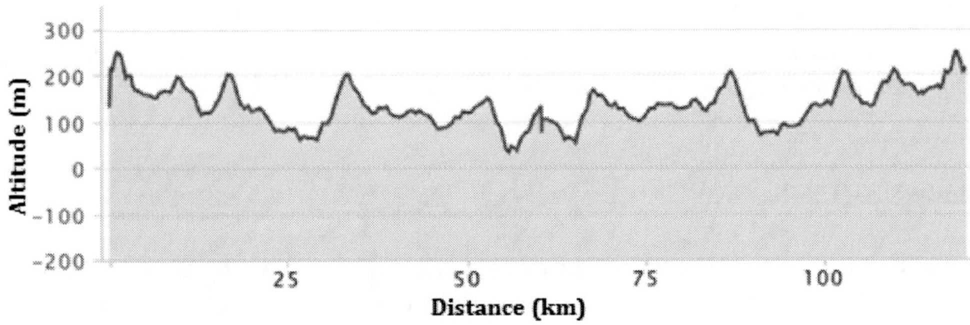

Figure 13. Altitude of the road trip.

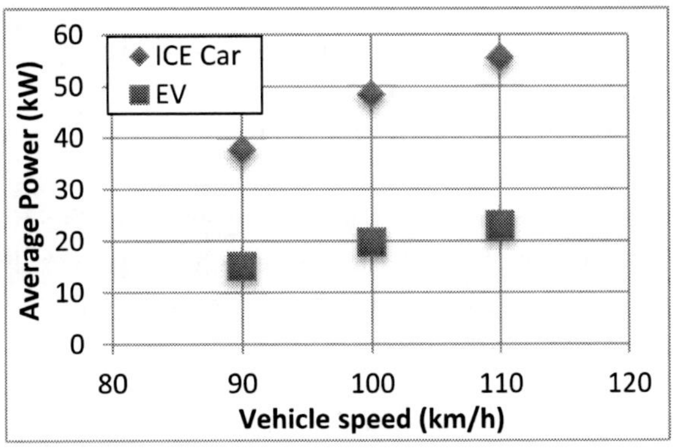

Figure 14. Average power for the long trip.

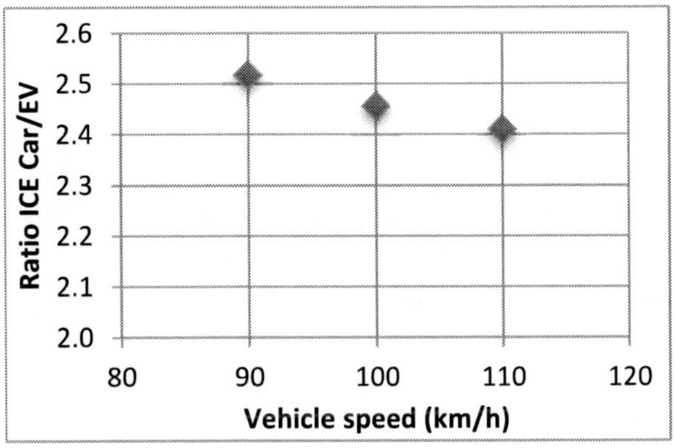

Figure 15. Relation of energy consumption (Diesel / Electric).

Only 3 speeds were evaluated (90, 100 and 110 km/h) for the ICE car, but for the EV speeds of 80 and 120 km/h were also tested. Speeds lower than 80 km/h are extremely slow for a motorway and speeds higher than 120 km/h are not legal (the legal limit in motorway in Portugal is 120 km/h).

The results are shown in Figure 14 for the Electric and for the Diesel powered cars. The values are an average of the energy consumption of the car, in terms of average power (in kW) in electricity and in Diesel fuel. As it can be seen, in terms of energy, the ICE car uses more than twice as much energy as the EV (Figure 15). These values, although different, are in line with the results for the short distance tests (Figure 9).

When plotting the data for short vs long tests (Figure 16 and Figure 17), as it would be expected, the values for the long trip are somehow higher, reflecting the altitude and level changes during the long trip, inexistent in the short tests.

Steady Up-Hill Tests

During the long trips the values for maximum consumption during the steepest sections of the road (6% gradient) were recorded and plotted (Figure 18 and Figure 19) for the EV and for the ICE car, respectively. In the figures, these values are plotted with the average consumption during the overall trip, for comparison. Note again the factor of 2 between the vertical axes of both plots.

Like for the previous results, Figure 20 shows the difference for energy consumption between both vehicles in the up-hill sections.

Again, the ratios are above 2, although lower than in the case for the leveled tests (Figure 9) and for the long trip tests (Figure 15).

As the vehicle required power is higher for this up-hill test, the efficiency of the ICE is here somehow higher than on the other conditions. This again proves the higher difference in efficiency between EV and ICE cars during low power requirements.

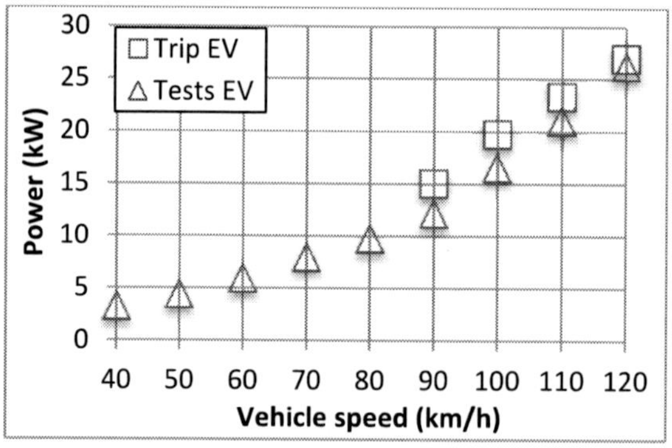

Figure 16. Average power for the short and long trip (EV).

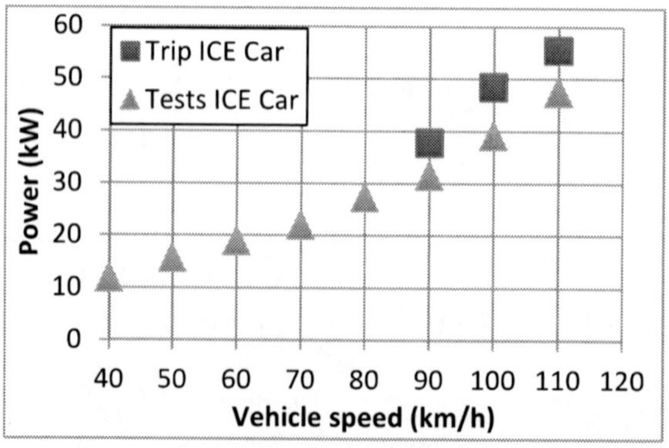

Figure 17. Average power for the short and long trip (ICE car).

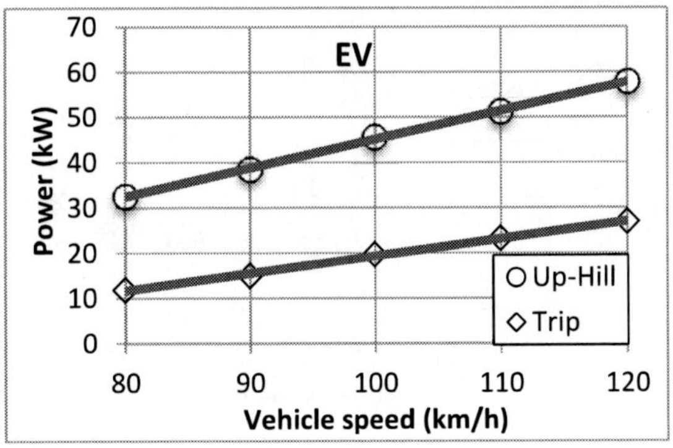

Figure 18. Road trips with EV (average and required power for up-hill).

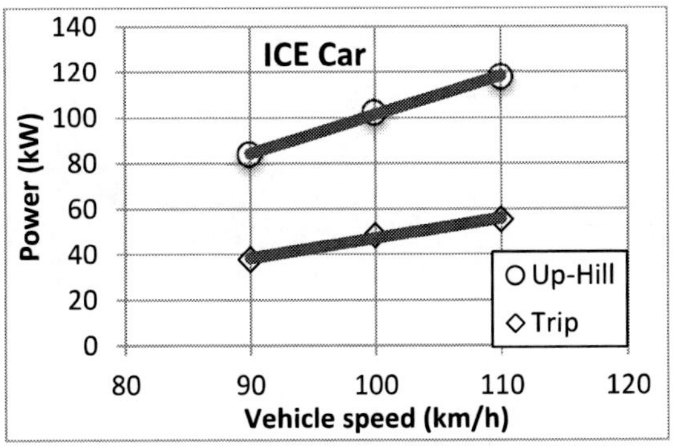

Figure 19. Road trips with ICE car (average and required power for up-hill).

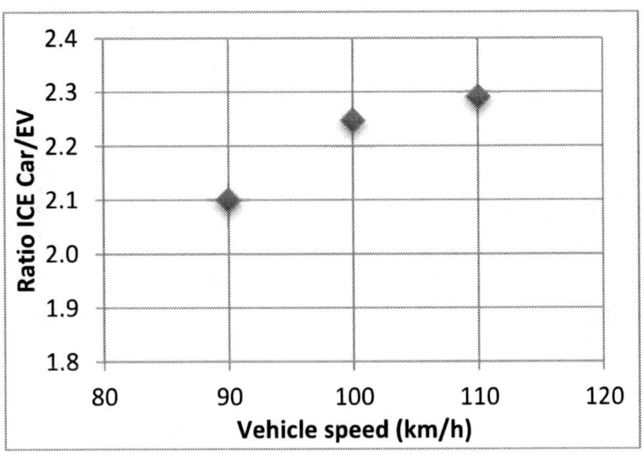

Figure 20. Relation of energy consumption (ICE Car / EV).

The overall results are summarized in Figure 21 and Figure 22 for the EV and for the ICE car, respectively, and for both cars in Figure 23.

When comparing the power used in both cars as a ratio for all the tested conditions (Figure 24), the ICE car almost always spends more than twice as much as the EV. However, this ratio is dependent on vehicle speed. For tests at constant speed in leveled ground the difference increases as the speed decreases. This shows the lower efficiency of the ICE towards low loads and specially using lower gear-box ratios.

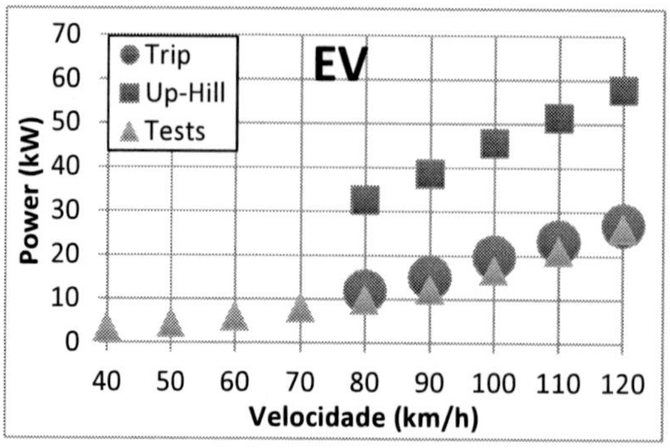

Figure 21. Various required power for EV.

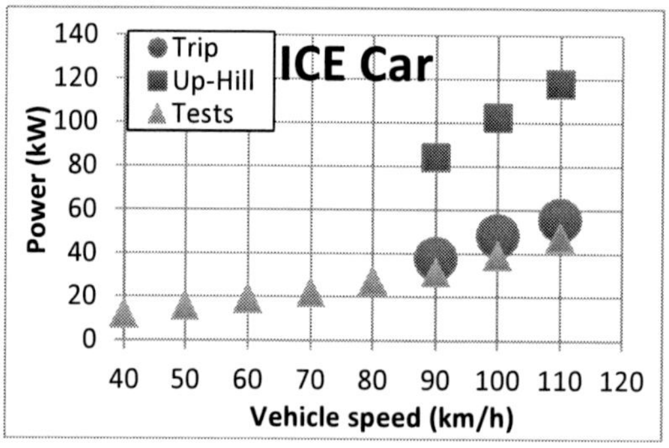

Figure 22. Various required power for ICE car.

The electric motor is much more efficient, as its efficiency does not plunge as much as with the thermal engine but mainly because it uses the same ratio for its entire speed range. When the comparison is for higher power levels (up-hill), the difference is lower, showing the improved efficiency of the thermal engine as the load increases.

Electric Vehicle Charging

The data presented above was for the energy "spent" for the locomotion, by the electric motor and controller in the case of the EV or by the diesel engine in the case of the ICE car. However, there is an efficiency for the electric charger and for the batteries during charging and discharging processes.

In this chapter, it was assumed that the ICE car had no inefficiencies of such kind, which is to say, that none of the diesel fuel supplied to the vehicle tank was "lost" during operation, as all the fuel delivered by the pump is effectively delivered to the ICE and burned in it.

The long trip presented earlier in this paper (Figure 10) is composed of two stages, departing (outward) from Guimaraes to Gaia and returning (inward) from Gaia to Guimaraes. As it can be seen in Figure 13, Guimaraes has an altitude of 200 m and Gaia (middle of the trip) an altitude of around 50 m.

Therefore the "outward" trip has a lower consumption (for the same speed) than the "inward" trip (Figure 25, for the EV). The difference between outward and inward consumption is roughly 20% higher for the trip returning to Guimaraes.

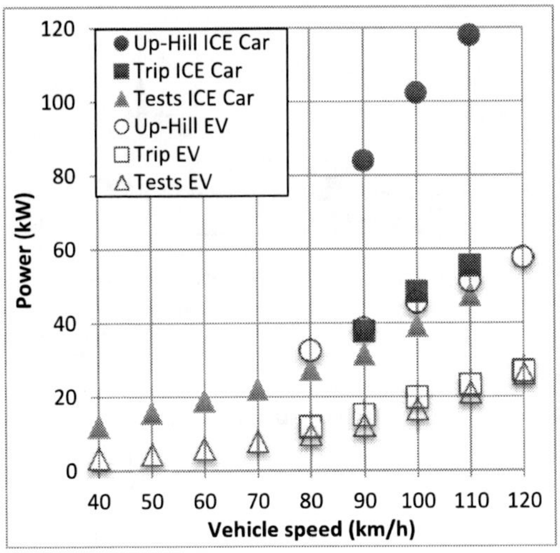

Figure 23. Comparison between required power between EV and ICE cars (in kWh/100 km).

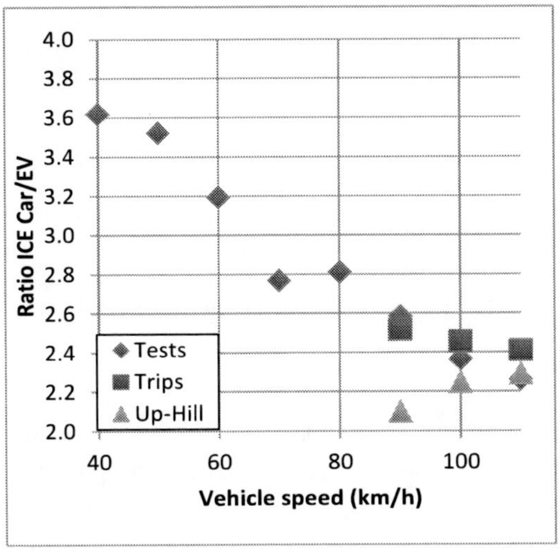

Figure 24. Ratio for power required in ICE and EV cars for the same conditions.

Before and after each stage of the trip the car was recharged to full SOC of the batteries and the energy supplied by the grid was measured.

It was possible to compare the charged (supplied) energy with the used (spent) energy for each of the trips' stages, which is plotted in Figure 25. The efficiency of the system (charger+battery) is displayed on Figure 26 showing the losses due to the operation of the charger and of the charging + discharging process in the battery.

Strangely in Figure 25, the values for energies charged for the outward trip at 110 and 120 km/h were higher than the values for the inward trips, although the energy spent for the inward trips was higher than that for the outward trips. It can be seen that the efficiencies of the charger-battery system corresponding to the outward trips at 110 and 120 km/h (which measured the energy for that trip) were much lower than those of the other tests. For the charging events for speeds up to 100 km/h, the charged energy was on average 25% in excess of the energy spent during the trip.

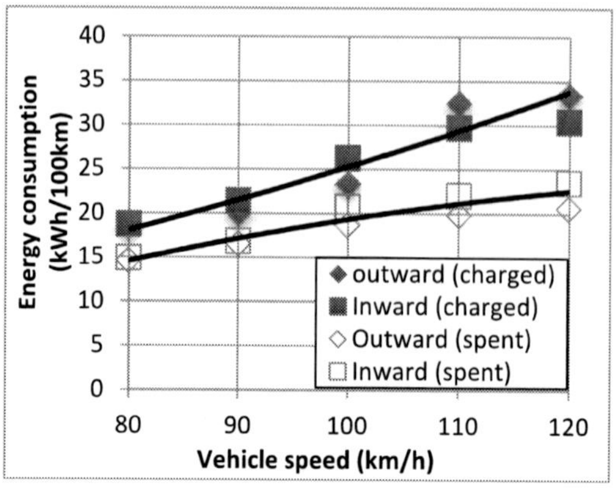

Figure 25. Energy spent during the trips and energy charged for those trips.

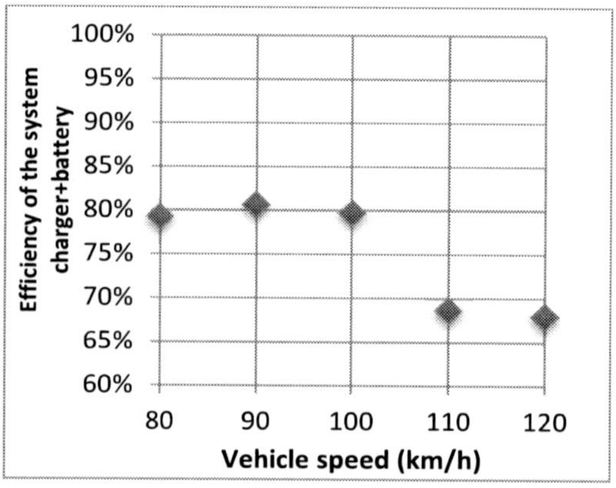

Figure 26. Efficiency of the charger-battery system.

For the outward trip at 110 km/h the charged energy was a massive 47% higher than the energy spent, showing the high level of energy lost with the discharging and subsequent charging of hot batteries.

With one exception, all the aforementioned charging events took place with the cooling fan running. As these results were peculiar, the return trips at 110 and 120 km/h were repeated, but the results were similar.

Surely the differences may not be due solely to the energy required for the operation of the cooling fan, even if it operated for several hours. In fact, this continuous operation of the cooling fan indicates that the batteries were hot, due to the high power requirement of those trips which heated the batteries up to a high temperature. At this higher temperature both the charging and the discharging efficiencies are probably worse. And if efficiency is lower, more energy will be lost as heat. This heat will further hinder the cooling of the batteries creating a loop and therefore preventing the system of ever achieving normal optimized operation. If confirmed, this behaviour would recommend the instalation of a more powerful battery cooling system.

The overall result for energy consumption for both cars can be seen in Figure 27. The ratio between the energy supplied to the EV (electricity charged) vs the energy supplied by the fuel to the ICE car ranged between 0.49 and 0.61 for the 90 km/h to 110 km/h range (Figure 28).

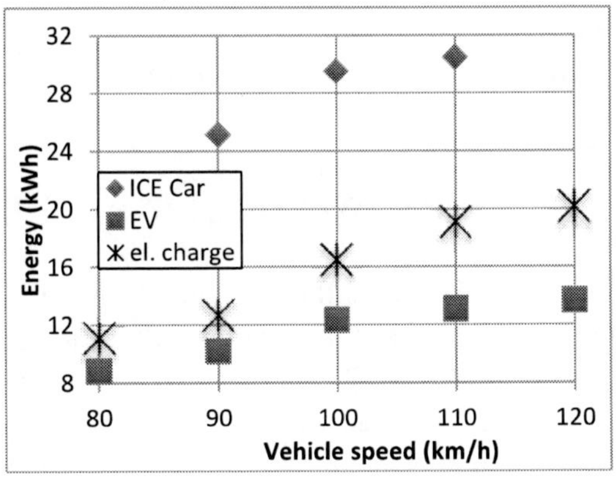

Figure 27. Energy spent in EV and ICE cars, compared with electricity charged on EV.

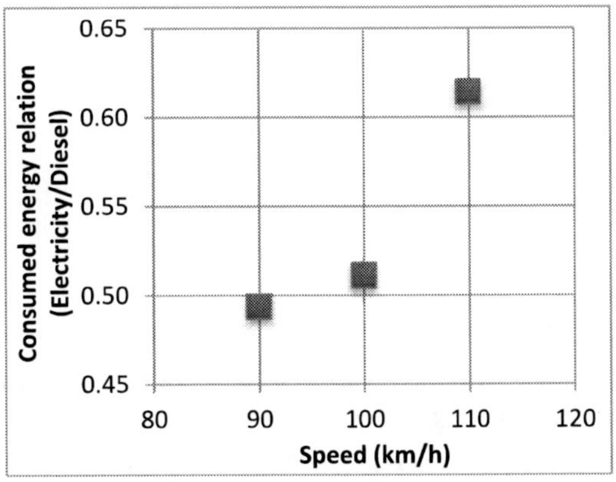

Figure 28. Relation between energy charged to the EV and the energy supplied in Diesel fuel.

CONCLUSION

As it was noticed that there is currently a lack of real life comparisons between the energy consumption of Electric Vehicles (EV) and Internal Combustion Engine (ICE) cars, a test program to assess this issue was envisaged and completed.

Short distance tests at levelled ground and 6% up-hill road as well as long trips were done for a broad range of fixed speeds. Measurements enabling the assessment of average energy consumption and required power were performed for both vehicles.

It was found that the electric powertrain presented much higher efficiency than the ICE (diesel engine) but the difference was reduced as the average speed and required power (e.g., at up-hill sections) increased. The ratio of ICE to EV power requirement varied between 2.1 (up-hill test) and 3.6 (minimum speed tested).

When considering total energy supplied to the EV the value was on average 25% higher than the energy supplied to the electric motor for most speeds (up to 100 km/h). For higher speeds (110 and 120 km/h) this ratio increased to 47%. Under these high speed conditions the higher required power caused the increase of battery temperature, which eventually reduced the discharging efficiency and probably also the subsequent charging efficiency.

Overall, the energy supplied by the electric grid to the EV ranged between 49% and 61% of the energy supplied by the fuel to the ICE (diesel) car for identical trips.

ACKNOWLEDGMENTS

The authors would like to thank Renault Portugal and Confiauto Braga for the lending of the test vehicles and also Braga - Capital Europeia da Cultura for the support. This research project was funded by Project MOBI-MPP (MIT-Pt/EDAM-SMS/0030/2008) supported by the MIT Portugal Program and FEDER funds through the Programa Operacional Factores de Competitividade - COMPETE and National Funds through FCT - Foundation for Science and Technology.

Francisco P. Brito benefited from post-doctoral grants SFRH/BPD/51048/2010 and SFRH/BPD/89553/2012 supported by the MIT Portugal EDAM and FCT. Delfim Pedrosa and Vítor Monteiro benefited from MOBI-MPP Project research grants and from FCT PhD grants SFRH/BD/86628/2012 and SFRH/BD/80155/2011, respectively.

REFERENCES

[1] J. Martins, F. Brito, *"Carros Elétricos,"* Publindústria, Porto, Portugal, Dez. 2011, ISBN 978-972-8692-64-3, http://www.engebook.com/2/7707/Carros-Eletricos.

[2] B. Ribeiro, F. Brito, J. Martins, *"A Survey on Electric/Hybrid Vehicles," Electric and Hybrid-Electric Vehicles - Overviews and Viewpoints*, pp. 9-22, Edited by Ronald, K. Jürgen, SAE International, Warrendale, US, 2010, (http://books.sae.org/book-pt-143/1/), ISBN 978-0768057171, DOI: doi-1000092221.

[3] Environmental Protection Agency (EPA), Department of Transportation (DOT), National Highway Traffic Safety Administration (NHTSA). Revisions and Additions to Motor Vehicle Fuel Economy Label; Final Rule. *Federal Register, Rules and Regulations,* vol. 76, 2011, http://www.gpo.gov/fdsys/pkg/FR-2011-07-06/pdf/2011-14291.pdf (accessed 22-05-2013).

[4] The European Parliament and the Council of the European Union, "Regulation (Ec) No 443/2009 of 23 April 2009- Setting emission performance standards for new passenger cars as part of the Community's integrated approach to reduce CO2 emissions from light-duty vehicles", *Official Journal of the European Union*, 05-06-2009, http://eur-lex.europa.eu/LexUriServ/LexUriServ.do?uri=OJ:L:2009:140:0063:008:en:PDF (accessed 22-05-2013).

[5] C. Silva, M. Ross, T. Farias, "Analysis and simulation of "low-cost" strategies to reduce fuel consumption and emissions in conventional gasoline light-duty vehicles," *Energy Conversion and Management,* pp. 215-222, 2009, ISSN 0196-8904 (http://www.sciencedirect.com/science/article/pii/S0196890408003920).

[6] J. Martins, K. Uzuneanu, B. Ribeiro, O. Jasansky, "Thermodynamic Analysis of an over-Expanded Engine," *SAE* 2004-01-0617, 2004.

[7] B. Ribeiro, J. Martins, "Direct Comparison of an Engine Working under Otto, Miller and Diesel cycles: Thermodynamic Analysis and Real Engine Performance," *SAE Technical Paper Series,* nº 2007-01-0261, 2007.

[8] F. Brito, J. Martins, R. Sousa, L. M. Gonçalves, "Temperature controlled Exhaust Heat Thermoelectric Generation," *SAE International Journal of Passenger Cars - Electronic and Electrical Systems*, vol. 5, 2012, DOI: 10.4271/2012-01-1214, http://saepcelec.sae journals.org/content/5/2/561.abstract.

[9] F. Brito, J. Martins, L. M. Goncalves, N. Antunes, D. Sousa, "Influence of Heat Pipe Operating Temperature on Exhaust Heat Thermoelectric Generation," *SAE International Journal of Passenger Cars - Mechanical Systems* 6(2), 2013, DOI: 10. 4271/2013-01-0559.

[10] C. Zdenek, M. Pavel, "Electric, hybrid electric and combustion engine driven cars and their impact on environment," *14th European Conference on Power Electronics and Applications,* pp. 1-5, Aug. 30 2011-Sept. 1 2011

[11] Porsche company, Porsche 918 Spyder: A unique combination of performance and efficiency, Press Release, 16-05-2013 http://www.porsche.com/uk/aboutporsche/ pressreleases/pcgb/?lang=noneandpool=international-deandid=2013-05-16 (accessed 22-05-2013)

[12] C. Silva, M. Ross, T. Farias, "Evaluation of energy consumption, emissions and cost of plug-in hybrid vehicles," *Energy Conversion and Management*, vol. 50, pp. 1635-1643, 2009.

[13] European Commission Joint Research Centre, Institute for Energy, "Well-to-wheels Analysis of Future Automotive Fuels and Power trains in the European Context - Report Version 3c", JRC Publication N° JRC65998, ISBN 978-92-79-21395-3 July 2011 http://publications.jrc.ec.europa.eu/repository/handle/111111111/22590 (accessed 31-05-2013).

[14] S. Eaves, J. Eaves, "A cost comparison of fuel-cell and battery electric vehicles," *Journal of Power Sources*, vol. 130, pp. 208-212, 2004.

[15] C. Camus, C. M. Silva, T. L. Farias, J. Esteves, "Impact of Plug-in Hybrid Electric Vehicles in the Portuguese Electric Utility System," *Power Engineering, Energy and Electrical Drives (POWERENG 2009)*, pp. 285-290 Lisbon, Portugal, Mar. 18-20, 2009.

[16] EDP Universal, "Rotulagem de energia elétrica EDP SU 2012", available online http:// www.edpsu.pt/pt/origemdaenergia/Folhetos%20de%20Rotulagem/Rotulagem%20de% 20energia%20el%C3%A9trica%20EDP%20SU%202012.pdf (accessed 29-05-2013).

[17] J. Ribau, C. Silva, F. Brito, J. Martins "Analysis of Four-Stroke, Wankel and Microturbine-Based Range Extenders for Electric Vehicles," Energy Conversion and Management, *Elsevier*, vol. 58, pp. 120-133, 2012, http://dx.doi.org/10.1016/j.encon man.2012.01.011.

[18] S. S. Williamson, S. M. Lukic, A. Emadi, "Comprehensive drive train efficiency analysis of hybrid electric and fuel cell vehicles based on motor-controller efficiency modeling," *IEEE Transactions on Power Electronics*, vol. 21, pp. 730-740, May 2006, DOI: 10.1109/TPEL.2006.872388.

[19] C. Alaoui, "Solid-State Thermal Management for Lithium-Ion EV Batteries," *IEEE Transactions on Vehicular Technology,* vol. 62, pp. 98-107, Jan. 2013.

[20] J. Patten, N. Christensen, S. Srivastava, G. Nola, "The Impact of Driving Conditions on PHEV Battery Performance," *Green Manufacturing Research Journal*, Paper 2, Jan. 2011.

[21] J. Martins, *"Motores de Combustão Interna"*, 4th edition (revised and extended, in Portuguese), Publindustria, Porto, Portugal, ISBN: 978-989-723-033-2, 2013.

[22] C. I. McClay, S. Williamson, "The variation of cage motor losses with skew," *IEEE Transactions on Industry Applications,* vol. 36, pp. 1563-1570, 2000, DOI: 10.1109/ 28.887207.

[23] T. A. Becker, I. Sidhu, B. Tenderich, *"Electric Vehicles in the United States A New Model with Forecasts to 2030,"* University of California, Berkeley, Center for Entrepreneurship and Technology (CET), v.2.0, Aug. 2009.

[24] K. Dyke, N. Schofield, M. Barnes, "The Impact of Transport Electrification on Electrical Networks," *IEEE Transactions on Industrial Electronics*, vol. 57, pp. 3917-3926, 2010.

[25] L. Jian, H. Xue, G. Xu, X. Zhu, D. Zhao, Z. Y. Shao, "Regulated Charging of Plug-in Hybrid Electric Vehicles for Minimizing Load Variance in Household Smart Micro-Grid," *IEEE Transactions on Industrial Electronics*, vol. 60, pp. 3218-3226, Aug. 2013.

[26] V. Monteiro, H. Gonçalves, João L. Afonso, "Impact of the Electric Vehicles on the Power Quality in a Smart Grid Context," *IEEE EQPU 11th International Electrical Power Quality and Utilization Conference*, Lisbon Portugal, pp. 1-6, Oct. 2011.

[27] J. McDowall, "Conventional battery technologies-present and future," *IEEE Power Engineering Society Summer Meeting,* pp. 1538-1540, 2000.

[28] A. Emadi, Y. J. Lee, K. Rajashekara, "Power Electronics and Motor Drives in Electric, Hybrid Electric, and Plug-In Hybrid Electric Vehicles," *IEEE Transactions on Industrial Electronics,* vol. 55, pp. 2237-2245, 2008.

[29] V. Monteiro, J. C. Ferreira, G. Pinto, D. Pedrosa, João L. Afonso, "iV2G Charging Platform," *IEEE ITSC 13th International Conference on Intelligent Transportation Systems*, Madeira Portugal, pp. 409-414, Sept. 2010.

[30] W. Kempton and J. Tomic, "Vehicle-to-grid power fundamentals: Calculating capacity and net revenue," *Journal of Power Sources*, vol. 144, pp. 268-279, 2005.

[31] B. Kramer, S. Chakraborty and B. Kroposki, "A review of plug-in vehicles and vehicle-to-grid capability," *IECON 2008 - 34th Annual Conference of IEEE Industrial Electronics*, pp. 2278-2283, 2008.

[32] V. Monteiro, H. Gonçalves, J. C. Ferreira, João L. Afonso. "Batteries Charging Systems for Electric and Plug-In Hybrid Electric Vehicles," *New Advances in Vehicular Technology and Automotive Engineering*, 1st ed., J. P. Carmo and J. E. Ribeiro, Ed. In Tech, 2012, pp. 149-168, ISBN 978-953-51-0698-2, http://dx.doi.org/10.5772/2617.

[33] Bimal, K. Bose, "Energy, Environment, and Advances in Power Electronics," *IEEE Transactions on Power Electronics*, vol. 15, pp. 688-701, July 2000.

[34] S. G. Wirasingha, R. Gremban, A. Emadi, "Source to Wheel (STW) Analysis of Plug in Hybrid," *IEEE Transactions on Smart Grids*, vol. 3, pp. 316-331, Mar. 2012.

In: Grid Electrified Vehicles
Editor: Carla Alexandra Monteiro da Silva

ISBN: 978-1-62808-839-7

Chapter 11

ON-ROAD MONITORING OF ELECTRIC BICYCLES AND ITS USE IN BIKE-SHARING SYSTEMS

Patrícia Baptista*

IDMEC - Instituto Superior Técnico, Universidade
Técnica de Lisboa, Lisboa, Portugal

ABSTRACT

Over the last decades there has been growing concern with increasing energy consumption accountable to the transportation sector. One possibility to reduce the impact of the transportation sector, particularly in urban environments, is to decrease the demand for energy intensive modes of transportation and by promoting alternatives that can provide a cheaper, less noisy and more sustainable alternative than a day-to-day car commute. One of those alternatives is bike-sharing, using either conventional or electric bicycles. This chapter presents a review of bike sharing systems (BSS) worldwide, including electric bycicle usage and the results of a research comprising almost 500 systems. From those, around 40 systems were identified as having electric bicycles. To better understand the impacts of using electric bicycles in bike-sharing, the monitoring of an electric and a conventional bicycle in an urban environment was promoted to perform an evaluation of speed, travel time and the use of electricity with a laboratory collecting second-by-second trip data (including, speed, location, altitude, energy use, etc.). Electric bicycles range (typically 40-60 km) appears to be adequate for use in electric BSS (≈15 km daily). In more detail, for the tour monitored, the average speed increase ranged between 8% and 26%. For sections with negative slopes the average speed decreased by 11% when moving from the conventional to electric bicycle, whereas it increased in neutral slopes by 6% and in positive slopes by 49%. Nonetheless, the full impact of the shift from conventional to electric bicycle can depend on the strategy chosen by the biker concerning how to use the electric support. In all, the planning of mixed bike sharing schemes, and in particular the definition of the ratio between electric and conventional bicycles, requires taking into consideration different factors such as the most common routes that will be served by the bicycles and the target population for the bicycles.

* Corresponding author: Patrícia Baptista. IDMEC - Instituto Superior Técnico, Universidade Técnica de Lisboa, Av. Rovisco Pais, 1 - 1049-001 Lisboa – Portugal.

Keywords: Energy; emissions; alternatives in transportation sector; bike-sharing; electric and conventional bicycles; urban mobility; system's planning

NOMENCLATURE

BSS	bike sharing system
CB	conventional bicycle
CC	credit card
CO_2	carbon dioxide
EB	electric bicycle
EV	fully electric vehicle
FC-HEV	fuel cell hybrid electric vehicles
FC-PHEV	fuel cell plug-in electric vehicles
HEV	hybrid electric vehicles
NO_x	nitrogen oxides
PEV	plug-in electric vehicles
PHEV	plug-in electric vehicles
PM	partivulate matter
TTW	tank-to-wheel

INTRODUCTION

Over the last decades there has been growing concern with increasing energy consumption accountable to the transportation sector, which has led to increasing challenges on how to decrease energy consumption and also local and global emissions.

In fact, the road transportation sector was responsible in 2010 for 35% of the final energy consumption (and 70% of the crude oil products consumption) [1].

Additionally, the transportation sector has lagged in improving its efficiency profile, as is presented in Figure 1 for several countries (in terms of emissions of CO_2 per unit of energy consumed). For total energy consumption, a 12% average reduction has been observed from 1995 to 2008 for the selected countries.

On the contrary, over the last decades or so, the road transportation sector faces a stable emissions of CO_2 per unit of energy consumed profile with an average 3% reduction from 1995 to 2008 for the selected countries in Figure 1.

Several steps have been taken to reduce emissions with technical progress leading to greater energy efficiency, but not enough to offset rising traffic volumes. The tightening of vehicle emission standards (with the application of the "Euro" standards) and improvements in fuel quality have significantly reduced transportation related emissions of local pollutants per vehicle per kilometer [2]. However, NO_x and PM have risen as a result of increasing diesel share in new vehicle sales.

Nevertheless, pollution levels still exceed legal limits in many urban areas meaning that further action must be taken to improve air quality.

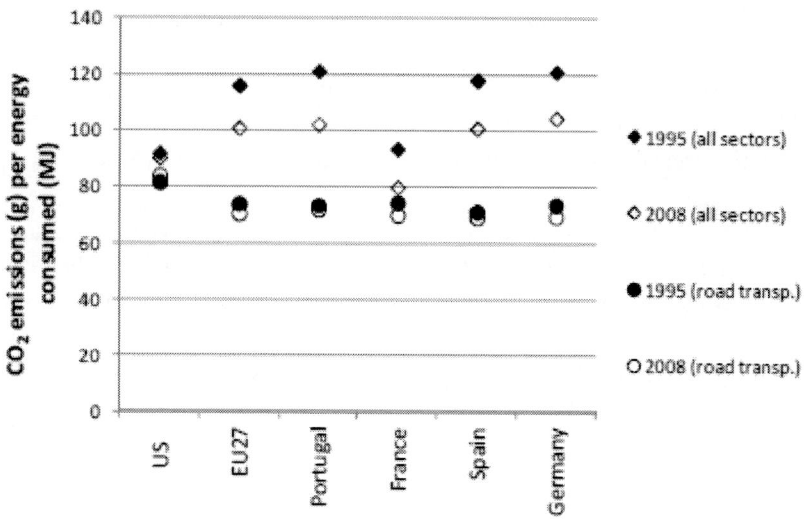

Figure 1. Emissions of CO_2 equivalent (g) per unit of energy consumed (MJ) from 1995 to 2008 considering all sectors and the road transportation sector for the US and several European countries [1].

For these reasons, it is expected in the near future that the transportation sector will face considerable changes, in the areas of vehicle technology, energy source and on the mobility management. It is likely that the market share of already established technologies such as hybrid electric vehicles (HEV) will raise and that, in the long term future, the market share of alternative vehicle technologies such as plug-in electric vehicles (PEV), covering plug-in hybrid (PHEV), fully electric (EV) or fuel cell vehicles (FC-HEV and FC-PHEV) will start increasing. Additionally, the energy pathways feeding the transportation sector may be diversified, with the incorporation of higher levels of biofuels and the introduction of alternative energy pathways such as electricity, hydrogen, natural gas, etc.

Moreover, addressing the reasons of why, where and when people use their vehicles, promoting the shift to other forms of transportation such as public transportation and promoting behavior changes regarding transportation usage (eco-driving, bike and car sharing and carpooling, etc.) can play a very important role where the previously mentioned challenges are concerned [3].

In more detail, electricity and hydrogen are alternative energy pathways that are regarded as an opportunity for significantly reducing the amount of CO_2 emitted by the transportation sector and increased renewable energy penetration [4].

Furthermore, the change to electricity or hydrogen would have a particularly beneficial impact for urban systems due to their zero local emissions (in their Tank-to-Wheel stage). However, such technologies still face weaknesses that prevent them from being true alternatives, mainly because they can only have a significant impact on a long-term scale due to the low renovation rates that fleets have [5].

Furthermore, this technology driven approach may not be completely successful unless a behavioral change happens that enables the users to be more efficient when using the different transportation modes. One possibility to reduce the impact of the transportation sector, particularly in urban environments, is to decrease the demand for energy intensive modes of transportation and by promoting alternatives that can provide a cheaper, less noisy and more sustainable alternative than a day-to-day car commute.

In general, three alternative transportation models can be identified [6]:

- Public transportation systems (bus, trains, subway systems and others);
- Vehicle sharing schemes (such as cars or bicycles); and
- Alternative transportation modes such as walking, private bicycles or others.

The promotion of each of these alternatives requires the development of diversified transportation policies that encourage people to use them, with each of them having their strengths and drawbacks. While most cities offer some sort of public transportation system, the promotion of vehicle sharing schemes and alternative transportation modes are more recent [7]. For these different alternatives, the use of bicycles is one of the more advantageous as it allows the users to move at significant speeds for short distances (typical in urban environments), resulting in no emissions and potentially having health benefits [8].

BICYCLES FOR A SUSTAINABLE URBAN MOBILITY

Bicycles allow people to travel longer, faster and with less effort than walking, while having a low impact on environment, consequently making it an efficient transportation mode for urban mobility. As a result, the importance of cycling has increased worldwide [9, 10], mainly in developing countries, namely in Asia, where two-wheelers are a first affordable measure towards individual mobility. A growing number of cities have been trying to integrate them in the daily mobility of their citizens, which for some countries has resulted in a significant share of trips being performed with bicycles, such as the Netherlands (26%), Denmark (18%) and Germany (10%) [11]. In the city of Amsterdam, 38% of all trips in 2008 were made using a bicycle, with 50% of Amsterdam's inhabitants riding a bike daily and 85% riding it at least once a week [12].

Promoting the use of bicycles calls for the development of specific policies that influence the trip at all levels, whether it is when riding it (through traffic calming and safe bikeways), parking it (by offering secure locations) or moving it around the city (through the integration of bicycles with public transportation systems).

In spite of its large acceptance, the use of bicycles still has several drawbacks that delay its widespread use. Some of the main problems identified by people when using conventional bicycles (CB) for urban transportation include the difficulty to travel very long distances and over hills, the possibility of arriving at a destination, such as work, sweaty or tired [13], being exposed to extreme cold or hot climates, among others. Nonetheless, Northern European countries have shown to have significantly high usages of bicycles despite their cold weathers. Furthermore, some cities with difficult topographies, such as Lisbon, have also begun promoting the use of bicycles through the creation of bike lanes and studying the possibility of having bike sharing schemes [14, 15].

Some of these issues can also be overcome through the use of electric bicycles (EB) [13]. Generally, it is expected that the use of EB can help reduce the effort required for performing trips as well as reduce travel time, though at a higher cost due to the electric system and the energy used.

There are usually two technological solutions regarding EB: power on demand or power assisted bicycles. Power on demand electric bicycles provide the user with the choice of different levels of electric assistance. Power assisted electric bicycles provide electric assistance as the pedal intensity increases.

Despite the high expectations for EB, few studies have tried to comprehend the real advantages of such bicycles in an urban context. Additionally, while previous studies addressed the estimated environmental impacts of EB compared to other transportation modes in China [16] and the users characterization and acceptance of this alternative technology in China [17] and in the United States [13], the experimental monitoring of bicycles has only been addressed for CB [18, 19]. Still, the results obtained by these types of studies are influenced by local variables such as the bikers' experience, road, traffic and the existence of bicycle lanes. Therefore, it is important to develop monitoring methodologies in different regions to obtain comparable situations.

This chapter analyses the current worldwide situation regarding bike-sharing systems (BSS) deployment, giving focus to electric BSS, and compares the on-road usage of EB and CB, in order to characterize their dynamic and energy usages profile and understand where the two technologies are more suited.

Bike-Sharing Systems: Definition and Benefits

The Public Bicycle System or Bike Sharing System (BSS) firstly appeared in Europe, approximately 50 years ago. After that, its popularity has grown considerably, and currently, it is possible to find BSS worldwide. Initially, the first systems were noticeably rudimentary and were used mainly for leisure activities. Afterwards, they have evolved into more modern schemes, representing an environmentally friendly and sustainable urban transportation solution. BSS are deployed essentially as a policy to reduce congestion and as a consequence to reduce pollutants emissions, to improve public health, to enhance the public transportation and to increase bike use [20].

BSS have been integrated in the public transportation system in order to connect commuters with other transports and address the issue of the "last mile" trip [21]. In more detail, bike sharing allows an extra mobility option for journeys which are too far to walk but too short to use a private vehicle or a public transportation. These systems have the potential to decrease private vehicle usage, to increase the number of cyclists in the streets and to promote modal shift [22]. However, BSS require the development of sustainable urban mobility policies mainly because they require infrastructure [23]. The implementation of these policies over a long-term period will not only achieve higher levels of cycling but will also allow the development of a cycling culture and therefore for it to be considered as a solution to address the urban mobility problem [22].

BSS are composed by a system of bicycles placed at docking stations placed at main locations around the city or particular areas. They allow one-way trips as well as round-trips at an instant access at the docking station by retrieving and parking bikes. Typically, members pay a membership fee to be admitted in the system, allowing them to use the system as many times as they want in that period. As an incentive to use of the bikes, the systems offer a free ride period (usually 30 minutes) after which additional fees are charged by intervals of time each 30 minutes extra, for example [24]. There are different types of ways users can join the system: annual, monthly, weekly, daily or per-trip basis, or 3 day passes. Depending on the type of system, users can access the bikes by using their credit cards (CC), membership card, by key or mobile phone. After using the bike, members can return the bike to any station they like and end their session when locking the bike at the dock.

In short-urban trips, bikes have significant advantages compared to other types of public transportation, mainly because they allow reaching destinations that other modes don't reach, they are less expensive to buy and maintain, they don't contribute to traffic congestion, they benefit the user in terms of exercising, etc. However, they also have disadvantages, such as, difficulty to use in some topography or weather conditions, can be viewed as unsafe by users and pedestrians, among others [25].

The typology of BSS deployed has varied along time and the following characterization can be made (Table 1) [26].

BSS have been implemented by municipalities as a sustainable urban mobility solution to reduce the environmental impact of the transportation sector and many BSS claim their environmental contribution [26]. Montreal's Bixi states that its program has saved over 1,360 ton of greenhouse gases since May 2009 [27]. Lyon states that its program, which was launched in 2005, has saved the equivalent of 8,437 ton of CO_2 from the atmosphere [28]. The Hangzhou system reveals that their bike sharing trips cover approximately 1.123,200 km, which would correspond to more than 200,000 kg of CO_2 if these km were travelled by a car [29].

Besides impacts in terms of reduced traffic congestion, BSS impacts in public health include environmental, social, economic and health benefits. Nevertheless, these benefits are not always equitable across socio-demographic groups. A study developed in London to examine the inequalities in uptake and usage of the city's scheme concluded that even though the system has an high usage level, usage is not equitably distributed, mainly because women and those living in deprived areas tend to not become members [30].

Regarding the transportation system, BSS reduce the dependence on private vehicle. Convenience, enhanced access, increased mobility, low cost, and urban space efficiency are also indicated as the main bike sharing advantages [31].

Additionally, these systems also have the potential to make people ride bicycles more and attract people who don't usually use a bike or to be a mobility option to business travelers to the city.

Table 1. Description of the different generations of BSS

1st Generation Distinguishing looking bikes (i.e. certain paint color) distributed around a city for free use, but subject to theft and poor organization.	2nd Generation Locking mechanism and checkout deposit payable at pickup. However, minimal deposit not enough to significantly reduce theft.
3rd Generation CC transactions and radio-frequency identification chips, allowing user identification and a security deposit needed to ensure accountability against theft and vandalism.	4th Generation Solar power and wireless communication, providing portable solutions (seasonal removals). The systems are modular and do not require excavation.

BIKE-SHARING WORLDWIDE

Early European public BSS were small-scale non-profit programs. One of the first examples was the Provos' White Bike, that started in 1965 in Amsterdam [20]. However, it rapidly failed due to theft and bicycle damage. As a result, the system evolved to coin-deposit systems (second-generation), which required a small coin deposit to unlock bikes from the docking stations. An example was Bycyklen in 1995 in Copenhagen, Denmark [26]. Additionally, France has been a promoter of IT-based systems (third-generation). The first 3rd generation system was deployed in 1998 in Rennes, France [26]. This BSS, named Smart Bike, was operated by Clear Channel Communications, Inc. and allowed free access for up to three hours for registered users, using a smartcard. JCDecaux launched Velo'v in Lyon in 2005. The Vélos Jaunes program in La Rochelle was replaced in 2006 with Yélo, a 3rd generation automated system. In 2007, Paris launched Vélib', which is today one of the world's largest public bike sharing programs [31].

In the last decade, the number of BSS has doubled. Nevertheless, since 2008 the increasing trends have decelerated, especially in Europe, due to the need to try to reduce the cost of BSS, and to mature individualized solutions to their particular problems (e.g. with a greater emphasis on either commuters or tourists). This results in slower decision-making processes, aggravated by the need for reliable and long term financing [32].

All European BSS have certain elements in common such as automated process for rental and return; easy and fast access; fixed stations; based on smart cards; customer registration using deposits, no anonymous use; one-way capability; high net density; and usually no charge for first 30 minutes. Currently, Taiwan, South Korea, China and Japan are some the world's fastest expanding markets for BSS.

These systems were initially launched in Asia in 1999, with Smart Bike (or TownBike) in Singapore, which was maintained until 2007.

In North America, even though bike sharing has been in action since the 1990s, the systems have developed quickly, initially with community-based and with second-generation BSS. The first US IT-based BSS, Tulsa Townies, began in Tulsa in 2007. It was the first worldwide solar-powered, completely automated docking-based system, and it does not charge for its service [33]. In the case of Canada, the first 2nd generation BSS, BIXI (BIcycle-TaXI), was initialized in 2009 in Montreal and expanded later to Ottawa and Toronto. Other example is Golden Community Bikeshare deployed in British Columbia in 2011.

Mexico City launched in February 2010 the program EcoBici, with 1200 bicycles at 90 stations, costing approximately US$25 per year. Other countries such as Argentina, Brazil, Colombia and Chile also have BSS programs. One of the first South America's BSS started in Santiago, Chile, in 2008 named B'easy. Rio de Janeiro, Brazil, developed the Samba program in 2009, which was later re-launched as BikeRio, with 60 stations, 600 bicycles and requiring that users make use of their mobile phone to unlock bicycles [31].

BIKE-SHARING SYSTEMS CHARACTERIZATION

BSS have been deployed around the world for nearly 40 years. Around 500 BSS were identified and their geographic distribution is presented in Table 2.

Table 2. Geographic distribution of bike-sharing systems and bicycle deployment

	Systems	Number of systems	Bikes	Number of bikes
Europe	354	71%	119333	24%
North America	45	9%	32791	7%
South America	18	4%	8065	2%
Asia	76	15%	329345	67%
Oceania	5	1%	2770	1%
Total	498	100%	492304	100%

Table 3. Top BSS (larger than 5000 bicycles)

Country	City	System	Number of bicycles
China	Wuhan	Public-Use Bicycle Program	70000
China	Hangzhou	Hangzhou Public Bicycle Service	65000
China	Shanghai	Rent me a Bicycle - Forever Bicycle Co.	28000
France	Paris	Vélib'	23500
China	Zhuzhou	Public Bicycle Rental	16000
China	Ninghai	Zhejiang Ninghai casual public bicycle system	10000
China	Taiyuan	New system - Electric	10000
China	Taizhou (Jiaojiang District)	Eversafe Public Bike System	10000
China	Taizhou (Linhai District)	Public Bike System	10000
UK	London	Barclays Cycle Hire	8000
China	Foshan – Chancheng	Chancheng Public bicycles	7600
US	New York	CitiBike - Alta Bike Share	7200
China	Nanchang	-	6000
China	Yantai	Yantai Publick Bike System	6000
Spain	Barcelona	Bicing	6000
Canada	Montreal	Bixi	5120
China	Foshan – Guicheng	Guicheng Public Bike Service	5000
China	Guangzhou	Public Bike Initiative	5000
China	Shenzhen – Yantian	Public Bike System	5000

They are mostly concentrated in Europe followed by Asia, and with North America if analyzing the number of systems. When analyzing the bicycles deployment around the world Asia presents a much higher value with 67% of bicycle deployment. Table 3 presents the larger BSS in the world. Excluding 5 systems in major world capitals, all the remaining large systems are implemented in China In order to better understand the city's typology where bike-sharing systems have been implemented a characterization in terms of area and population was performed, as presented in Table 4, for the 482 cities studied.

Table 4. Characterization of cities in terms of population, area, density and existence of a bikeway infrastructure

	Cities	Average number of inhab. (thousand inhab.)	Average area (km^2)	Density ($inhab/km^2$)	Bikeways		
					Cities identified	Deployment (km)	Inhab. per km of Bikeway
Europe	349	302	325	2248	58	140	7103
North America	41	924	2645	1923	25	278	5095
South America	15	3987	1025	5632	6	74	40980
Asia	72	3322	2721	3575	7	157	10370
Oceania	5	1922	6280	1973	-	-	-
Total	482	974	997	2546	96	173	8935

Table 5. Distribution between different types of system owners

	Type of system owners	
Agency	2	0.6%
Consortium	4	1.1%
Government	3	0.8%
Municipal Company	2	0.6%
Municipality	295	81.5%
Private	33	9.1%
Project	2	0.6%
Transport company	16	4.4%
University	5	1.4%
Total	362	100%

Bike-sharing systems are present in two different types of cities across the world: while European and American cities present similar population density (approximately 2000 inhabitants per km^2) and present a bikeways infrastructure (140-280 km), South American and Asian cities present much denser cities' profiles with a population density higher than 3500-5000 inhabitants per km^2.

Bikeways infrastructures are widely spread in Asia, but its presence is still scarce in South America, resulting in a high number of inhabitants served by each kilometer of bikeway.

The analysis of the different bike-sharing systems allows the identification of the main stakeholders responsible for their implementations. Table 5 demonstrates that most of BSS are owned by the city's municipality (around 80%). Other options include private companies, the city's transportation company and universities or consortiums.

In terms of the promoter of the BSS, Table 6 shows that private companies (≈80 %) own the main share as promoters of these systems, followed by the municipalities (8.3 %) and transport companies (5.6 %).

Table 6. Distribution between different types of promoters

	Type of system operator	
Agency	11	2.9 %
Consortium	11	2.9 %
Government	1	0.3 %
Municipal Company	10	2.7 %
Municipality	21	5.6 %
Private	293	78.6 %
Project	1	0.3 %
Transport company	21	5.6 %
University	4	1.1 %
Total	373	100 %

Table 7. Private promoters' distribution by number of cities and number of bicycles

System operators		Number of systems			Number of bicycles	
		For all systems		For systems >2	For all systems	
Bicicinttà System	Italy	67	13.5%	31.3%	4681	1.0%
next bike	Germany	38	7.6%	17.8%	4958	1.0%
JCDecaux	France	24	4.8%	11.2%	42342	8.6%
Adshell / Clear Channel Communications	US	12	2.4%	5.6%	11435	2.3%
Velopass	Switzerland	11	2.2%	5.1%	926	0.2%
metropolradruhr	Germany	9	1.8%	4.2%	2190	0.4%
callabike	Germany	8	1.6%	3.7%	4115	0.8%
B-cycle	US	5	1.0%	2.3%	883	0.2%
Shanghai Forever Bicycle Co.	China	5	1.0%	2.3%	35320	7.2%
Alta Bicycle Share	US	4	0.8%	1.9%	6325	1.3%
MIBISI - MOVUS	Spain	4	0.8%	1.9%	859	0.2%
Bike Nation	US	3	0.6%	1.4%	6600	1.3%
U-Bike	Taiwan	3	0.6%	1.4%	2130	0.4%
Veolia Transport / Veloway	France	3	0.6%	1.4%	1924	0.4%
CEMUSA	Spain	2	0.4%	0.9%	251	0.1%
Decobike	US	2	0.4%	0.9%	1000	0.2%
Effia	France	2	0.4%	0.9%	4640	0.9%
Impursa	Spain	2	0.4%	0.9%	390	0.1%
ISUDA	China	2	0.4%	0.9%	0	0.0%
Keolis	French	2	0.4%	0.9%	2445	0.5%
PBSC	US/Canada	2	0.4%	0.9%	330	0.1%
Smoove	French	2	0.4%	0.9%	1200	0.2%
ViaCycle	US	2	0.4%	0.9%	55	0.0%
Others (only 1 place)		284	57.0%	-	357305	72.6%
Total		498	100%	100%	492304	100%

From these private promoters, the most relevant in terms of market share were identified and are presented in Table 7. If analyzing the private promoters for the number of bike-sharing systems, the Bicicintà System present a considerable share. However, if we look to the deployed number of bicycles, the JCDecaux systems largely prevails.

Bike-sharing systems have been available in the last 4 decades, but in the last 5 years they have observed a very high deployment rate mainly in Europe, as is presented in Figure 2.

Analyzing the number of bicycles deployed and the number of docking stations available, a good comparison between the types of systems deployed worldwide is obtained (Table 8).

The average number of bicycles deployed per 1000 inhabitants is of approximately 2, with around 5 docking station per 10 km^2, with both North America and Europe following this trend. South America presents a low deployment of both bicycles and stations, while Asia presents a large deployment of bicycles but with few stations.

When analyzing the bike sharing stations deployment, it is observed that the average distance between stations is ≈400 m (Table 9).

Systems with a large number of bicycles tend to have lower distance between stations, while small systems tend to have higher distances.

In order to analyze the acceptance of these kinds of services, Table 10 presents the average number of members and compares it with the population and deployed bike profiles.

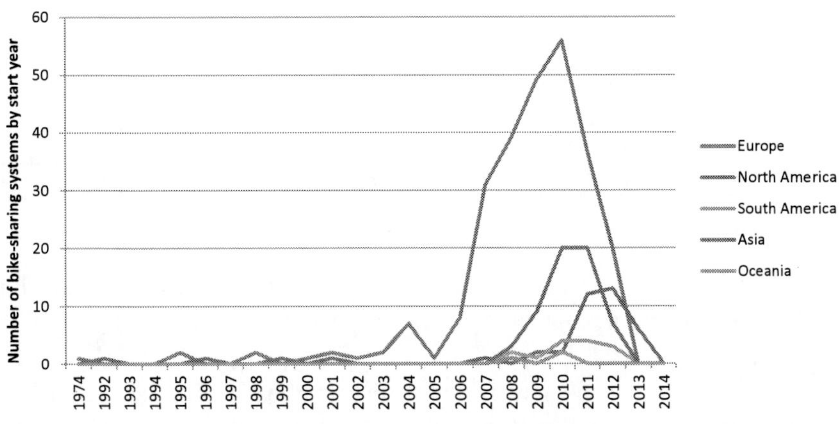

Figure 2. Bike-sharing systems' starting year by geographic region.

Table 8. Bike-sharing systems characterization in terms of number of bikes and stations

	Average number of bikes	Bikes per 1000 inhabitants	Average number of stations	Stations per 10 km^2
Europe	425	2.3	35	6.0
North America	781	1.4	73	3.3
South America	504	0.3	19	1.4
Asia	4916	3.2	161	2.5
Oceania	923	0.4	58	13.4
Total	1204	2.3	57	5.1

Table 9. Average distance (m) between stations

	Total systems	Systems size distribution		
		Small (<100 bicycles)	Medium (100-500 bicycles)	Large (>500 bicycles)
Europe	424	2000	388	329
North America	400	-	-	400
South America	650	-	1000	300
Asia	408	-	500	390
Oceania	500	-	-	500
Total	431	2000	439	364

Table 10. Membership characterization of the considered bike-sharing system

	Sampling	Average number of members	Members/Population (%)	Members per bike
Europe	25%	13753	4.7%	23
North America	36%	46617	11.5%	139
South America	33%	13127	1.4%	-
Asia	20%	39876	3.4%	5
Oceania	40%	7795	0.2%	20
Total	25%	20912	5.2%	80

European and North American systems present a significant adhesion to bike-sharing services (6-8%), while more populated areas like South America and Asia have lower adhesion values (≈1%).

In terms of number of member weighted to the number of bicycles, Europe and North America present similar values of ≈30 members for each bicycle, only surpassed by South America with ≈40 members for each bicycle.

A daily trip characterization was also performed as is presented in Table 11. In average, 5.7 trips per bike per day are performed lasting 18 minutes and 2.6 km. Each bicycle rides in average 15 km per day.

When analyzing the utilization of the BSS (see Table 12 and Table 13 for data per inhabitant and per BSS member), the Asian systems present the higher utilization rates with higher trips per inhabitant or member (9.62 trips per month compared to the average 2.123). However the comparison for annual data is not possible due to lack of data. Europe and North America present similar profiles with ≈20 trips performed annually and 2 monthly per each enrolled member.

In terms of pricing, three different types of bicycle usage are defined: regular, usage of bike-sharing system regularly in a monthly or annual basis; casual: few and uncommon usage of bike-sharing system in a daily or weekly basis; and EB, usage of an electric bike-sharing system.

Table 11. Daily trip statistics

	Trips/bike/day	Average time/trip (min)	Average distance per trip (km)	Average distance per day per bike (km)
Europe	6.3	19	2.3	14
North America	3.4	15	3.3	11
South America	-	-	-	-
Asia	5.9	-	-	-
Oceania	0.4	-	-	-
Total	5.7	18	2.6	15

Table 12. Usage characterization in terms of trips per inhabitant for the considered bike-sharing system

	Trips per inhabitant											
	Annual				Monthly				Daily			
	Total	Small	Medium	Large	Total	Small	Medium	Large	Total	Small	Medium	Large
Europe	0.708	0.155	0.334	2.544	0.018	0.013	0.020	-	0.043	0.000	0.090	0.023
North America	0.696	-	0.710	0.689	0.001	0.001	-	-	0.001	0.000	0.000	0.006
South America	0.015	-	0.023	0.006	0.000	0.000	-	0.000	0.011	0.016	-	0.001
Asia	-	-	-	-	0.574	-	0.006	1.143	0.029	-	0.000	0.033
Oceania	0.001	-	-	0.001	0.001	-	-	0.001	-	-	-	-
Total	0.652	0.155	0.361	1.502	0.082	0.006	0.018	0.381	0.030	0.004	0.063	0.026

Table 13. Usage characterization in terms of trips per member for the considered bike-sharing system

	Trips per member											
	Annual				Monthly				Daily			
	Total	Small	Medium	Large	Total	Small	Medium	Large	Total	Small	Medium	Large
Europe	27.367	14.072	23.370	55.573	1.501	-	1.501	-	0.100	0.023	0.036	0.263
North America	16.039	-	4.000	19.049	2.500	2.500	-	-	0.030	0.049	0.022	0.007
South America	3.116	-	2.656	3.575	0.348	0.348	-	-	0.091	0.128	-	0.053
Asia	-	-	-	-	9.615	-	-	9.615	0.374	-	-	0.374
Oceania	0.417	-	-	0.417	0.495	-	-	0.495	-	-	-	-
Total	23.239	14.072	20.506	32.550	2.133	1.424	1.501	5.055	0.120	0.045	0.031	0.248

Table 14 presents the share distribution for the free first minutes of bicycle usage. In a Regular usage most systems have 30 minutes free, and in a Casual usage no free minutes and up to 30 minutes free are almost equally divided.

Systems having EB provide the user with up to 30 minutes free (63%), however, 38% of the systems do not allow their users to have any free minutes.

Table 14. Free first minutes according to the type of bicycle usage

	Free (first x min)		
	Regular	Casual	EB
0	22%	31%	38%
0>t>=30	45%	40%	63%
30>t>=60	28%	22%	0%
t>60	6%	7%	0%

Table 15. Pricing mechanism and average values according to the type of bicycle usage

			Average value (€)
Regular	subscription		2.33
	annual fee		24.53
	monthly fee	CB	31.75
		EB	42.50
Casual	subscription	Casual	2.41
	7 days fee	CB	11.79
		EB	15.00
	3 days fee	CB	1.91
	daily fee	CB	4.55
		EB	17.00

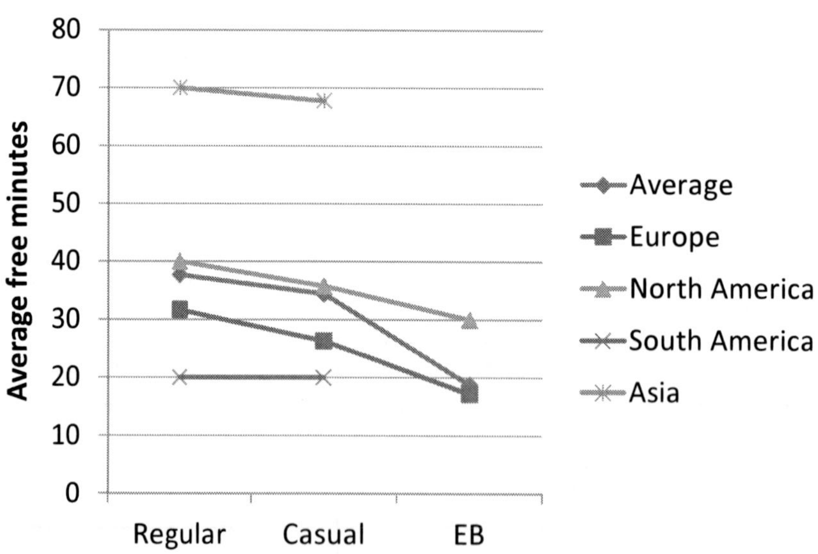

Figure 3. Average free minutes according to the type of bicycle usage.

For all the analyzed systems, Figure 3 presents the average free minutes associated to each type of bicycle usage, confirming that a Regular use presents higher free minutes (in average 38 minutes), followed by Casual (34 minutes) and by Electric bicycles (19 minutes). Asian systems present much higher free minutes values with up to 4 hours of free bicycle usage. Table 15 presents the pricing mechanisms associated to each bike-sharing system and the average values associated to each kind of pricing mechanisms. 8% of the analyzed systems present a subscription fee with an average value of ≈2€. 34% of the systems have an annual fee (≈25 €) and 14% have a monthly fee (≈32 €) applicable to Regular users.

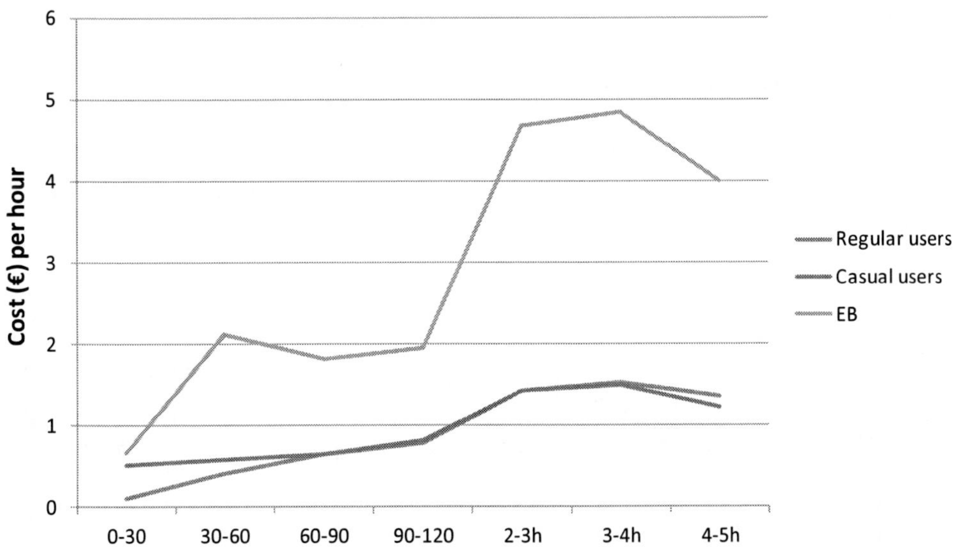

Figure 4. Cost per hour of renting the bicycles according to the type of bicycle usage.

For Causal users, 24% of the systems present a daily fee (≈5€), 16% present a weekly fee (≈12€), and 2% have a 3 days fee. For EB, the most used mechanism is a daily fee (≈17 €), followed by a monthly fee (≈43 €) and by a weekly fee (≈15 €).

Finally, the cost per hour of renting the bicycles is presented in Figure 4 a) according to the different types of usages. EB present a much higher profile, while the Casual usage profile also presents a slightly higher profile when compared to the Regular usage profile.

ELECTRIC BIKE-SHARING SYSTEMS

Over 40 electric BSS were identified, as is presented in Table 16. For the analyzed BSS, 68% of the systems are mixed bicycle sharing systems, while 32% correspond to only EB. For the totality of the systems, a 52% ratio between EB and the total number of bikes, while for mixed systems, that ratio corresponds to 22%. Additionally, the average number of EB being deployed per system is of 56 bicycles (excluding the Taiyuan system). Most of the systems are newly deployed systems (66%), while 34% correspond to conventional systems where EB were added. Little information is available on the type of EB used. Most EB seem to be bicycle adaptations and the differentiation between power-assist and power on-demand is difficult. The BSS service is not for free, with 88% of the systems charging for the use of electric bikes, while 12% provides the service for free.

On-Road Monitoring of Conventional and Electric Bicycles: Overview and Methodology

The use of EB in BSS presents an increasing trend. EB are expected to help reduce the human effort required as well as reduce travel time, though at a higher cost due to the electric system and the energy used.

Table 16. Summary of electric BSS and their characteristics

Electric Bike sharing system			Start	Promoter	Only EB		Bikes		Ratio EB/	System		Free	
Country	City	System			Y	N	CB	EB	(EB+CB)	New	Adapt	Y	N
Portugal	Águeda	Beagueda	2011		x			10	100%	x		x	
Spain	[1]	Bizimeta	2009	Rezikleta		x	93	70	43%	x			x
Spain	[2]	Lozocleta	2011			x	75	25	25%	x			x
Spain	Getxo	Getxo en bici/Getxo bizi	2009			x	57	10	15%				x
Spain	Guadalajara	Guay-bici	2010	Zero Vehículos Eléctricos		x	50	60	55%		x		x
Spain	Leioa	BiziLeioaBiziz	2011			x	67	6	8%	x			x
Spain	Palencia	Palencia te presta la bici		Urbaser		x	45	5	10%				x
Spain	Paterna (Valencia)	Bicipaterna	2008	MIBISI – MOVUS		x	290	10	3%		x		x
Spain	Rivas Vaciamadrid	Bicinrivas	2011			x	120	30	20%	x			x
Spain	Segovia	Segovia de BICIo	2009	Empresa Bicicletas de Castilla y León		x	98	97	50%	x			x
Spain	Torrent (Valencia)	Torrentbici	2009	MIBISI – MOVUS		x	390	10	3%	x			x
Spain	Villaquilambre	Villaquilambre te presta la bici /Biciquilambre	2008	Ingenia Soluciones		x	111	5	4%	x		x	
France	Grenoble	Métrovélo	2006	Semitag/Transdev		x	3200	?					x
France	Montpellier	Vélomagg'	2007	Smoove		x	1000	?			x		x
France	Orléans	Vélo'+	2007	Effia		x	350	?			x		x
France	Poitiers	Cap'Vélo	1998			x	221	76	26%		x		x
France	St Etienne	Vélivert	2010			x	760	130	15%		x		
Italy	Ascoli Piceno	Ascoli by bike	2012	Bicincittà system	x			20	100%	x			x
Italy	Arezzo	ARBike	2011			x	35	10	22%		x		x
Italy	Cantù	BICIamo		Bicincittà system		x	12	6	33%	x		x	
Italy	Capo D'orlando	Biciclando	2012	TMR S.r.l.		x	10	30	75%	x			x
Italy	Cascina	Cascina inBici	2011		x			8	100%	x		x	
Italy	Genova	Mobike	2009	Bicincittà system		x	60	?		x			
Italy	Jesi	Jesiinbici	2011	Bicincittà system	x			20	100%	x			x
Italy	Sestri Levante	TigullionBike	2012	Bicincittà system		x	18	12	40%				x
Italy	Siracusa	GoBike	2009	Bicincittà system		x	200	50	20%	x			x
Italy	Varese	Gimme Bike	2008	Bicincittà system		x	66	?			x		
Netherlands		OV-fiets				x	?	?			x		x
Germany	Aachen	Callabike	2011	Callabike	x		0	15	100%				x

[1] Berango, Sopelana, Urduliz, Barrika, Gorliz, Plentzia and Lemoiz.
[2] Gargantilla de Lozoya, Pinilla de Buitrago, Navarredonda, Villavieja de Lozoya and Buitrago de Lozoya.

Electric Bike sharing system			Start	Promoter	Only EB		Bikes		Ratio EB/	System		Free	
Country	City	System			Y	N	CB	EB	(EB+CB)	New	Adapt	Y	N
Germany	Stuttgart	Callabike	2004	Callabike		x	400	?	0%		x		x
Swithzerland	Several locations	Velopass		Velopass		x	88	?	0%		x		x
US	San Francisco	City Car Share - E-Bikes	2012	City CarShare		x	45	?	0%	x			
China	Taiyuan	New system – Electric	2012		x			10000	100%				
China	Xi'a n	Public Bike System - Forever Bicycle Co.	2011	Xi'an Jingjian Public Bicycle Co.		x	800	?		x			x
Singapore	Singapore	Smoove			x			?	100%	x			x
Japan	Awaji, Sumoto and Minami-Awaji			Awajishima tourism association	x			40	100%				
Japan	Fukutsu			Town Mobile Network Kitakyushu	x			30	100%				
Japan	Kitakyushu	City Bike	0		x		0	116	100%	x			x
Japan	Naha		2006	Non-Profit Org.	x			50	100%				
Japan	Sakai	Sakai Community bikes	2010		x			450	100%	x			x
Japan	Setagaya		2010	Non-Profit Org.	x			108	100%	x			x

Table 17. Characteristics of the bikers

Biker	Age	Sex	Biker Weight (kg)	Conventional bicycle use
B1	30	Male	70	Commuting and leisure
B2	32	Male	65	Short trips
B3	23	Male	75	Leisure
B4	27	Male	67	Commuting and leisure
B5	27	Male	76	Commuting

However, very few studies have quantified the real world benefits of EB in an urban environment. Therefore, it is important to develop monitoring methodologies in different regions to obtain comparable situations. The evaluation of EB and CB was performed through the monitoring of trips performed by 5 different bikers (B1 to B5), with each biker performing the same urban tour with both bicycles. The bikers used the EB first and the CB after, with a minimum resting period of 1 hour in-between.

The bicycles used by all bikers were the same, in order to enable a fairer evaluation. The specifications of the two bicycles used are the following:

- Conventional bicycle (CB) (Orbita Aluminio): weight of 15 kg, 21 gears; and

- Electric bicycle (EB) (QWIC Trend2): power-on-demand electric bicycle with 25.7 kg, 6 electric gears, 7 mechanical gears and a detachable Li-ion battery with a 360 Wh capacity, provided by Prio. Energy [34]

Regarding the bikers, they were all using an EB for the first time. To enable them to adjust to the use of a different bicycle, they were allowed to have brief test periods to understand how the electrical support operated. The characteristics of the bikers are shown in Table 17.

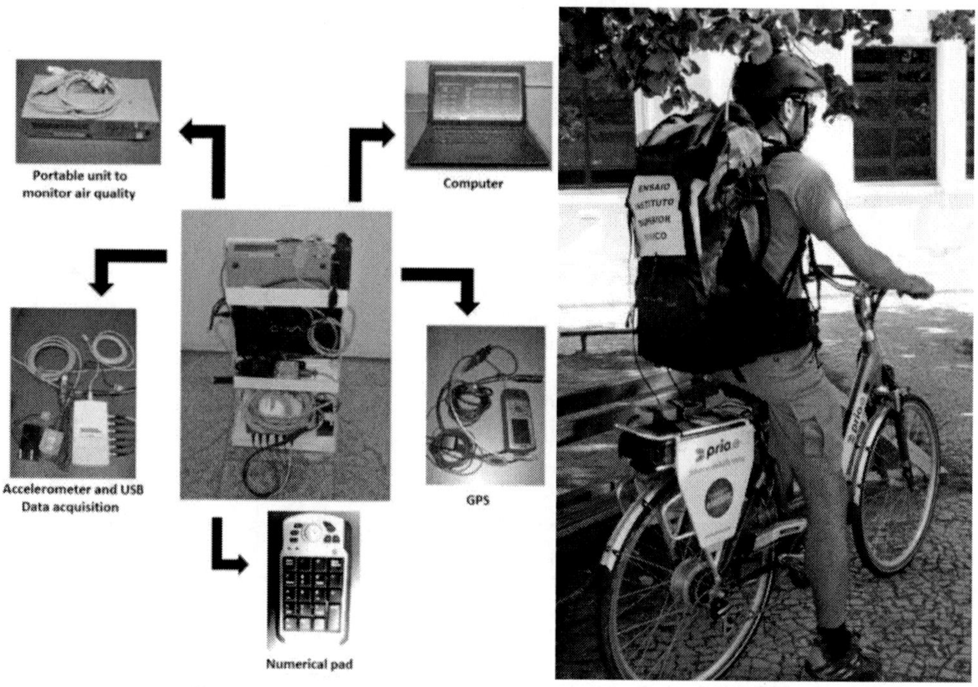

a b

Figure 5. MoveLab components and experimental apparatus used for the real time monitoring (a) and electric bicycle and one of the bikers of the performed study (b).

Table 18. Technical description of the equipment in MoveLab

Monitoring equipment	Data acquired	Temporal resolution of data
GPS (Garmin GPS map 76CSx)	Speed (km/h), altitude (m), location	1 s
Voltage and current probes (Fluke i1010)	Voltage, current	0.5 s

Each trip was monitored during the ride using a monitoring laboratory designed to assess energy and environmental impacts associated to non-motorized modes (MoveLab) and corresponds to a backpack weighting 12 kg that the user (pedestrian or biker) carries, as shown in Figure 5. MoveLab is equipped with a GPS to record the dynamic profile of the trip (including location, altitude and speed), a particle matter measuring equipment, voltage and current probes to assess the levels of electric assistance, and biometric sensors (recording heart rate and breathing intensity). All this equipment was carried by the rider in the backpack, however, for the analysis performed in this work, the data collected by the biometric sensors and air quality measurements was not considered. When asked to carry the MoveLab backpack, the bikers saw no inconvenience since in their daily routines they already carry backpacks weighting around 5 to 8 kg.

All the equipment is connected to a laptop to record the data throughout the travel. The technical description of the equipment used is presented in Table 18. Since the equipment used has different temporal resolution the data was processed into a second by second time basis.

The GPS allows collecting speed, location and also altitude information via an integrated barometric altimeter. The altimeter was adequately installed inside the backpack, avoiding that pressure fluctuations due to movement affected the readings. An external antenna was used to avoid GPS signal losses.

Voltage probes were installed directly in the bicycle battery terminals, while current measurements were done on the circuit that connects the battery to the electric motor. The signals provided by the probes were collected by a National Instruments DAQ board installed also on the backpack. For battery voltage signal a voltage divider circuit was placed before the DAQ board to account for the 0-10 V limit of the acquisition device. Both GPS and battery data were collected in a PC using a program developed in LabView by the authors to integrate the different communication protocols (serial port and NMEA protocol for GPS and analog data via USB port for the voltage and current collected in the DAQ board) that allows to synchronize the data, capturing all the equipment readings in a 1 Hz basis. A solid-state disk PC was used to avoid data loss while in motion.

GPS readings of speed were used to post process distance travelled, acceleration and road grade. Altitude and distance were used to determine the trip road grade using an algorithm that, for each point of the trip, finds the points 50 m before and after and uses this information to establish a second order polynomial fit based on three points of distance and altitude. The derivative of the polynomial fit in the studied point allows determining road grade, which is presented in rad. Battery data was used to determine, at each point of the trip, the power provided by the electric motor, according to the biker demands, and integrate this data along the trip to find the cumulative energy spent on the predefined tour.

With the data collected it is possible to understand and quantify how riders changed their use profile (in terms of speed and acceleration), changing from a conventional bicycle to an electric one and the strategies (regarding electric assistance) adopted by the five riders when operating the EB. To compare the use of conventional and EB, a round-trip tour of approximately 8.5 km in Lisbon, Portugal, was performed by each biker with both bicycles. The tour consisted on going from Instituto Superior Técnico (IST) main campus (point A and H of Figure 6 a) to downtown Lisbon and back, passing through the top of the Parque Eduardo VII (sections C and F) and Avenida da Liberdade (sections D and E) on both ways. With this tour, the bikers crossed different parts of the city of Lisbon including traffic intensive avenues, side roads with very little traffic and a street with a bike lane.

In terms of topography, the tour had significant slopes, as shown in Figure 6 which presents the altitude profile of the tour. This tour enabled the evaluation of the different bicycles in different slope types as well as the testing of the bikers' strategy for using the EB under different conditions.

On-Road Monitoring of Conventional and Electric Bicycles: Results

The trips performed with the electric and conventional bicycles by each biker were compared by using the data collected by the GPS system and the electric energy usage profiles. That corresponded to more 8 hours of data, which, for statistical significance, had the outliers with sampling lower than 10 data points removed.

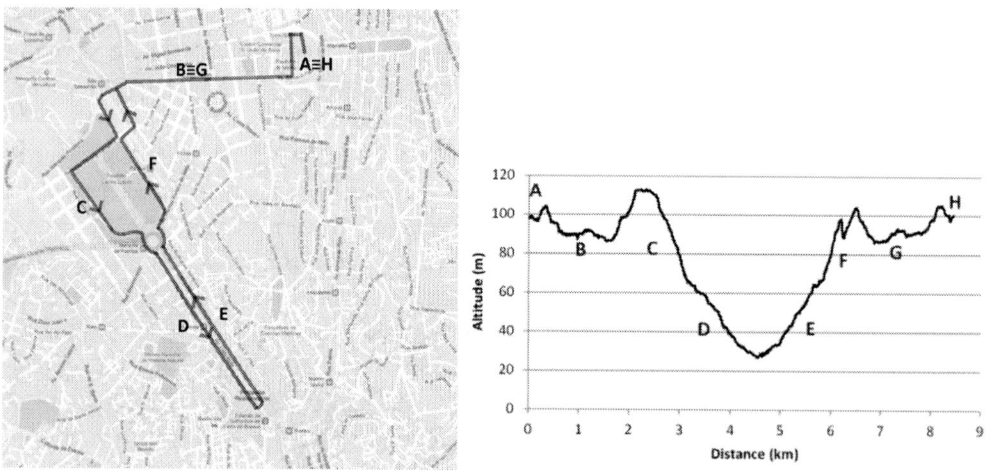

a b

Figure 6. Route map a) (source: Google Maps) and typical topography profile of the tour performed by the bikers b). Sections of tour: A and H – IST main campus; B and G – Plain avenue with bike lane; C – Park (descendant); D – Av. Liberdade (descendant); E – Av. Liberdade (ascendant); and F – Park (ascendant).

Table 19. Time, distance, mean speed and maximum speed by trip

Parameters	B1		B2		B3		B4		B5		Average	
	EB	CB	EB	CB	EB	EB	CB	CB	EB	CB	EB	CB
Time (min)	39	45	69	66	54	59	43	44	31	35	47	50
Distance (km)	9.3	9.2	8.8	7.8	9.1	7.8	8.8	8.0	8.4	8.5	8.9	8.3
Average speed (km/h)	14.2	12.2	7.7	7.1	10.0	7.9	12.3	10.8	16.6	14.5	12.2	10.5
Max speed (km/h)	45.4	40.7	26.2	34.6	38.9	37.5	42.2	46.2	42.9	45.8	39.1	41.0

The tour performed by the bikers was pre-defined, however some adjustments were made during the trip itself by the bikers based on obstacles, traffic and crossing roads positioning, which led to slightly differences on the distances travelled. The total amount of time required by each biker in each bicycle to perform the desired route, the total distance traveled, the average speed and the maximum speed can be seen in Table 19.

The changes in route made by the bikers resulted in a deviation of distance travelled between 0.8 km and -0.7 km in comparison with the 8.5 km reference value for the trip. These changes represent approximately a 10% change in total distance travelled. Nonetheless, based on the analysis of the data collected by the GPS system, the overall directions of the route were maintained by all bikers.

The average speed of the tour varied between 7.7 km/h and 16.6 km/h when using the EB and 7.1 km/h and 14.5 km/h when using the CB. For all the bikers, the use of the EB resulted in an increase of average speed during the tour. This increase ranged from 8%, corresponding to a 0.6 km/h higher average speed, to 26%, corresponding to a 2.1 km/h higher average speed.

The maximum speed reached by the bikers ranged from 26.2 km/h to 45.4 km/h when using the EB and from 34.6 km/h to 46.2 km/h when using the CB. Three of the bikers reached a higher maximum speed when using the CB.

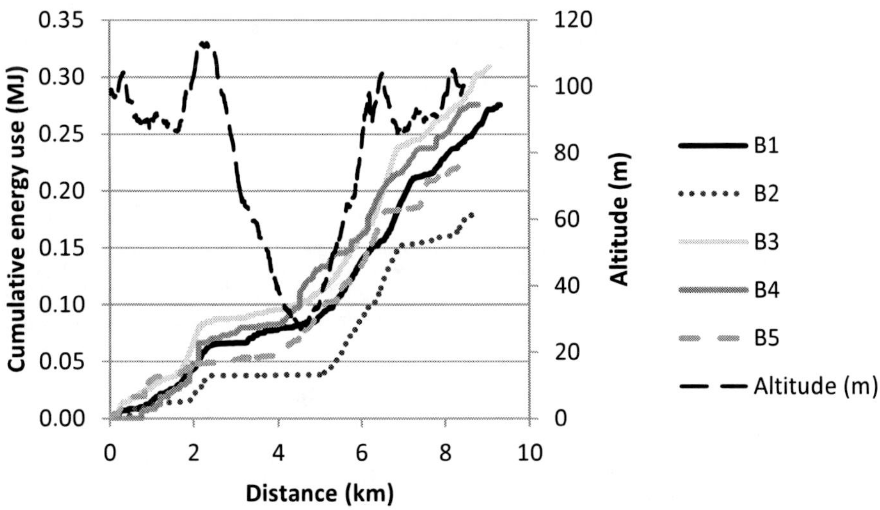

Figure 7. EB cumulative electric energy use per biker during the trips.

Table 20. TTW energy consumption when using electric bicycles

	B1	B2	B3	B4	B5	Average
Total energy (MJ)	0.276	0.179	0.309	0.276	0.228	0.254
Energy per kilometer (MJ / km)	0.030	0.020	0.034	0.031	0.027	0.028
Average speed (km/h)	14.2	7.7	10.0	12.3	16.6	12.2

Generally, the use of electric energy assistance in the EB occurred in the same parts of the tour, as shown in Figure 7. However, it is possible to see that some bikers used the electric support system more constantly throughout the journey, even if at lower levels of electric assistance.

The total electricity used by each biker is shown in Table 20, as well as the total amount of energy used in the trip divided by the total amount of kilometers made. These values regard the Tank-to-Wheel stage (TTW), which is related to the required energy consumption associated to driving the vehicle, in this case, the bicycle. It should be noted that this energy consumption is calculated over the total of the trip, which includes kilometers that were done both with and without the use of the electric support. The results do not show a clear relation between the increase in speed when moving from CB to EB and the amount of energy used.

This can be attributed to several different factors such as the low experience with EB of each biker, traffic conditions, and others.

The EB energy consumption per kilometer has an average of 0.028 MJ/km. When comparing the EB with the typical CB energy use (which rounds 0.06 MJ/km [35-37]), the electric energy share may range from 33 to 56% of the total energy use for the five bikers.

The EB energy consumption per kilometer is also much lower than the typical electric vehicle, with around 0.6 MJ/km, or a typical conventional vehicle, with around 2 MJ/km [5, 38], even considering that these are able to provide other types of mobility services. This proves that EB, even when considering an occupancy factor of 4 for vehicles, continue to be more efficient than cars. When analyzing the speed profiles, using the data collected by the GPS system, the average speed of each biker in each bicycle was calculated for different grade values, as shown in Figure 8 for the average of the bikers.

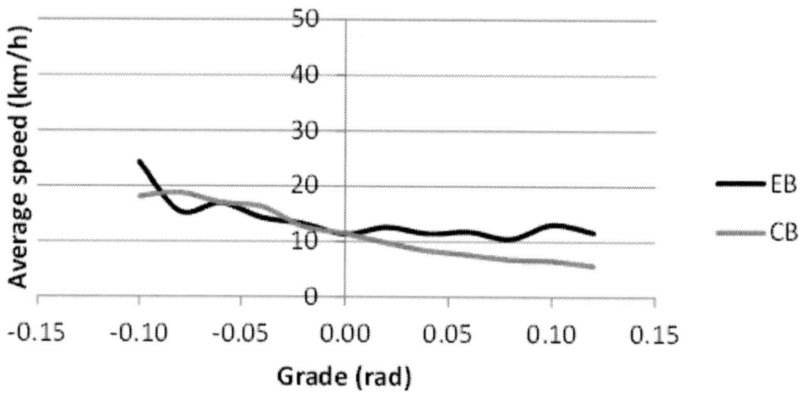

Figure 8. Average speed as function of road grade for the average of the bikers.

Table 21. Average speed increase by each biker in different slopes, when comparing the EB with the CB

	Negative slope	Neutral slope	Positive slope
B1	-15%	15%	36%
B2	-11%	7%	25%
B3	-12%	16%	86%
B4	-6%	5%	52%
B5	-11%	-7%	49%
Average	-11%	6%	49%

The analysis showed that the use of the EB led to an increase on the average speed obtained by the bikers in the parts of the tour with positive grades. While the average speed tended to decrease with the increase of the grade when using the CB, the bikers were able to stabilize or even increase their average speed on increasing positive grades when using the EB. On the other hand, when the grade was negative or near zero, the differences between the speeds obtained by the bikers with the different bicycles were small.

The average speed obtained by each cyclist in negative, neutral and positive slopes is shown in Table 21. The portions of the route with grades between -0.03 rad and 0.03 rad were considered to have neutral slopes. As mentioned before, the use of the EB enabled significant increases in speed in portions with positive slopes, which ranged from 25% to 86%, with an average of 49%.

In negative slopes, all bikers were faster with the CB. Overall, the average difference was of an 11% lower speed when using the EB. In neutral slopes, only one biker did not increase his speed when using the EB. Nonetheless, the average increase was of only 6%. These low changes in speed in negative and neutral slopes can be due to the lack of feeling safe when using the EB. Also, when the speed was already high with the CB, the electric option did not allow the bikers to gain much from its use.

Moreover, it is possible to observe different bikers' strategies when using EB. One of the most important factors when assessing the impact of using an EB concerns the user behavior when driving it. When analyzing the share of kilometers made with electricity backup for different slopes, as is presented in Figure 9.

Biker 2 uses electric backup mainly for positive slopes. In more detail, he used the electricity backup almost constantly in positive slopes with high energy use but only 43% in negative slopes and only 61% in neutral slopes with low energy uses.

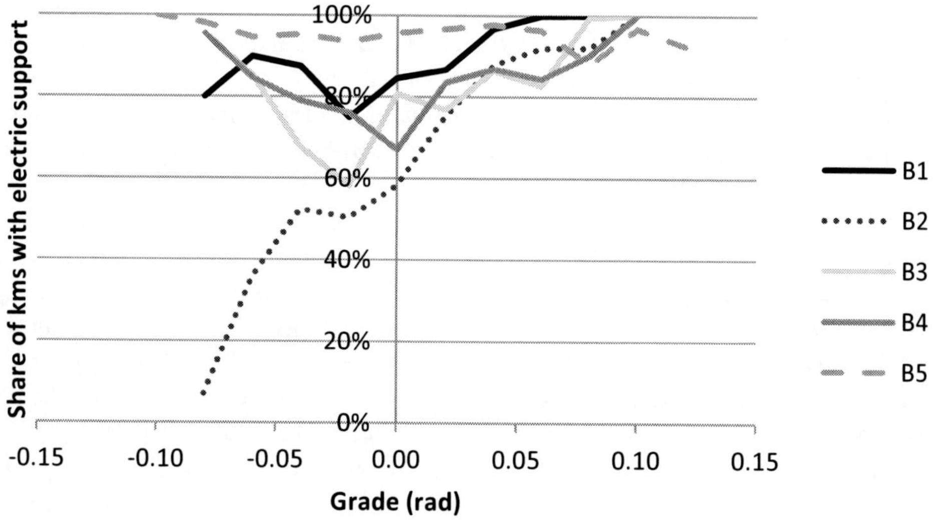

Figure 9. Share of kilometers made with electric backup in different slopes.

Bikers 1, 3 and 4 used the electric support mainly in parts of the tour with negative or positive slopes, but with the amount of energy required from the EB increasing with the slope. Finally, biker 5 used the electric backup constantly throughout the tour, with a share of kilometers made with electric support of 95%, but having a more balanced use of energy per kilometer than the other bikers (lower than other bikers in positive slopes and higher in negative slopes). Hence, medium electric assistance levels were used throughout.

CONCLUSION

This chapter presents a detailed characterization of an alternative transportation service being deployed worldwide, bike-sharing. Conventional and electric BSS were characterized, with almost 500 systems identified and characterized. From those, around 40 systems were identified as having EB. Europe has been a pioneer in the deployment of BSS, and currently prevails in terms of number of BSS. However, when considering bicycle deployment, Asia is clearly ahead.

The average number of bikes per 1000 inhabitants is ≈2 and the average number of docking stations per 10 km^2 is ≈5. The average distance between stations is ≈400 m. The estimated average trips per bike per day was of 5.7, corresponding to a 18 minute trip, covering 2.6 km, which represents an average distance per bike of 15 km. Most systems do not charge the first 30 minutes of usage, but these free first minutes tend to be lower for electric bikes. The average fees of BSS were estimated. Casual users' present higher subscription and usage cost when compared to Regular users. EB users also have an average 112 % increase in price.

In terms of electric BSS, the average number of EB being deployed per system is of 56 bicycles. 68% of the systems being deployed are mixed bicycle sharing systems, while 32% correspond to only EB. Considering the totality of the systems, a 52% ratio between EB and the total number of bikes was observed, while for mixed systems that ratio corresponds to 22%. In order to better understand the impacts of using EB in bike-sharing, the monitoring of an electric and a CB in an urban environment was promoted.

An evaluation of speed, travel time and the use of electricity for a conventional and an EB was performed with a laboratory collecting second-by-second trip data (including, speed, location, altitude, energy use, etc.).

Both bicycles have strengths and weaknesses that need to be considered when planning the introduction of bicycles in a city context. EB electric range (40-60 km) appears to be adequate for use in electric BSS (≈15 km daily). Secondly, EB are able to provide higher speeds in routes that require the biker to go through sections with high slopes due to the support that the electric backup can provide.

However, when the electric support is not used, whether by choice or because the battery is exhausted, the average speeds obtained are lower than the ones obtained with CB, which can be justifiable due to their much higher weight and lower number of mechanical gears.

CB have shown to be able to provide the same average speeds as EB in sections with negative and neutral slopes, even for bikers that use the electric backup heavily. The main drawback of CB concerns the sections of the route with positive slopes, in which the average speed obtained is much lower than the one achieved with an EB, even despite their much lower weight and higher number of mechanical gears.

In more detail, for the tour monitored, the average speed increase ranged between 8% and 26%. For sections with negative slopes the average speed decreased by 11% when moving from the CB to EB, whereas it increased in neutral slopes by 6% and in positive slopes by 49%. Nonetheless, the full impact of the shift from conventional to EB can depend on the strategy chosen by the biker concerning how to use the electric support.

As a result, the planning of mixed bike sharing schemes, and in particular the definition of the ratio between electric and CB, requires taking into consideration different factors such as the most common routes that will be served by the bicycles and the target population for the bicycles.

The routes that will most commonly be performed will determine the different slopes that will be required to overcome. The target population will provide information on the ability of the bikers to achieve higher speeds in CB and might influence the strategy that the bikers will take when using EB. The model of the bicycle is also an important aspect to take into consideration when planning such schemes.

In brief, the planning of mixed conventional and electric BSS should take into consideration the topography of the main routes that are going to be served in addition to the target population that will use the bicycles, in order to have more effective policies to encourage the use of BSS.

Acknowledgments

Thanks are due to the researchers that collaborated in these studies and to the five bikers that accepted to perform the monitored tour. The author acknowledges Fundação para a Ciência e Technologic for the Post-Doctoral financial support (SFRH/BPD/79684/2011). The author would also like to acknowledge PRIO.E for providing the electric bicycle.

References

[1] EUROSTAT, Environment and Energy, EUROPA Eurostat – Data Navigation Tree, 2012 [cited September 2012].

[2] European Environment Agency, National emissions reported to the Convention on Long-range Transboundary Air Pollution (LRTAP Convention), 2010.

[3] IEA, Energy Technology Perspectives, scenarios and strategies to 2050 by International Energy Agency, 2010.

[4] Pina, A., Ioakimidis, C., Ferrão, P., Introduction of electric vehicles in an island as a driver to increase renewable energy penetration, In: *IEEE International Conference on Sustainable Energy Technologies* 2008, 2008, Singapore.

[5] Baptista, P., C. Silva, T. Farias, and J. Heywood, Energy and environmental impacts of alternative pathways for the Portuguese road transportation sector, *Journal of Energy Policy*, 2012, 51, p. 802–815.

[6] Wang, R., Autos, transit and bicycles: Comparing the costs in large Chinese cities, *Transport Policy*, 2011, 18, p. 139–146.

[7] Shaheen, S. and A. Cohen, *Worldwide Carsharing Growth: An International Comparison,* U.D. Institute of Transportation Studies, 2008.

[8] Lindsay, G., A. Macmillan and A. Woodward, Moving urban trips from cars to bicycles: impact on health and emissions, *Australian and New Zealand Journal of Public Health,* 2011, 35(1).

[9] Urban Audit, Urban Audit: Welcome to the Urban Audit web site, [cited October 2012; Available from: http://www.urbanaudit.org/.

[10] Freemark, Y., Transit Mode Share Trends Looking Steady; Rail Appears to Encourage Non-Automobile Commutes, *The Transport Politic*, 2010 [cited October 2012]; Available from: http://www.thetransportpolitic.com/2010/10/13/transit-mode-share-trends-looking-steady-rail-appears-to-encourage-non-automobile-commutes/.

[11] Buehler, R. and J. Pucher, Walking and Cycling in Western Europe and the United States - Trends, Policies, and Lessons, *TR News* May-June 2012: Active Transportation: Implementing the Benefits, 2012.

[12] Gardner, G., Power to the Pedals, *World Watch Magazine*, 2010, 23(4).

[13] Dill, J. and G. Rose, E-bikes and Transportation Policy: Insights from Early Adopters, In: *91th Annual Meeting of the Transportation Research Board*, 2012.

[14] Galp Energia, Ciclovias e percursos pedonais, 2012 [cited September 2012]; Available from: http://www.galpenergia.com/PT/sustentabilidade/eficiencia-energetica/na-sociedade/o-desafio-da-mobilidade-sustentavel/Paginas/Ciclovias-e-percursos-pedonais.aspx.

[15] Martinez, L., L. Caetano, T. Eiró, and F. Cruz, An optimisation algorithm to establish the location of stations of a mixed fleet biking system: an application to the city of Lisbon, In: *15th Edition of the Euro Working Group on Transportation*, 2012, Paris.

[16] Cherry, C., J. Weinert and Y. Xinmiao, Comparative environmental impacts of electric bikes in China, *Transportation Research Part D,* 2009, 14, p. 281–290.

[17] Cherry, C. and R. Cervero, Use characteristics and mode choice behavior of electric bike users in China, *Transport Policy*, 2007, 14, p. 247-257.

[18] Capelli, C., et al., Energy cost and efficiency of riding aerodynamic bicycles, *European Journal of Applied Physiology* 1993, 67, p. 144-149.

[19] Parkin, J. and J. Rotheram, Design speeds and acceleration characteristics of bicycle traffic for use in planning, design and appraisal, *Transport Policy*, 2010, 17, p. 335–341.

[20] Shaheen, S., S. Guzman and H. Zhang, Bikesharing in Europe. the Americas. and Asia: Past. Present and Future, *Transportation Research Record* 1659, 2010.

[21] Liu, Z., X. Jia and W. Cheng, Solving the Last Mile Problem: Ensure the Success of Public Bicycle System in Beijing, *Procedia - Social and Behavioral Sciences*, 2012, 43, p. 73-78.

[22] Kumar, A., K. Teo and A. Odoni, *A Systems Perspective of Cycling and Bike-sharing Systems in Urban Mobility*, 2012.

[23] Global Transport Knowledge Partnership – International Road Federation, Bicycle-sharing schemes: enhancing sustainable mobility in urban areas, *Background Paper No. 8* CSD19/2011/BP8, 2011.

[24] Daddio, D. W., Maximizing bicycle sharing: an empirical analysis of capital bikeshare usage, *C.H. Master's Project*, Editor, 2012.

[25] DeMaio, P. and J. Gifford, Will Smart Bikes Succeed as Public Transportation in the United States?, *Journal of Public Transportation*, 2004, 7(2).

[26] DeMaio, P., Bike-sharing: History, Impacts, Models of Provision, and Future, *Journal of Public Transportation*, 2009, 12(4).

[27] BIXI, [cited September 2012]; Available from: http://www.bixisystem.com/.

[28] Velo'v, [cited September 2012]; Available from: http://www.velov.grandlyon.com/?L=1.

[29] ICLEI – Local Governments for Sustainability, Hangzhou, China: The world's largest bike sharing program, 2011.

[30] Ogilvie, F. and A. Goodman, Inequalities in usage of a public bicycle sharing scheme: Socio-demographic predictors of uptake and usage of the London (UK) cycle hire scheme, *Preventive Medicine*, 2012, 55, p. 40-45.

[31] Mineta Transportation Research, Public Bikesharing in North America: Early Operator and User Understanding, *MTI Report II-26*, Editor, 2012.

[32] Spicycle, Cycling on the rise - Public Bicycle and Other European Experience, 2009.

[33] Tulsa Townies, [cited Ocotober 2012]; Available from: http://www.tulsa-townies.com/default.aspx.

[34] Prio Energy, Prio Energy - Electric Mobility, 2012 [cited October 2012]; Available from: http://www.prioenergy.com/.

[35] Ulrich, K., *The Environmental Paradox of Bicycling, Working Paper* — Department of Operations and Information Management, The Wharton School, 2006.

[36] Lemire-Elmore, J., *The Energy Cost of Electric and Human-Powered Bicycles*, in Term Paper 81781999, 2004.

[37] Bouwman, M., *An environmental assessment of the bicycle and other transport systems*, 2000, Center for Energy and Environmental Studies of University of Groningen, Faculty of Spatial Sciences, The Netherlands.

[38] Baptista, P., Tomás, M., Silva, C., Plug-in hybrid fuel cell vehicles market penetration scenarios, *International Journal of Hydrogen Energy*, 2010, 35(18), p. 10024-10030.

CONTRIBUTORS

Andrew A Frank (Chapter 1) is CTO at Efficient Drivetrains Incorporated and Prof Emeritus of Mechanical and Aeronautical Engineering University of California-Davis. Education: BS and MS Mechanical Engineering Univ. of California Berkeley 1955, 1958MS and Phd Electrical Engineering, Univ. of Southern California 1965, 1967. Experience: (1). 15 years at Rockwell International work on Helicopters, Military missiles, Apollo Rocket to the Moon. (2). 18 years at Univ. of Wisconsin, Electrical and Computer Engineering Hybrid Computing Lab Director with Research on Hybrid Electric and Flywheel cars. (3). 25 Years at the Univ. of California-Davis. Research on the Design of Energy Storage Vehicles and Plug-In Hybrid cars and trucks using CVT's and advanced powertrain concepts. (4). 6 Years CTO of Efficient Drivetrains, Commercializing his patent portfolio on Plug-In Hybrid Electric and Electric cars. Prof Frank's objective now is to provide the technology necessary for the world to transition off of Fossil Fuel to Renewable Energy in the shortest possible time frame and with the minimum cost but with winners in all segments of society from industry to the common man.

Alexander Hanzlik (Chapter 2) holds an Msc degree in Computer Science and a PhD degree in Computer Engineering from the Vienna University of Technology as well as an

MBA degree from the PEF Private University of Management, Vienna. His main research interests are Simulation, Fault-tolerant Computing and Distributed Real-Time Systems. He has about 20 years professional experience in software development, system design and project management. He is author and co-author of more than 20 scientific papers. Currently, Alexander Hanzlik is under contract at the AIT Austrian Institute of Technology, where he is concerned with design and simulation of distributed electronic control systems for automotive applications.

Erwin Kristen (Chapter 2) studied Electronics and Economy at the University of Applied Science Technikum Vienna. He worked as a hardware developer at ALCATEL and as system engineer in railway safety technology. For more than 15 years he has been working on time triggered bus systems, like TTP/C and FlexRay. Mr. Kristen gained experience in the automotive industry as FAE at DECOMSYS. Since 2003 he is the research Project leader at AIT for automotive and industrial networks.

Adi Wolfson (Chapter 3) is the head of the Chemical Engineering department at Sami Shamoon College of Engineering (SCE) and one of the founders of SCE's Green Processes Center. Prof. Wolfson earned his PhD summa cum laude in chemical engineering at Ben-Gurion University of the Negev in Beer-Sheva, Israel, and continued his post-doctoral studies at the University of Leuven, Belgium. Prof. Wolfson has worked in the field of green chemistry and engineering for the last 10 years, mainly researching catalytic applications in organic reactions and the development of alternative, environmentally friendly solvents. Prof.

Wolfson is also working with national and local authorities and he collaborates with industry to promote sustainability and sustainable development in Israel.

Dorith Tavor (Chapter 3) is the Dean of Chemical Engineering, chair of the Green Processes Center and chair of the Center of Learning Promotion at Sami Shamoon College of Engineering (SCE). Prof Tavor earned her PhD in chemical engineering and MBA at Ben-Gurion University of the Negev in Beer-Sheva, Israel. Prof. Tavor founded the chemical engineering department and the Green Processes Center at Sami Shamoon College of Engineering. Prof. Tavor's work in the field of green engineering focuses mainly on catalytic applications in organic reactions and the development of alternative, environmentally friendly solvents and the recycling of industrial wastewater. Through her connections in academia, the community, and industry, Prof. Tavor is working to promote sustainability and sustainable development in Israel. In addition, Prof. Tavor is the director at SCE of an incentive program to encourage the greater involvement of women and girls in science and technology.

Shlomo Mark (Chapter 3) is the chair of the Software Engineering Department and the head of the Negev Monte Carlo Research Center (NMCRC) at Shamoon College of Engineering (SCE). Dr. Mark earned his Ph.D. in nuclear engineering and an M.Sc. in biomedical engineering and in management and safety engineering. His main research interests are service science, agile programming, scientific programming, computational modeling for physical, environmental and medical applications, and Monte Carlo simulations.

Scott Peterson (Chapter 4) recently earned a PhD from Carnegie Mellon University in Engineering and Public Policy where he studied battery degradation, emissions and integration of plug-in vehicles with the grid. He also holds a Masters in Environmental Engineering from Texas A&M University and a certificate in Advanced International Affairs from the Bush School. He graduated Phi Beta Kappa with a degree in Chemistry from Texas A&M University. He is interested in technical, political, and economic analysis of energy, transportation and sustainability issues.

Jamie Davies (Chapter 5) is a consumer research analyst at the UC Davis Plug-in Hybrid & Electric Vehicle Research Center. He is engaged in the evaluation of consumers' use of and experience with alternative fueled vehicles. Emphasis is on the design and practice of qualitative and quantitative research to measure consumers experience with PEVs, understand market potential, and assess impacts of vehicle attributes and charging infrastructure on driving experiences. Broader goals and motives of his research projects are to understand vehicle environmental impacts, consumer and fleet purchase trends and behavior, how to build value for potential alternative fueled vehicle customers, and other strategies to increase market share.

Ken Kurani (Chapter 5) is an Associate Researcher at the University of California, Davis' Institute of Transportation Studies; he has a PhD in Civil and Environmental Engineering. Within the interdisciplinary setting of the Institute, his research focus on the role of human behavior: he works at the intersection of lifestyle, automobility, energy, and the environment. He is especially interested in enriching the behavioral approaches to understanding consumers' responses to new transportation technologies such as new propulsion systems and fuels for automobiles. His research also explores how citizen/consumers can use such new technologies to shape both their own lives as well as efforts to market and regulate transportation and communication networks according to the characteristics of energy efficiency, air quality, safety, and social equity. His present research includes household response to hybrid, plug-in hybrid, and electric vehicles as well as consumer/citizen valuation of energy use and its private, social, and environmental consequences. This work focuses on interfaces between people and their energy use that help to translate energy from an abstraction into actions.

Åshild Vatne (Chapter 6) received her B.Sc Electrical Engineering from Høgskolen i Bergen (College of Bergen, Norway) in 2010, and M.Sc in Electrical Engineering from the Norwegian University of Science and Technology (NTNU) in 2012. This chapter is based on her final project and master thesis.

Marta Molinas (Chapter 6) received her M. Eng from University of the Ryukyus (Japan) in 1997 and Dr. Eng from Tokyo Institute of Technology (Japan). She is currently Professor in Department of Electric Power Engineering, Norwegian University of Science and Technology.

Jan A. Foonæs (Chapter 6) received his M.Eng from Norges Tekniske Høgskole (NTH) in 1974, and Dr. Eng from NTH in 1978. He is currently Senior Adviser in Nord-Trøndelag Elektrisitetsverk AS (NTE) and Adjunct Associate Professor in Department of Electric Power Engineering, Norwegian University of Science and Technology.

Filipe J. Soares (Chapter 7) received the Physics degree (five-year course) from the Faculty of Sciences and an Electrical Engineering (Renewable Energies) Postgrad from Porto University, in 2004 and 2007, respectively. He also received the Ph.D. degree in Sustainable Energy Systems, in the MIT|Portugal Program, from Porto University, in 2012. Currently he is a researcher in the Power Systems Unit of INESC Porto and his research activity is directed towards the integration of renewable energy sources and electric vehicles in electric distribution grids, as well as to the development of advanced functionalities for the management of electric vehicles' charging and their participation in the electricity markets. In these areas, he has actively participated in the European project MERGE and in the Portuguese project REIVE, both focusing the integration of EV in the electric power system. He is author or co-author of more than 40 papers.

J. A. Peças Lopes (Chapter 7) obtained a degree in Electrical Engineering at the Faculty of Engineering of Porto University (FEUP) in 1981. In 1988 he completed his Ph.D. and in 1996 he obtained his Aggregation degree both at FEUP. He is presently Full Professor at FEUP and member of the Board of Directors of INESC Porto. He was responsible by INESC activities in EU research projects MicroGrids and MoreMicroGrids and was the Scientific Coordinator of the EU-financed MERGE project. He has been responsible for a large number of consultancy and research projects in Portugal, Cape Verde and Brazil under contracts with the industry and research national projects. He was also responsible by the participation of INESC in the Green Islands project. Prof. Peças Lopes has published more than 200 papers and supervised more than 20 post graduation students.

Pedro R. Almeida (Chapter 7) received the Electrical Engineering degree (five-year course) from the Faculty of Engineering of Universidade do Porto in 2007. He also received the Ph.D. degree in Sustainable Energy Systems, in the MIT|Portugal Program, with Universidade do Porto, in 2012. Currently he is a Power System Engineer in the Power Systems Consulting Unit of Tractebel Engineering (GDFSUEZ) where he conducts consulting activity for different players in power systems develops research activity directed towards the integration of distributed energy resources in electric distribution grids, in particular planning activities and integration of distributed energy resources in balancing and wholesale electricity markets, Belgian project GREDOR. Previously, he had actively participated in the European project MERGE and in the Portuguese project REIVE, both focusing the integration of EV in the electric power system. He is author or co-author of more than 40 papers.

Carla M. Silva (Chapter 8) received the Mechanical Engineering degree (five year course) in 2000 and the PhD degree in Mechanical Engineering in 2005, both at Instituto Superior Técnico, Technical University of Lisbon, Lisboa Portugal. She went for a post doc at both IST and University of Michigan working on CO2 mitigation in road vehicles. Currently she is a research assistant at IDMEC-Institute of Mechanical Engineering at IST. She coordinates the Vehicle simulation and Life cycle research group of IDMEC. She has been awarded 7 prizes of recognition. She is author/co-author of more than 20 per-reviewed Isi Web of knowledge papers and has more than 50 papers in International conferences. Areas of interest are new fuels; alternative vehicles simulation; Energy and environment.

Sara Marques (Chapter 8) completed her degree in September 2004 in Mechatronics Engineering. Her professional life began with an internship at Siemens S.A. – Electrical Switchboard factory in Corroios, Portugal. Afterwards she joined ATEC – Training Academy for Industry as assistant coordinator from Industrial Maintenance Technician (Mechatronics) and as trainer in the field of electricity and electronics. In 2007 she was part of the preparing course to enter Volkswagen Autoeuropa (New Engineers Course), and at that year-end she integrated the Strategic Industrial Engineering department in Volkswagen Autoeuropa. With the purpose of deepening her knowledge in the industrial field area, she entered MITPortugal

Engineering Design and Advanced Manufacturing (EDAM) area, in Advanced Studies Course Technology Management Enterprise (TME). From September 2009 is a member of IDMEC, in Energy, Environment and Transports Research Group (DTEA) in Instituto Superior Técnico, as PhD student (MITPortugal – Leaders for Technical Industries), performing research in the fields of life-cycle analysis and automotive energy and environmental management.

Teresa Batista (Chapter 8) received the Environmental Engineering degree (five year course) in 2007 in Universidade de Coimbra, Portugal.

She collaborated with a Portuguese NGO, Quercus ANCN in a program of energetic efficiency in the domestic sector consumption and after that she cooperated with EDP in the implementation of an Environmental Management System in a power station. Then she integrated an internship programme, "Inov Contacto" in Italy.

She is now collaborating with IDMEC-Institute of Mechanical Engineering at Instituto Superior Técnico, Technical University of Lisbon and finishing her master degree thesis about methodologies of environmental ranking vehicles, theme in which she had already worked as a researcher of MIT Portugal.

Alexandre Lucas (Chapter 8) received the Electrical Engineering degree in 2007 in Instituto Superior de Engenharia de Lisboa and in 2009 Masters in Business and Industrial Strategy by Instituto Superior de Economia e Gestão. From 2009 to 2013 he pursued his PhD degree in Sustainable Energy Systems by the MIT Portugal Program (IST/MIT). He developed research at MIT during 2011/12 academic year while attending complementary studies at Harvard E. School in Energy Management. Since 2005 he has worked with EFACEC Systems Engineering form base project development and tendering for international markets, to project management. Currently, he is a team coordinator in the Energy Transmission and Distribution area. He is a published author in peer-reviewed International journals in the Energy area and has focused his research on energy supply infrastructures. Areas of interest are: Life cycle analysis, energy conversion technologies; alternative vehicles; energy management and environment.

Brandon M. Graver (Chapter 9) was born in New York's picturesque Hudson Valley in April 1986. He received his Bachelor of Science in Environmental Engineering and Science, Technology & Society from the oldest technological university in the English-speaking world – Rensselaer Polytechnic Institute – in May 2008. He joined the Department of Civil, Construction and Environmental Engineering at North Carolina State University in August 2008 to pursue a Master of Science degree. His research focuses on the estimation and measurement of energy use and emissions in the transportation sector.

H. Christopher Frey (Chapter 9) is Distinguished University Professor of Environmental Engineering in the Department of Civil, Construction, and Environmental Engineering at NC State. He leads research on measurement and modeling of real-world fuel use and emissions of onroad and nonroad vehicles and on exposure and risk analysis, and has conducted research on quantification of sensitivity, uncertainty, and variability in environmental and risk models, quantification of uncertainty in emission factors and inventories, and modeling and evaluation of power generation and environmental control systems. He has published over 100 journal papers, 160 conference papers, and 60 technical reports, and has authored or co-authored 7 book chapters and 1 book.

He is Chair of the U.S. Environmental Protection Agency's Clean Air Scientific Advisory Committee (CASAC), which advises the EPA Administrator on the National Ambient Air Quality Standards. He is chairing CASAC review panels for ozone, lead, and oxides of nitrogen. He is a member of the EPA Science Advisory Board. He was a lead author for guidance on uncertainty in greenhouse gas emission inventories for the Intergovernmental Panel on Climate Change. He has also served on expert panels of the National Research Council and World Health Organization. He is chair of the Air & Waste Management Association's ET-1 technical committee on mobile source emissions and advisor of the NC State student chapter of AWMA.

He is a fellow of A&WMA and a fellow and past president of the Society for Risk Analysis. He received the 2008 NCSU Alumni Association Outstanding Research Award, the 1999 Chauncey Starr Award of SRA, and the 2012 Lyman A. Ripperton Environmental Educator Award of the A&WMA. Dr. Frey has a B.S. Mechanical Engineering from the University of Virginia, and from Carnegie Mellon University he has a Master of Engineering in Mechanical Engineering and PhD in Engineering and Public Policy.

Delfim Pedrosa (Chapter 10) was born in Leiria, Portugal on July, 1986. He is a PhD student at Centro Algoritmi of University of Minho. He finished his Integrated Master Course in Industrial Electronic Engineering and Computers in the Minho University in 2010. After that he was a Research Fellow at the Department of Industrial Electronics of the University of Minho, Portugal. His main researching interests are in Industrial Electronics and Electrical Vehicles.

Francisco P. Brito (Chapter 10) is a Post-doctoral Researcher at the University of Minho (Guimarães, Portugal) and Invited Assistant Professor at the Mechanical Engineering Department of ISEP (Porto, Portugal).He has been involved in the coordination team of MIT-Portugal Project MOBI-MPP on the Assessment and Development of Integrated Systems for

Electric Vehicles and has published around 20 articles in Scientific books, Journals and Conferences, including the book “Carros Elétricos” (Publindústria, 2012). His current research interests include vehicle energy efficiency and waste heat recovery as well as internal combustion and electric mobility powertrain Systems.

Jorge Martins (Chapter 10) got his PhD in the area of Internal Combustion Engines from the University of Birmingham, UK. Currently he is Associate Professor at the University of Minho, Portugal, where he lectures on Thermodynamics and I. C. Engines, and where he is the head of the Thermal Engines Laboratory. He has been an Invited Professor at various universities, mainly in Brazil. For the last 5 years Jorge Martins has been working on the Development of Electric Vehicles. He leads a large Portuguese project on that subject, with the participation of three major Portuguese universities (Minho, Porto and Technical of Lisbon).He has authored two books: Internal combustion Engines (in Portuguese), now on the 4th edition (2013), and Electric Cars (in Portuguese), edited in 2012, and over 70 technical and scientific papers. He was International Expert of FP5, FP6 and FP7 for DG XII and DG TREN in various calls for the evaluation of European Projects.

João Luiz Afonso (Chapter 10) was born in Rio de Janeiro, Brazil, in 1963. He is Associate Professor at the Department of Industrial Electronics of the University of Minho, Portugal, where he works since 1993. He received the degree in Electrical Engineering and the M.Sc. degree in Electrical Engineering from the Federal University of Rio de Janeiro, Brazil, in 1986 and 1991, respectively. In 2000 he obtained his PhD in Industrial Electronics from the University of Minho, Portugal. He lectures the subjects of Electrical Machines,

Complements of Power Electronics, Electrical Power Quality, Active Power Filters, and Renewable Energy. His researching activities are related with the development of Active Power Filters, Power Quality Analyzers, Power Electronics for Renewable Energy Sources and Electric Vehicles, and with studies of Power Quality and Energy Efficiency.

Vítor Monteiro (Chapter 10) is a researcher at the University of Minho, Centro Algorimi, 4800-059 Guimarães, Portugal. Currently he is a PhD student in the Group of Energy and Power Electronics (GEPE). His research interests are related with Power Electronic Converters, Digital Control, Power Quality, Smart Grids, and Power Electronics towards Electric Vehicles.

Patrícia Baptista (Chapter 11) received the Chemistry degree in 2006 at Instituto Superior Técnico, Technical University of Lisbon, Portugal, with her final essay focused on the development of biodiesel quality control methods. She obtained her PhD in Sustainable Energy Systems within the MIT Portugal Program at Instituto Superior Técnico in 2011, in the area of the evaluation of the impacts of alternative vehicle technologies and alternative energy sources in the road transportation sector. Patricia is currently a Postdoctoral researcher in the field of the impacts of electric mobility, driver behavior and ICTs in urban mobility.

EDITOR CONTACT INFORMATION

Dr. Carla Alexandra Monteiro da Silva
Institute of Mechanical Engineering-IDMEC
Insituto Superior Técnico
carla.silva@ist.utl.pt

INDEX

A

B

C

D

F

G

H

I

K

L

Q

R

S